Sustainable Structural Materials

From Fundamentals to Manufacturing, Properties and Applications

Editors

Elango Natarajan

Faculty of Engineering, Technology and Built Environment
UCSI University, Kuala Lumpur, Malaysia
and
Department of Mechanical Engineering
PSG Institute of Technology and Applied Research
Coimbatore, Tamil Nadu, India

Kalaimani Markandan

Department of Chemical & Petroleum Engineering
Faculty of Engineering, Technology and Built Environment
UCSI University, Malaysia

Cik Suhana Binti Hassan

Department of Mechanical & Mechatronics Engineering
Faculty of Engineering
Technology and Built Environment
UCSI University, Kuala Lumpur, Malaysia

Praveennath G. Koppad

National Institute of Technology Karnataka
Mangaluru, Karnataka, India

CRC Press is an imprint of the
Taylor & Francis Group, an **informa** business
A SCIENCE PUBLISHERS BOOK

Cover Credit: Image by Freepik

First edition published 2025
by CRC Press
2385 NW Executive Center Drive, Suite 320, Boca Raton FL 33431

and by CRC Press
4 Park Square, Milton Park, Abingdon, Oxon, OX14 4RN

CRC Press is an imprint of Taylor & Francis Group, LLC

Library of Congress Cataloging-in-Publication Data (applied for)

ISBN: 978-1-032-42313-5 (hbk)
ISBN: 978-1-032-42316-6 (pbk)
ISBN: 978-1-003-36222-7 (ebk)

DOI: 10.1201/9781003362227

Typeset in Times New Roman
by Prime Publishing Services

Preface

In an era where sustainability is no longer just a trend but a necessity, the construction and materials industry is at a pivotal crossroads. As we face the daunting challenges of climate change, resource depletion, and environmental degradation, the call for innovative solutions has never been more urgent. This book, *Sustainable Structural Materials: From Fundamentals to Manufacturing, Properties and Applications,* aims to bridge the gap between theory and practice, offering a comprehensive exploration of sustainable materials and their role in shaping a more resilient future.

The journey toward sustainability in structural materials begins with a deep understanding of the fundamental principles that govern their behavior and performance. In this book, we delve into the science behind various materials, emphasizing their environmental impact, lifecycle assessments, and the technological advancements that pave the way for greener alternatives. We specifically explore innovative solutions such as biomass materials, which offer renewable and carbon-neutral options for construction, and additive manufacturing techniques that minimize waste while allowing for complex geometries and tailored properties.

Moreover, we examine the role of biogas and energy harvesting technologies in enhancing the sustainability of structural materials. By integrating these approaches into the lifecycle of construction materials, we can reduce reliance on fossil fuels, promote circular economies, and optimize energy use throughout the built environment. Each chapter is designed to provide readers with both theoretical knowledge and practical insights into the manufacturing processes, properties, and real-world applications of these sustainable materials.

Additionally, we examine the promising potential of recycled plastics as structural materials. By transforming plastic waste into viable building components, we not only reduce landfill burdens but also create durable and versatile materials for construction. This integration of recycled materials is a key step in promoting a circular economy within the industry. Moreover, we discuss the role of biogas and energy harvesting technologies in enhancing the sustainability of structural materials. By integrating these approaches into the lifecycle of construction materials, we can reduce reliance on fossil fuels, promote circular economies, and optimize energy use throughout the built environment.

As we navigate through the complexities of sustainable engineering, we invite architects, engineers, researchers, and students alike to engage with the content and consider how they can contribute to a more sustainable built environment. The case studies and examples presented throughout this book illustrate the exciting possibilities that arise when innovation meets sustainability, from using agricultural byproducts as structural components to harnessing energy through innovative biogas systems.

It is our hope that this book serves as a valuable resource for those seeking to make informed decisions about materials and construction practices. Together, we can build a future that prioritizes not only the needs of today but also the well-being of generations to come. Thank you for joining us on this journey towards a more sustainable future in structural materials.

Editors

Contents

Chapter **1**

Additive Manufacturing of Polymer Composites: Processing and Structural Design

Kalaimani Markandan
Department of Chemical and Petroleum Engineering, Faculty of Engineering, Technology and Built Environment, UCSI University, 56000, Malaysia
Email: kalaimani@ucsiuniversity.edu.my

1.1 INTRODUCTION

Thermoplastic polymeric materials such as acrylonitrile butadiene styrene (ABS), polylactic acid (PLA), polymethylmethacrylate (PMMA), polyamide (PA) and polycarbonate (PC), as well as thermosetting polymeric materials such as epoxy resins have been commonly used in various additive manufacturing (AM) technologies. 3D printed polymeric materials have found potential applications in aerospace, automotive industry and sports equipment, owing to the light weight, high stiffness and strength of the structures, which maximizes fuel efficiency while retaining structural integrity. The realization of large specific stiffness and strength of 3D printed polymeric materials today has been made possible by intentional lattice designs to prevent 'soft' bending and buckling deformation mode while ensuring even distribution of load throughout the structure [1,2]. Whereas the relative stiffness and the strength of structures vary according to the solid fraction in power law relationship (i.e., exponent depends on the design topology), it should be noted that the absolute values of mechanical properties are proportional to the intrinsic modulus and the strength of the material [3].

Despite the recent advancements in 3D printing technology, most 3D printed polymeric structures are vastly conceptual prototypes instead of functional structures or components, due to the lack of strength and stiffness, which hampers the functional and load bearing capabilities. To this end, one possibility to enhance the specific stiffness and strength of the material is via the addition of reinforcement particles. Examples of these particles include graphene [4], iron and copper [5,6], tungsten [7], aluminium [8], alumina [8–11], barium and calcium titanate [12], diamond microparticles [13], carbon fibers [14] and glass beads [15]. Many studies have consistently reported on the manufacturing process and performance of 3D printed polymer composites, with an emphasis on process flexibility and geometrical precision to fabricate complex structures while attaining excellent mechanical properties. Figure 1.1 shows the relationship between the principle, design, materials and process of 3D printing.

Today, more than a thousand 3D printers are available in the market, which rely on various printing technologies. Nevertheless, the International Organization for Standardization (ISO) and American Society for Testing and Materials (ASTM) 52900:2015 standards have categorized additive manufacturing into seven major classes such as binder jetting (e.g., 3D inkjet technology), directed energy deposition (e.g., laser deposition, laser engineered netshaping, electron beam, plasma arc melting), material extrusion (e.g., fused deposition modeling, fused filament fabrication, fused layer modelling), material jetting (e.g., 3D inkjet technology, direct ink writing), powder bed fusion (e.g., electron beam melting, direct metal laser sintering, selective laser sintering/melting), sheet lamination (e.g., laminated object manufacturing, ultrasound consolidation or ultrasound additive manufacturing) and vat polymerization (e.g., stereolithography, digital light processing) [16, 17]. The technological advancements in the aforementioned 3D printing techniques have enabled the capability to produce complex products which are real, innovative as well as robust. However, despite the existence of these seven major classes, only vat polymerization, material extrusion and material jetting have been emphasized in literature till date to additively manufacture polymer composites.

It is worth noting that research in the past decade has focused on the additive manufacturing of filler-reinforced polymer composites which can strengthen the 3D printed parts for use as functional models instead of non-functional prototypes. Most 3D printed polymer composites have exhibited anisotropic properties attributed to the alignment of fillers in the matrix. In fact, many studies have consistently established that the highest modulus in these composites was attainable when fillers adopted the isostrain configuration of Voight composite (i.e., fillers aligned perfectly along the loading direction), whereas the lowest possible modulus in the polymer composite was achieved when the fillers adopted an isostress configuration of Reuss composite (i.e. fillers aligned perpendicularly to the lateral load direction). In many instances, alignment of fillers is highly desirable for uniaxial loads, which has led to the realization of various filler alignment techniques in polymer matrices, such as electric field induced alignment, magnetic induced alignment, fiber spinning with drawing, selection of naturally aligned fillers, as well as lining up of fillers along the print direction due to the intense shear stresses caused by nozzle walls.

In fact, the emphasis in 3D printing has been on technique, design topology and an extensive profiling of the mechanical properties (and mechanical anisotropy) of polymer composites under quasistatic and dynamic conditions. In this chapter, the authors make an attempt to gather information on two principal topics and discuss them in depth: (i) strengthening mechanism associated with filler alignment and theoretical models for prediction of mechanical properties, and (ii) functional properties and applications of polymer composites with anisotropic behavior. In particular, the chapter focusses on the work done in the last decade to showcase the progress in this field.

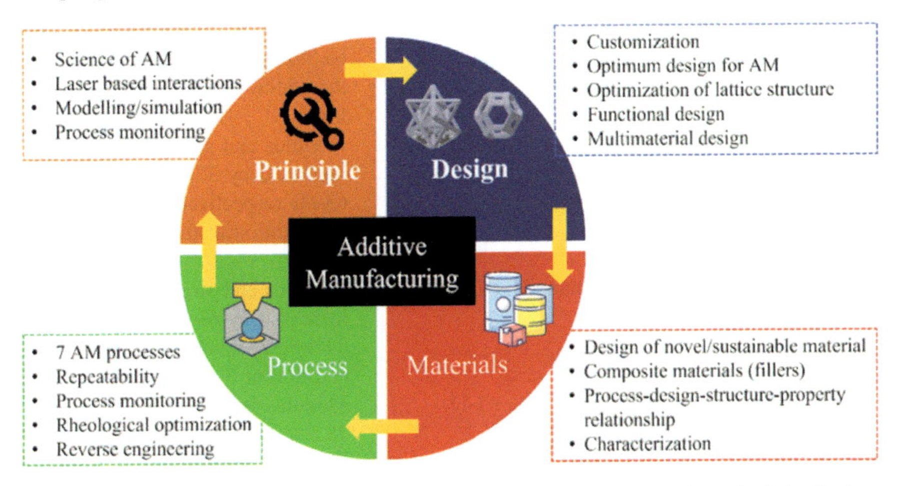

Figure 1.1 Schematic diagram illustrating the relationship between the principle, design, materials and process of additive manufacturing.

1.2 ADDITIVE MANUFACTURING OF POLYMER COMPOSITES

Addition of fillers has been reported to enhance the specific stiffness and strength of the additively manufactured composites. Several common fillers that have been reported in literature till date are carbon fibers (CF), boron nitride, graphene, carbon nanotubes (CNT), zinc ferrite ($ZnFe_2O_4$) and barium titanate ($BaTiO_3$), among others. CF have been commonly explored in FDM-based AM since the resultant structural properties of polymer composites improve with the length of the fibers, due to the improved stress transfer, load bearing capabilities and better fiber-matrix adhesion which improves bending and the tensile strength of the composite even at lower fill densities (i.e. lower weight) [18]. Some studies have used the commercial MarkOne® and MarkTwo® 3D printers by Markforged, which allow the continuous fiber 3D printing of polymer composites through the use of pre-impregnated CF, Kevlar fiber [19] and glass fiber [20].

Table 1.1 Fillers used in the AM of polymer composites and the key findings reported.

AM Technique	Matrix Filler	Filler Morphology	Key Findings	Ref.
SLA	Photocurable resin-TiO_2 particles	0-D	Modulus and strength increased by >1000% with the addition of 1 wt.% filler	28
SLA	Photocurable resin-TiO_2 particles	0-D	Modulus increased by 20.7% with the addition of 0.5 wt.% filler; no enhancement in strength	29
SLA	Photocurable resin-Ag-TiO_2 particles	0-D	Modulus and strength increased by 53.3% and 60.8% respectively with the addition of 1.2 wt.% filler	30
SLA	Photocurable resin-softwood lignin	3-D	Modulus and strength increased by 37.4% and 63.7% respectively with the addition of 0.8 wt.% filler	31
FFF	PLA-cellulose nanofiber	1-D	Modulus and strength increased by 63.2% and 84% respectively with the addition of 3 wt.% filler	32
FDM	PLA-CNT	1-D	Modulus increased by 28.3% with the addition of 2.5 wt.% filler; no improvement in strength	33
FDM	ABS-ZnO_2 particles	0-D	Modulus and strength increased by 16.7% and 92.3% respectively with the addition of 11 wt.% filler	34
FDM	PC-ABS-graphene	2-D	Modulus and strength increased by 59.3% and 52% respectively with the addition of 11 wt.% filler	35
FDM	ABS-CF	1-D	Strength increased by 42% with the addition of 2 wt.% filler	36

On the other hand, graphene—a two-dimensional sheet of hexagonally arranged carbon atoms—is another popular choice of reinforcement in the 3D printing of polymer composites, given its high Young's modulus of 1 TPa [21], breaking strength of 42 N/m [21] and low mass density of approximately 2 g cm [22]. To date, several techniques have been reported for the 3D printing of graphene-based polymer composites such as DIW, FDM, SLA and SLS.

Several 0-dimensional fillers have also been used as reinforcements in the additive manufacturing of polymer composites. For example, selective heat melting (SHM) technique was developed by a group of authors at *Nanyang Technological University, Singapore* to fabricate 3D polyethylene (PE)-copper (Cu) and polyethylene-iron (Fe) composites [23]. The strength of PE-Cu composite was reportedly comparable to that of engineering plastics such as polycarbonate; besides, the composite was electrically conductive (0.152 ± 0.28 S/m) at par with physically cross-linked graphene assemblies. Other 0-D fillers that have been used as reinforcing fillers in 3D printed polymer composites include $BaTiO_3$, which has shown great promise as a functional dielectric material for capacitors or light-weight passive antennas [24], and carbon black, which exhibits rich active sites, high specific area, perfect electrical conductivity and great chemical stability with good conductive loss, polarization loss and suitable impedance matching as a microwave absorbing material [25–27]. Table 1.1 summarizes some of the fillers used in the AM of polymer composites and the key findings reported in these studies.

1.3 TECHNIQUES OF CONTROLLING FILLER ALIGNMENT

Controlling filler alignment in 3D printing can unleash additional benefits such as producing composite materials with advanced and anisotropic mechanical, electrical and thermal properties which are highly advantageous to a myriad of engineering applications. To this end, various techniques have been utilized in 3D printing to spatially arrange, align or orient the filler material. Many studies have reported the use of external fields such as electric field, magnetic field, shear force field and ultrasound wave field for the alignment of fillers in polymer matrices.

For example, ultrasound wave field spatially arranges or orients fillers via acoustic radiation force with low attenuation in low viscosity fluids which aids in dimensional scalability [37]. Alignment of fillers by ultrasound wave filed has been reported to be material agnostic, i.e., independent of material properties or shape, except for density or the difference in the compressibility of the filler and the polymer matrix [38]. In a shear force field, flow (liquid material) or strain (solid material) is created to orient the filler material in the direction of the shear force. Niendorf and Raeymaekers reported that ultrasound wave field assisted alignment of carbon microfibers decreases with increasing filler concentration since microfibers entangle with each other, which physically deters filler alignment in a desired orientation [39]. Besides, alignment is greatly influenced by the polymer resin viscosity since changes in rheology during extrusion generates viscous drag forces that oppose alignment. Other studies have reported that shear field to achieve filler alignment can create streamlines which result in undesired alignment (randomized fiber architecture) due to Jeffrey's orbits phenomenon during extrusion through the convergent channel [40].

For electric and magnetic field assisted filler orientation, the fillers should be electrically conductive and ferromagnetic respectively. Besides, ultrahigh electric [41] and magnetic field [42] strengths are required ($\approx$20 kVm^{-1} and $\approx$8000 mT respectively), which restricts the dimensional scalability of the samples. Unlike the quasi-instantaneous filler alignment via mechanical control, an external field such as an electric or magnetic field allows the alignment of fillers over tens of mm, owing to the physical limitations posed by the external field [43, 44]. As such, electric or magnetic fields may limit the degree of filler alignment since it is not capable of overcoming the viscous drag force or filler entanglement. Nevertheless, recent advances in inkjet printing, such as electrohydrodynamic printing (e-jet), have improved spatial resolution and unleashed possibilities of depositing microspheres that form arrays post deposition as well as ordered nanowires with the application of electric field [45]. Similarly, magnetic field has assisted in the deposition of ordered nanowire patterns and micropillars—all of which provide opportunities for multi-material printing when used with multiple printheads configuration [46].

In another interesting study, Lai et al. proposed for the first time the Selection of Naturally Aligned Graphene (SNAG) in SLA process where graphene platelets were aligned along the print axis [4]. The natural selection of graphene was due to graphene reducing the transmission of laser light by more than half the wavelength (i.e., 405 nm) commonly used in SLA 3D printers. In particular, the horizontal graphene platelets create a 'shadowing effect' which blocks laser light and prevents polymerization. On the other hand, the vertically aligned platelets are naturally selected for 3D printing and securely incorporated into the print layer due to the unique 2D morphology of graphene (large area and small thickness). The natural selection of graphene platelets aligned to the print axis enables the composite to adopt the isostrain (Voigt) arrangement which improves the material modulus with 10 times less graphene concentration in comparison with other techniques reported in literature. The SNAG process has been ascertained by polarized light microscopy experiments where image intensity varies with the polarizer angle, indicating the polarization effect of graphene platelets on the transmitted light [47].

1.4 ALIGNMENT OF FILLERS AND STRENGTHENING MECHANISM

Carbon fiber has been commonly explored as a filler in many 3D printed composites and it has been consistently shown that the resulting composite exhibit anisotropic properties. A summary of mechanical properties of several 3D printed polymer matrices reinforced with carbon fiber is shown in Figure 1.2, where it can be seen that fibers aligned along the loading direction (0°) exhibit higher stiffness and tensile strength compared to the fibers aligned in other orientation angles. The main theoretical model that can be used to predict the effect of fiber orientation on the elastic moduli of 3D printed polymer composites is the isostress and isostrain limits. The isostrain limit has been used extensively to predict the upper limit of composite modulus. The isostrain configuration of a Voight polymer composite is attained when the anisotropic fillers are aligned perfectly along the loading direction. It can be described using the equation below:

$$E_c = E_f V_f + E_m V_m \tag{1}$$

On the contrary, when load is applied laterally (i.e. perpendicular to filler alignment), an isostress configuration of Reuss composite is obtained. In the Reuss configuration, the constituents are arranged such that each constituent is subjected to the same stress and can be described using the equation below:

$$E_c = \frac{E_f E_m}{V_f E_m + V_m E_f} \tag{2}$$

Where E_c, E_f and E_m represent the modulus of the composite, the fiber and of the polymer matrix respectively, and V_f represents the volume fraction of the fiber in the composite. The isostrain and isostress configuration of fibers can yield the highest and the lowest possible modulus respectively for a composite. This is because, from equation (1), it can be seen that the modulus for a composite with fillers in isostrain configuration varies linearly with the volume fraction of fibers (i.e. average moduli of polymer matrix and fiber). In this configuration, the fibers as reinforcement will be the main load bearing component of the system, thereby contributing significantly to enhance E_c. On the other hand, when fibers are in isostress configuration, E_c is dominated by E_m unless a high volume fraction of fibers is reinforced in the matrix.

In a similar vein, Markandan et al. used the modified version of Halpin-Tsai equation[48] which takes into account the length, width and thickness of graphene to compare the predicted and the measured stiffness values of stereolithographically fabricated polymer graphene composites in different print orientations [47]. The ratio of composite modulus (E_c) to neat polymer matrix (E_m) is predicted using the equation below:

$$\frac{E_c}{E_m} = \frac{3}{8}\left[\frac{1+((w+l)/t)\left(\dfrac{(E_g/E_m)-1}{(E_g/E_m)+((w+l)/t)}\right)V_g}{1-\left(\dfrac{(E_g/E_m)-1}{(E_g/E_m)+((w+l)/t)}\right)V_g}\right] + \frac{5}{8}\left[\frac{1+2\left(\dfrac{(E_g/E_m)-1}{(E_g/E_m)+2}\right)V_g}{1-\left(\dfrac{(E_g/E_m)-1}{(E_g/E_m)+2}\right)V_g}\right] \tag{3}$$

Similarly, experimental strength values of 3D printed graphene nanocomposites in different orientations can be compared with the model developed by Pukanszky [49], which can be given as:

$$\sigma_R = \frac{1-\varphi_f}{1+2.5\varphi_f}\exp(B\varphi_f) \tag{4}$$

where σ_R is the relative strength defined as σ_c/σ_m; here, σ_c and σ_m are the strength of the composite and the matrix respectively, φ_f is the volume fraction of nanofillers, whereas "*B*" is a parameter that determines the quantitative level of polymer-graphene interfacial adhesion and can be expressed as:

$$B = (1 + A_c \rho_f t_i) \ln\left(\frac{\sigma_i}{\sigma_m}\right) \tag{5}$$

where σ_i is the strength of the interphase, t_i is the thickness of the interphase, and ρ_f and A_c are the density and the specific area of the filler. A_c for composites containing plate-like nanofillers such as graphene can be expressed as:

$$A_c = \frac{2l^2 + 4lt}{\rho_f l^2 t} \tag{6}$$

where l and t represent the length and the thickness of filler. The modulus and strength anisotropy, A, across different print orientations of the polymer-graphene composite are calculated from the equation below:

$$A = \frac{\text{Property in Print Axis} - \text{Property in Perpendicular Axis}}{\text{Property in Perpendicular Axis}} \tag{7}$$

From Figure 1.2(C–F), the anisotropic behavior across all polymer-graphene composites at various print orientations were observed—an average anisotropy of ≈15.2 ± 4.6% and ≈27.2 ± 3.7% in modulus and strength respectively for composites reinforced with 0.02–0.05 wt.% graphene. Markandan et al. (2021) established SNAG process to induce graphene alignment in SLA 3D printing [47]. However, significant deviation can be seen between experimental anisotropy and theoretical anisotropy originating from the isostrain and isostress model. This deviation explains the limitation of the SNAG process where the tendency of horizontally aligned graphene platelets to block UV light is higher at higher graphene concentrations. As such, the graphene alignment no longer obeys the isostrain configuration of the SNAG process, which results in significantly reduced anisotropy.

1.5 STRUCTURAL DESIGNS

Various 3D printing techniques enable the possibility to engineer materials and structures across various scales such as nanoscale, microscale, mesoscale and macroscale, as shown in Figure 1.3. The possibility of designing materials and structures across various scales can be useful to realize many corresponding functionalities. For example, AM in nanostructure scale can influence electron and phonon transportation, which in turn affects thermal/electrical conductivity and magnetic properties. On mesoscale, structural design involves path planning of continuous fibers in the polymer matrix or by controlling the cellular or lattice topology. When the structure is topologized, it is possible to align the fibers in the direction of force transfer and thus strengthen the stress-intensive regions [57].

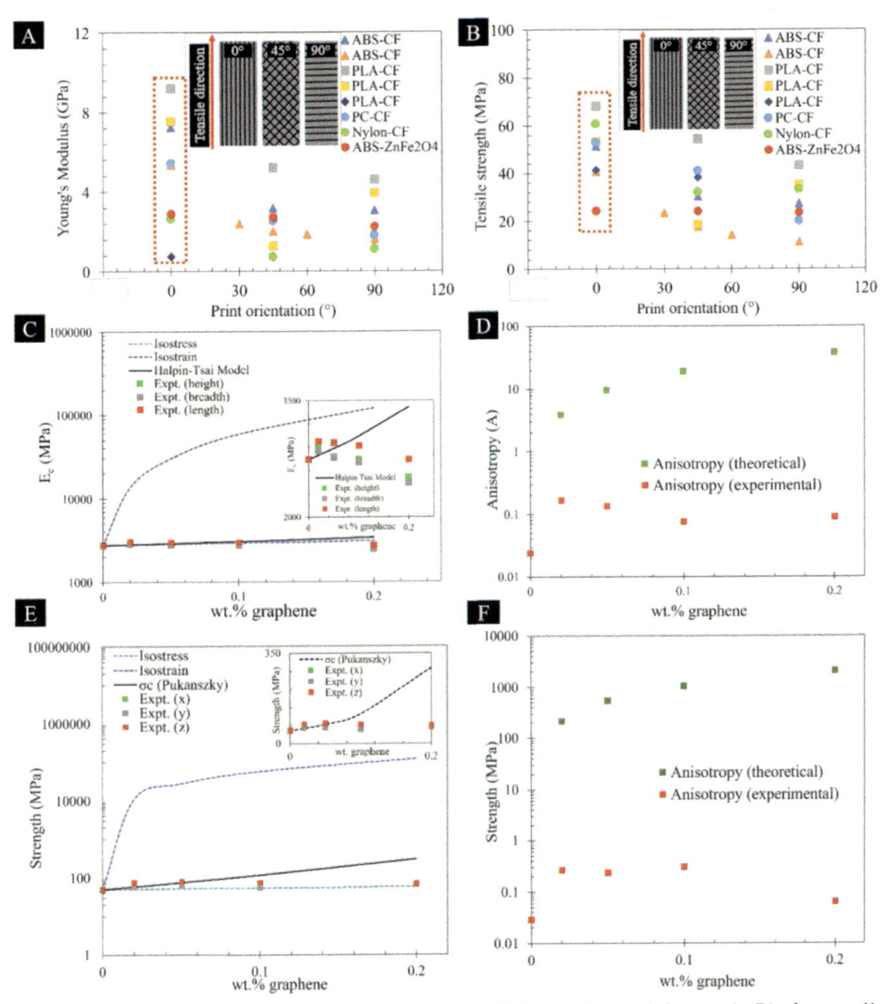

Figure 1.2 Effect of print orientation on the (A) Young's modulus and (B) the tensile strength of various 3D printed polymer composites (ABS–CF [50, 51], PLA–CF [50, 52, 53], PC-CF [54], nylon-CF [55] and ABS–Zn Fe_2O_4 [56]) (C) experimental and theoretical (Halpin-Tsai, isostress and isostrain) modulus values at different print orientations [47], (D) experimental and theoretical (i.e. isostrain and isostress) anisotropy in modulus [47], (E) experimental and theoretical (isostress, isostrain and Pukanszky34) strength values at different print orientations [47], (F) experimental and theoretical (i.e. isostrain and isostress) anisotropy in strength [47].

Stiffness and strength can be considered two primary design factors while designing any mechanical part or component via AM. Lattice structures which are typically made of repeating unit cells and assembled into 3D patterns are excellent materials for lightweight and energy absorbing applications. Two common lattice structures that have been investigated for 3D printing of polymer composites are the stretch or bending dominated lattice structures. Lattices with stretch

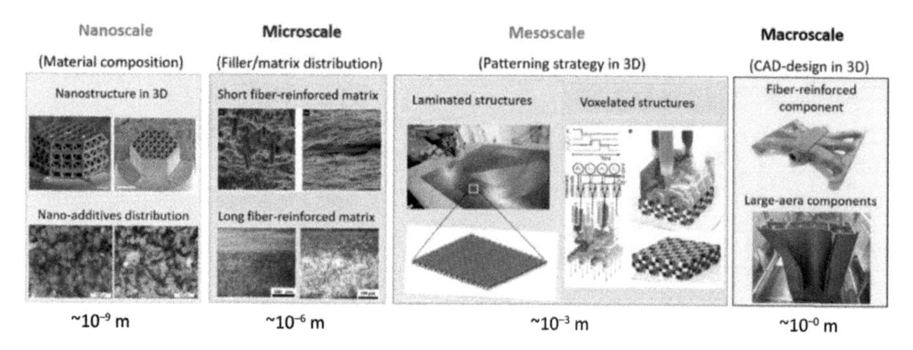

Figure 1.3 Structural design in multiple scales such as nanoscale, microscale, mesoscale and macroscale, and the phases of composite with spatial distribution in 2D and 3D [58].

dominated structures exhibit higher stiffness compared to bending dominated lattices. On the other hand, bend-dominated deformation provides relatively higher energy absorption capabilities than stretch dominated structures. The bend-dominated structure has been reported to enhance energy absorption during external compression. Generally, the energy absorption or damping characteristics of lattices arises from the hard phase retaining the mechanical strength while the relatively soft phase dissipates the mechanical energy when subjected to external cyclic loading or impact. As such, the spatial arrangement of the hard and soft phases can assist in force distribution which in turn induces force dissipation. For example, in a study by Yu et al. the effects of relative density and loading rate on the energy absorption behavior of Kelvin foams were investigated and it was reported that the deformation mode of Kelvin foams was dominated by relative density [59]. However, there exists a critical relative density below which the lattice structure is dominated by cell wall bending as the main deformation mechanism that results in cell collapse (Figure 1.4). Above the critical relative density, the cell wall of the Kelvin lattice is dominated by stretching deformation which leads to an X-based deformation shape.

Although octet truss, a stretch dominated lattice, has been consistently reported to exhibit better stiffness compared to bending dominated lattices, some studies have also reported the enhanced energy absorption behavior of the octet truss lattice, in particular when reinforced with nanofillers. For example, in a study by Lai et al. octet truss lattices with a truss width of 1 mm and relative density of 0.31, and different graphene concentrations and post-print heat treatments, showed evident improvement in the dynamic strength and impact energy absorption capability of the lattice (Figure 1.5) [4]. Besides, strength and energy absorption at high strain rates of the lattice significantly outperformed solid Al alloys and foams, and porous ceramic foams (such as A&T's Eco-core), on a per unit weight basis. The octet truss lattice with 0.02 wt.% of graphene content showed energy absorption per unit mass of 38.9 kJ/kg, which was approximately twice that of Balsa wood (i.e. 22 kJ/kg) – another well-known porous material [41].

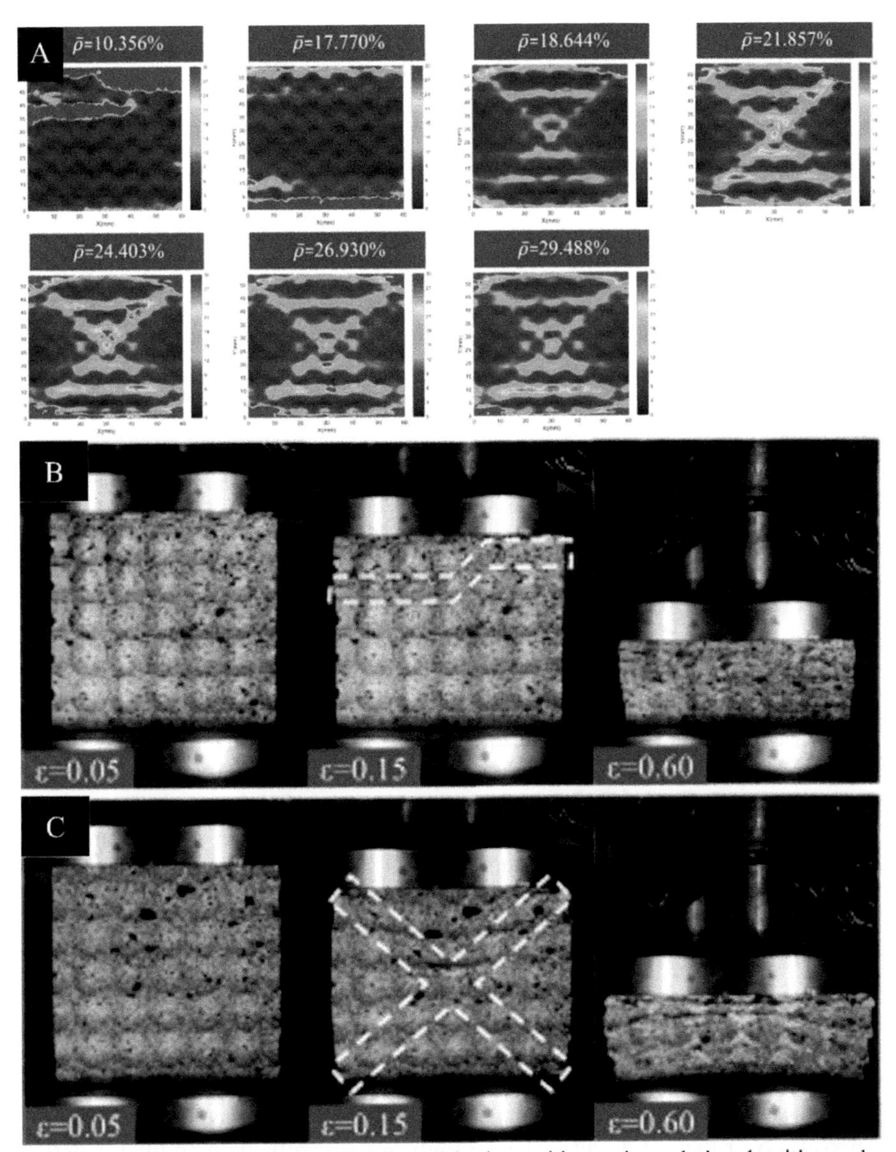

Figure 1.4 (A) Deformation of Kelvin cell lattices with varying relative densities under quasi-static loading and nominal strain of 10%; (B, C) Deformation images at relative densities of 10.356% and 24.403% respectively at different nominal strains [59].

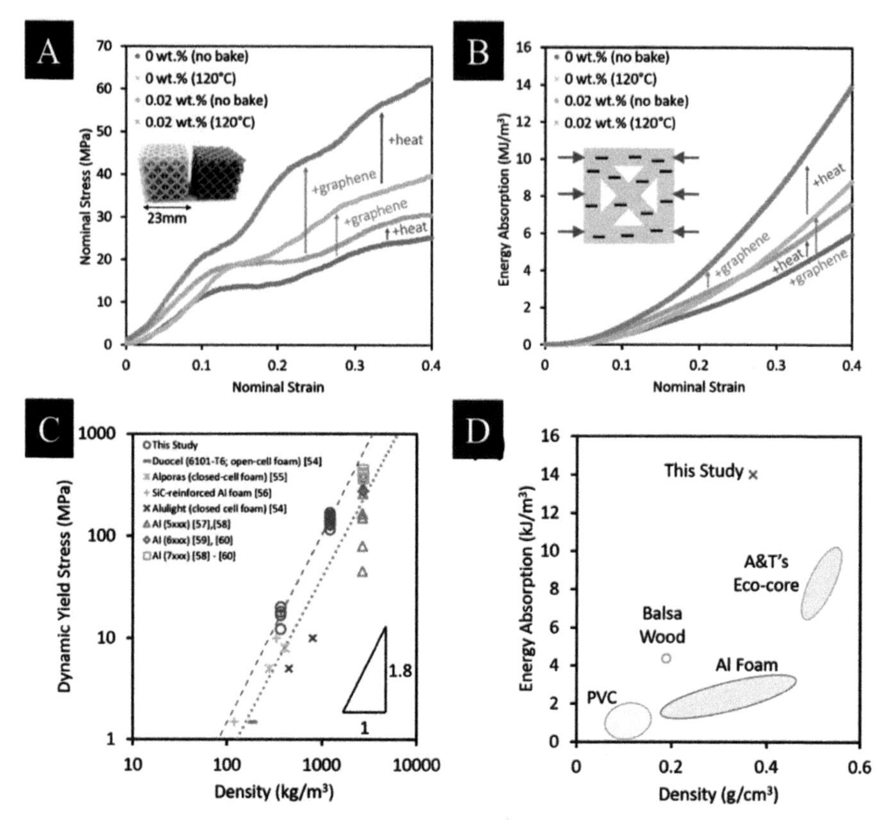

Figure 1.5 (A) Stress-strain response of octet truss lattices with different graphene concentrations and post-print heat treatments (truss width of 1 mm and relative density of 0.31) (B) energy absorption per unit volume of octet truss lattices (C) comparison of dynamic strength vs. density, and (D) energy absorption vs. density of 3D printed lattice reinforced with graphene nanocomposites against other lightweight engineering materials [4].

1.6 CONCLUSION

In conclusion, progress has been made in the additive manufacturing of polymer composites over the last decade. The possibility of spatially aligning filler material renders a huge potential for various applications ranging from biomimetics and tissue engineering to smart materials and aerospace engineering. In fact, several techniques to control filler alignment have been investigated with increased degree of control, while exploring newer materials and designs for 3D printing. To date, CF has been most commonly reported as reinforcement in 3D printing, owing to the significant enhancement in its properties and close-to-theoretical values attainable in a highly aligned CF composite. Regardless of the type of reinforcement, studies have consistently reported that the alignment of filler renders composites with designer properties that were previously unattainable. Despite the extensive studies reported on filler alignment via stimulus such as shear, electric, magnetic and ultrasound, it would be interesting to explore newer techniques with sensitive tuning which addresses concerns such as dimensional scalability, repeatability and wider material choice that can cater towards mass manufacturing. Furthermore, studies exploring various fillers in structural designs of 3D printed lattices remain limited; they can be investigated in detail in future studies.

REFERENCES

[1] do Rosário, J.J., Berger, J.B., Lilleodden, E.T., McMeeking, R.M. and Schneider, G.A. 2017. The stiffness and strength of metamaterials based on the inverse opal architecture. Extrem. Mech. Lett. 12: 86–96. https://doi.org/10.1016/j.eml.2016.07.006.

[2] Lal Lazar, P.J., Subramanian, J., Natarajan, E., Markandan, K. and Ramesh, S. 2023. Anisotropic structure-property relations of FDM printed short glass fiber reinforced polyamide TPMS structures under quasi-static compression. J. Mater. Res. Technol. 24: 9562–9579. https://doi.org/10.1016/j.jmrt.2023.05.167.

[3] Ashby, M.F. 2006. The properties of foams and lattices. Philos. Trans. R. Soc. A Math. Phys. Eng. Sci. 364(1838): 15–30. https://doi.org/10.1098/rsta.2005.1678.

[4] Lai, C.Q., Markandan, K., Luo, B., Lam, Y.C., Chung, W.C. and Chidambaram, A. 2021. Viscoelastic and high strain rate response of anisotropic graphene-polymer nanocomposites fabricated with stereolithographic 3D printing. Addit. Manuf. 37: 101721. https://doi.org/10.1016/j.addma.2020.101721.

[5] Nikzad, M., Masood, S.H. and Sbarski, I. 2011. Thermo-mechanical properties of a highly filled polymeric composites for fused deposition modeling. Mater. Des. 32(6): 3448–3456. https://doi.org/10.1016/j.matdes.2011.01.056.

[6] Hwang, S., Reyes, E.I., Moon, K.-sik, Rumpf, R.C. and Kim, N.S. 2015. Thermo-mechanical characterization of metal/polymer composite filaments and printing parameter study for fused deposition modeling in the 3D printing process. J. Electron. Mater. 44(3): 771–777. https://doi.org/10.1007/s11664-014-3425-6.

[7] Shemelya, C.M., Rivera, A., Perez, A.T., Rocha, C., Liang, M., Yu, X. et al. 2015. Mechanical, electromagnetic, and X-ray shielding characterization of a 3D printable tungsten–polycarbonate polymer matrix composite for space-based applications. J. Electron. Mater. 44(8): 2598–2607. https://doi.org/10.1007/s11664-015-3687-7.

[8] Boparai, K., Singh, R. and Singh, H. 2015. Comparison of tribological behaviour for Nylon6-Al-Al_2O_3 and ABS parts fabricated by fused deposition modelling: this paper reports a low cost composite material that is more wear-resistant than conventional ABS. Virtual Phys. Prototyp. 10(2): 59–66. https://doi.org/10.1080/17452759.2015.10 37402.

[9] Kurimoto, M., Yamashita, Y., Ozaki, H., Kato, T., Funabashi, T. and Suzuoki, Y. 2015. 3D printing of conical insulating spacer using alumina /UV-Cured-Resin composite. Annu. Rep. Conf. Electr. Insul. Dielectr. Phenomena, CEIDP. 2015-Decem. 463–466. https://doi.org/10.1109/CEIDP.2015.7352047.

[10] Martin, J.J., Fiore, B.E. and Erb, R.M. 2015. Designing bioinspired composite reinforcement architectures via 3D magnetic printing. Nat. Commun. 6: 1–7. https:// doi.org/10.1038/ncomms9641.

[11] Kokkinis, D., Schaffner, M. and Studart, A.R. 2015. Multimaterial magnetically assisted 3D printing of composite materials. Nat. Commun. 6(1): 8643. https://doi.org/ 10.1038/ncomms9643.

[12] Isakov, D.V., Lei, Q., Castles, F., Stevens, C.J., Grovenor, C.R.M. and Grant, P.S. 2016. 3D printed anisotropic dielectric composite with meta-material features. Mater. Des. 93: 423–430. https://doi.org/10.1016/j.matdes.2015.12.176.

[13] Kalsoom, U., Peristyy, A., Nesterenko, P.N. and Paull, B. 2016. A 3D printable diamond polymer composite: a novel material for fabrication of low cost thermally conducting devices. RSC Adv. 6(44): 38140–38147. https://doi.org/10.1039/c6ra05261d.

[14] Kaur, M., Yun, T.G., Han, S.M., Thomas, E.L. and Kim, W.S. 2017. 3D printed stretching-dominated micro-trusses. Mater. Des. 134: 272–280. https://doi.org/10.1016/j.matdes.2017.08.061.

[15] Chung, H. and Das, S. 2006. Processing and properties of glass bead particulate-filled functionally graded Nylon-11 composites produced by selective laser sintering. Mater. Sci. Eng. A. 437(2): 226–234. https://doi.org/10.1016/j.msea.2006.07.112.

[16] Tofail, S.A.M., Koumoulos, E.P., Bandyopadhyay, A., Bose, S., O'Donoghue, L. and Charitidis, C. 2018. Additive manufacturing: scientific and technological challenges, market uptake and opportunities. Mater. Today. 21(1): 22–37. https://doi.org/10.1016/j.mattod.2017.07.001.

[17] Mahamood, R.M., Akinlabi, S.A., Shatalov, M., Murashkin, E.V. and Akinlabi, E.T. 2019. Additive manufacturing/3D printing technology: a review. Ann. Dunarea Jos Univ. Galati Fascicle XII Weld. Equip. Technol. 30: 51–58. https://doi.org/10.35219/awet.2019.07.

[18] Heidari-Rarani, M., Rafiee-Afarani, M. and Zahedi, A.M. 2019. Mechanical characterization of FDM 3D printing of continuous carbon fiber reinforced PLA composites. Compos. Part B Eng. 175: 107147. https://doi.org/10.1016/j.compositesb.2019.107147.

[19] Melenka, G.W., Cheung, B.K.O., Schofield, J.S., Dawson, M.R. and Carey, J.P. 2016. Evaluation and prediction of the tensile properties of continuous fiber-reinforced 3D printed structures. Compos. Struct. 153: 866–875. https://doi.org/10.1016/j.compstruct.2016.07.018.

[20] Justo, J., Távara, L., García-Guzmán, L. and París, F. 2018. Characterization of 3D printed long fibre reinforced composites. Compos. Struct. 185: 537–548. https://doi.org/10.1016/j.compstruct.2017.11.052.

[21] Lee, C., Wei, X., Kysar, J.W. and Hone, J. 2008. Measurement of the elastic properties and intrinsic strength of monolayer graphene. Science. 321(5887): 385–388. https://doi.org/10.1126/science.1157996.

[22] Liu, Y., Xie, B., Zhang, Z., Zheng, Q. and Xu, Z. 2011. Mechanical properties of graphene papers. J. Mech. Phys. Solids. 60(4): 591–605. https://doi.org/10.1016/j.jmps.2012.01.002.

[23] Markandan, K., Lim, R., Kumar Kanaujia, P., Seetoh, I., bin Mohd Rosdi, M.R., Tey, Z.H., et al. 2020. Additive manufacturing of composite materials and functionally graded structures using selective heat melting technique. J. Mater. Sci. Technol. 47: 243–252. https://doi.org/10.1016/j.jmst.2019.12.016.

[24] Khatri, B., Lappe, K., Habedank, M., Mueller, T., Megnin, C. and Hanemann, T. 2018. Fused deposition modeling of ABS-Barium titanate composites: a simple route towards tailored dielectric devices. Polymers (Basel). 10(6): 666. https://doi.org/10.3390/polym10060666.

[25] Rabbi, M.F. and Chalivendra, V. 2020. Strain and damage sensing in additively manufactured CB/ABS polymer composites. Polym. Test. 90: 106688. https://doi.org/10.1016/j.polymertesting.2020.106688.

[26] Garcia Rosales, C.A., Garcia Duarte, M.F., Kim, H., Chavez, L., Hodges, D., Mandal, P., et al. 2018. 3D printing of shape memory polymer (SMP)/carbon black (CB) nanocomposites with electro-responsive toughness enhancement. Mater. Res. Express. 5(6): 065704. https://doi.org/10.1088/2053-1591/aacd53.

[27] Dawoud, M., Taha, I. and Ebeid, S.J. 2018. Strain sensing behaviour of 3D printed carbon black filled ABS. J. Manuf. Process. 35: 337–342. https://doi.org/10.1016/j.jmapro.2018.08.012.

[28] Mubarak, S., Dhamodharan, D., Divakaran, N., Kale, M.B., Senthil, T., Wu, L., et al. 2020. Enhanced mechanical and thermal properties of stereolithography 3D printed structures by the effects of incorporated controllably annealed anatase TiO_2 nanoparticles. Nanomaterials. 10(1): 1–24. https://doi.org/10.3390/nano10010079.

[29] Aktitiz, İ., Aydın, K. and Topcu, A. 2021. Characterization of TiO_2 nanoparticle–reinforced polymer nanocomposite materials printed by stereolithography method. J. Mater. Eng. Perform. 30(7): 4975–4980. https://doi.org/10.1007/s11665-021-05574-x.

[30] Mubarak, S., Dhamodharan, D., Kale, M.B., Divakaran, N., Senthil, T., Sathiyanathan P., et al. 2020. A novel approach to enhance mechanical and thermal properties of SLA 3D printed structure by incorporation of metal–metal oxide nanoparticles. Nanomaterials 10(2): 217. https://doi.org/10.3390/nano10020217.

[31] Zhang, S., Li, M., Hao, N. and Ragauskas, A.J. 2019. Stereolithography 3D printing of lignin-reinforced composites with enhanced mechanical properties. ACS Omega. 4(23): 20197–20204. https://doi.org/10.1021/acsomega.9b02455.

[32] Ambone, T., Torris, A. and Shanmuganathan, K. 2020. Enhancing the mechanical properties of 3D printed polylactic acid using nanocellulose. Polym. Eng. Sci. 60(8): 1842–1855. https://doi.org/10.1002/pen.25421.

[33] Patanwala, H.S., Hong, D., Vora, S.R., Bognet, B. and Ma, A.W.K. 2018. The microstructure and mechanical properties of 3D printed carbon nanotube-polylactic acid composites. Polym. Compos. 39(S2): E1060–E1071. https://doi.org/10.1002/pc.24494.

[34] Aw, Y.Y., Yeoh, C.K., Idris, M.A., Teh, P.L., Elyne, W.N., Hamzah, K.A. et al. 2018. Influence of filler precoating and printing parameter on mechanical properties of 3D printed acrylonitrile butadiene styrene/zinc oxide composite. Polym.—Plast. Technol. Eng. 58(1): 1–13. https://doi.org/10.1080/03602559.2018.1455861.

[35] Tambrallimath, V., Keshavamurthy, R., Bavan, S.D., Patil, A.Y., Yunus Khan, T.M., Badruddin, I.A., et al. 2021. Mechanical properties of Pc-Abs-based graphene-reinforced polymer nanocomposites fabricated by FDM process. Polymers (Basel). 13(17): 2951. https://doi.org/10.3390/polym13172951.

[36] Bilkar, D., Keshavamurthy, R. and Tambrallimath, V. 2019. Influence of carbon nanofiber reinforcement on mechanical properties of polymer composites developed by FDM. Mater. Today Proc. 46(10): 4559–4562. https://doi.org/10.1016/j.matpr.2020.09.707.

[37] Friedrich, L., Collino, R., Ray, T. and Begley, M. 2017. Acoustic control of microstructures during direct ink writing of two-phase materials. Sens. Actuators, A Phys. 268: 213–221. https://doi.org/10.1016/j.sna.2017.06.016.

[38] Wadsworth, P., Nelson, I., Porter, D.L., Raeymaekers, B. and Naleway, S.E. 2020. Manufacturing bioinspired flexible materials using ultrasound directed self-assembly and 3D printing. Mater. Des. 185: 108243. https://doi.org/10.1016/j.matdes.2019.108243.

[39] Niendorf, K. and Raeymaekers, B. 2021. Additive manufacturing of polymer matrix composite materials with aligned or organized filler material: a review. Adv. Eng. Mater. 23(4): 1–18. https://doi.org/10.1002/adem.202001002.

[40] Roy, M., Tran, P., Dickens, T. and Quaife, B.D. 2020. Effects of geometry constraints and fiber orientation in field assisted extrusion-based processing. Addit. Manuf. 32: 101022. https://doi.org/10.1016/j.addma.2019.101022.

[41] Kamat, P.V., Thomas, K.G., Barazzouk, S., Girishkumar, G., Vinodgopal, K. and Meisel, D. 2004. Self-assembled linear bundles of single wall carbon nanotubes and their alignment and deposition as a film in a Dc field. J. Am. Chem. Soc. 126(34): 10757–10762. https://doi.org/10.1021/ja0479888.

[42] Fujiwara, M., Oki, E., Hamada, M., Tanimoto, Y., Mukouda, I. and Shimomura, Y. 2001. Magnetic orientation and magnetic properties of a single carbon nanotube. J. Phys. Chem. A. 105(18): 4385–4386. https://doi.org/10.1021/jp004620y.

[43] Wu, S., Zhang, J., Ladani, R.B., Ghorbani, K., Mouritz, A.P., Kinloch, A.J. et al. 2016. A novel route for tethering graphene with iron oxide and its magnetic field alignment in polymer nanocomposites. Polymer (Guildf). 97: 273–284. https://doi.org/10.1016/j.polymer.2016.05.024.

[44] Pothnis, J.R., Kalyanasundaram, D. and Gururaja, S. 2021. Enhancement of open hole tensile strength via alignment of carbon nanotubes infused in glass fiber-epoxy-CNT multi-scale composites. Compos. Part A Appl. Sci. Manuf. 140: 106155. https://doi.org/10.1016/j.compositesa.2020.106155.

[45] Korkut, S., Saville, D.A. and Aksay, I.A. 2008. Collodial cluster arrays by electrohydrodynamic printing. Langmuir. 24(21): 12196–12201. https://doi.org/10.1021/la8023327.

[46] Ahn, T., Kim, H.J., Lee, J., Choi, D.G., Jung, J.Y., Choi, J.H., et al. 2015. A facile patterning of silver nanowires using a magnetic printing method. Nanotechnology. 26(34): 345301. https://doi.org/10.1088/0957-4484/26/34/345301.

[47] Markandan, K., Seetoh, I.P. and Lai, C.Q. 2021. Mechanical anisotropy of graphene nanocomposites induced by graphene alignment during stereolithography 3D printing. J. Mater. Res. 36(21): 4262–4274. https://doi.org/10.1557/s43578-021-00400-5.

[48] Affdl, J.C.H. and Kardos, J.L. 1976. The Halpin-Tsai equations: a review. Polym. Eng. Sci. 16(5): 344–352. https://doi.org/10.1002/pen.760160512.

[49] Pukánszky, B. 1990. Influence of interface interaction on the ultimate tensile properties of polymer composites. Composites. 21(3): 255–262. https://doi.org/10.1016/0010-4361(90)90240-W.

[50] Jiang, D. and Smith, D.E. 2017. Anisotropic mechanical properties of oriented carbon fiber filled polymer composites produced with fused filament fabrication. Addit. Manuf. 18: 84–94. https://doi.org/10.1016/j.addma.2017.08.006.

[51] Somireddy, M., Singh, C.V. and Czekanski, A. 2020. Mechanical behaviour of 3D printed composite parts with short carbon fiber reinforcements. Eng. Fail. Anal. 107: 104232. https://doi.org/10.1016/j.engfailanal.2019.104232.

[52] Ferreira, R.T.L., Amatte, I.C., Dutra, T.A. and Bürger, D. 2017. Experimental characterization and micrography of 3D printed PLA and PLA reinforced with short carbon fibers. Compos. Part B Eng. 124: 88–100. https://doi.org/10.1016/j.compositesb.2017.05.013.

[53] Liu, Z., Lei, Q. and Xing, S. 2019. Mechanical characteristics of wood, ceramic, metal and carbon fiber-based PLA composites fabricated by FDM. J. Mater. Res. Technol. 8(5): 3743–3753. https://doi.org/10.1016/j.jmrt.2019.06.034.

[54] Gupta, A., Fidan, I., Hasanov, S. and Nasirov, A. 2020. Processing, mechanical characterization, and micrography of 3D-printed short carbon fiber reinforced polycarbonate polymer matrix composite material. Int. J. Adv. Manuf. Technol. 107(7–8): 3185–3205. https://doi.org/10.1007/s00170-020-05195-z.

[55] Kubota, M., Hayakawa, K. and Todoroki, A. 2022. Effect of build-up orientations and process parameters on the tensile strength of 3D printed short carbon fiber/PA-6 composites. Adv. Compos. Mater. 31(2): 119–136. https://doi.org/10.1080/09243046.2021.1930497.

[56] Hamzah, K.A., Yeoh, C.K., Mohd Noor, M., Teh, P.L., Sazali, S.A., Aw, Y.Y., et al. 2019. Effect of the printing orientation on the mechanical properties and thermal and electrical conductivity of ABS-$ZnFe_2O_4$ composites. J. Mater. Eng. Perform. 28(9): 5860–5868. https://doi.org/10.1007/s11665-019-04313-7.

[57] Li, N., Link, G., Wang, T., Ramopoulos, V., Neumaier, D., Hofele, J., et al. 2020. Path-designed 3D printing for topological optimized continuous carbon fibre reinforced composite structures. Compos. Part B Eng. 182: 107612. https://doi.org/10.1016/j.compositesb.2019.107612.

[58] Yuan, S., Li, S., Zhu, J. and Tang, Y. 2021. Additive manufacturing of polymeric composites from material processing to structural design. Compos. Part B Eng. 219: 108903. https://doi.org/10.1016/j.compositesb.2021.108903.

[59] Duan, Y., Du, B., Shi, X., Hou, B. and Li, Y. 2019. Quasi-static and dynamic compressive properties and deformation mechanisms of 3D printed polymeric cellular structures with Kelvin cells. Int. J. Impact Eng. 132: 103303. https://doi.org/10.1016/j.ijimpeng.2019.05.017.

Chapter 2

Application of Nanomaterial Technology in Sustainable Biogas Production

Yong Wei Tiong[1, 2]

[1]NUS Environmental Research Institute, National University of Singapore, #02-01, T-Lab Building, 5A Engineering Drive 1, Singapore 117411, Singapore

[2]Bioprocessing Technology Institute (BTI), Agency for Science, Technology and Research (A*STAR), 20 Biopolis Way, #06-01 Centros, Singapore 138668, Singapore

Email: tiong_yong_wei@bti.a-star.edu.sg

2.1 INTRODUCTION

Agriculture and food processing industries are the two biggest waste producers globally. Two critical waste-related challenges faced by these industries are emission of odors and gases, and manure management [1]. These gas emissions include greenhouse gases, ammonia (NH_3), and hydrogen sulfide (H_2S), which destroy environmental resources [2,3]. Rising energy prices, more restrictive regulatory requirements, and increasing concern over greenhouse gas emissions have prompted the usage of waste biomass to generate nonconventional energy, through anaerobic digestion (AD) technology for wastes treatment for instance [4,5]. AD consists of a series of microbial biological conversions that convert the organic portion of the waste biomass into bioenergy and bioresources in the absence of oxygen [6]. The spontaneous series of reactions involved therein are hydrolysis, acetogenesis, acidogenesis and methanogenesis, as detailed in Figure 2.1. In brief,

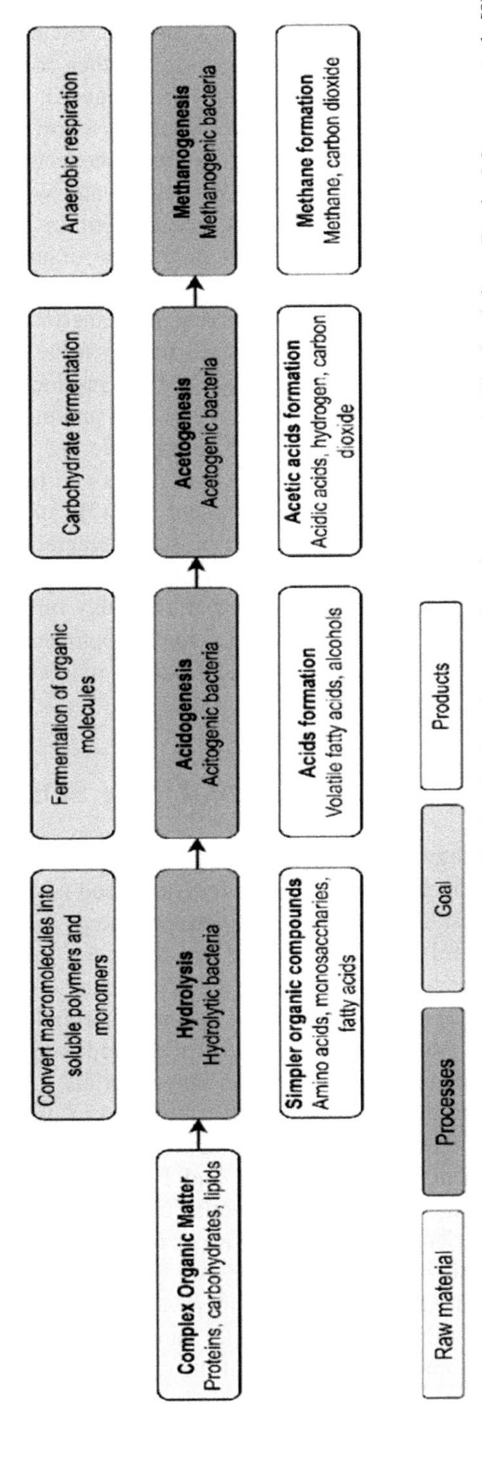

Figure 2.1 Phases of AD processes with the respective feedstock, functions, and products generated. Obtained from Rocha-Meneses et al. [8].

the complex organic matter of the waste biomass (cellulose) is converted into simpler monomers (sugars and alcohols) by hydrolytic bacteria (hydrolysis). The molecules obtained from the hydrolysis step are further converted into organic acids and alcohols by acidogenic bacteria (acidogenesis). These products are subsequently converted into acetic acid (acetogenesis), followed by getting further metabolized into methane and carbon dioxide (methanogenesis) [7].

AD aims for cleaner renewable energy recovery and volume reduction, i.e., to produce biogas (methane), which is a great benefit in resolving the global warming issue, as well as the consequent production of nutrient-rich digestate (a valuable biofertilizer for cultivating plants) simultaneously [9, 10]. To date, significant research has gone into concurrent methane-rich biogas generation and nutrient rich digestates production, based on a definite volume of biomass. Nevertheless, the AD process for maximum biogas production still encounters certain drawbacks, including low methane yields, high amounts of carbon dioxide, high concentrations of ammonia nitrogen, poor buffering capacity, as well as production of inhibitory substances [11]. These issues can be addressed through different strategies such as waste pretreatment [12, 13], anaerobic co-digestion [14, 15], and enhancement via nanomaterial supplementation [16, 17]. Among these strategies, the addition of nanomaterials is of high interest due to the diverse possibilities of fabricated materials that inevitably further enhance the AD performance notably [17]—in terms of electron acceptors/donors and cofactors of key enzymatic activities (e.g., pyruvate-ferrodoxin oxidoreductase) that play a major role in biogas formation [18].

2.2 NANOMATERIAL TECHNOLOGY SOLUTIONS

Nanomaterial technology has been successfully deployed in many applications, such as nanoelectronics, nanomedicine, nanodevices, and food nanotechnology [19, 20]. Nevertheless, the latest breakthroughs in renewable energy generation application, particularly for the oil and gas industry are the most recent breakthroughs. Nanomaterials consist of particles with a size less than 100 μm in at least one dimension (0D, 1D, 2D or 3D). Depending on their morphology, dimensions and chemical properties, nanomaterials are further divided into zero-valent metals, metal oxides, and carbon-based conductive nanomaterials.

AD relies on the microbial consortium conversion of organic waste, making the concentration of inorganic nutrients at appropriate levels in AD crucial. For instance, macronutrients (C, H, O, N, S and P) aid in microbial growth and metabolism through the synthesis of essential carbohydrates, proteins and fats. Meanwhile, micronutrients (Fe, Ni and Co) act as co-factors of many enzymes [21]. The effects of the utilization of nanomaterials on biogas production have been recognized as a promising solution to improve the biological as well as the chemical process of AD, enhance the buffer capacity process stability, increase the biogas yield, and improve the quality of the consequent production of effluent sludge (anaerobic digestates)—a valuable biofertilizer [22]. Supplying microbial communities with trace metals (nanomaterials) can enhance the microbial genera responsible for acidogenesis and hydrolysis stages [23], followed by methanogenesis for the production of biogas (methane) [24]. Methanogenesis is the most significant pathway for optimum biogas (methane) production. It can occur via three different pathways, depending on the type of substrate, i.e., acetoclastic (acetates), hydrogenotrophic (H_2/CO_2), and methylotrophic (reducing methyl groups), with the former two pathways considered as dominant [25]. Recent research has suggested that direct interspecies electron transfer (DIET) has occurred between organisms or in conjunction with nanomaterials [26]. Specifically, DIET can be stimulated in engineered systems to improve the desired treatment goals and energy recovery in systems such as AD and microbial technologies. Figure 2.2 illustrates the application of nanomaterials for electron transfer between oxidizing bacteria and methanogens, which can be further categorized into 3 groups—zero-valent, metal oxides, and carbon-based.

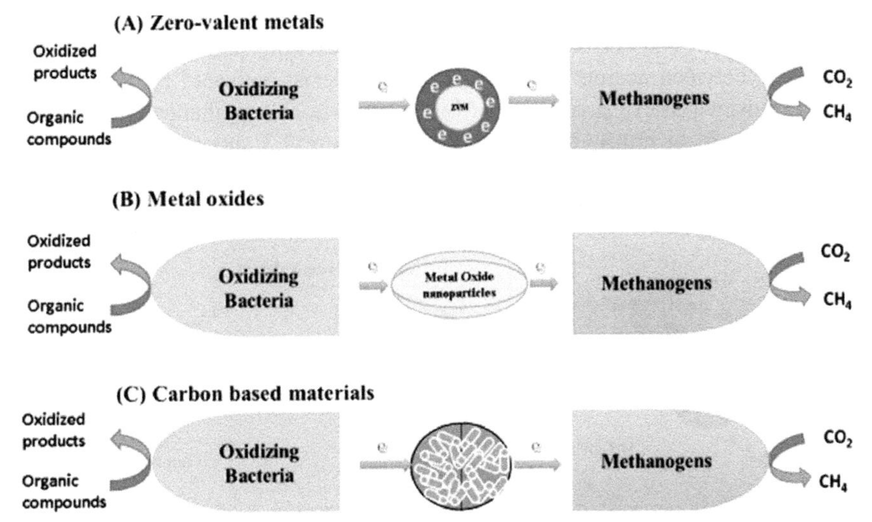

Figure 2.2 Application of nanomaterials for electron transfer between oxidizing bacteria and methanogens (A) electron through zero-valent metals, (B) electron transfer through metal oxides, (C) electron transfer through carbon-based materials. Obtained from Jadhava et al. [27].

This chapter evaluates the recent developments in the utilization of various types of nanomaterials as catalysts in the AD process. Furthermore, both economic and environmental impacts on the usage of nanomaterials have been further addressed, along with the prospective opportunities for the utilization of nanomaterials in AD.

2.3 VARIOUS NANOMATERIALS IN AD ENHANCEMENT

Nanomaterials play a significant role in AD processes, leveraging their unique properties such as dimension, composition, surface area, decomposition capacity, and catalytic nature. Their nano-sized structures and specific physicochemical properties can significantly influence the rate of AD. Moreover, nanomaterials interact with substrates and microorganisms in AD systems, further enhancing their efficacy in facilitating efficient digestion processes. Nanomaterials can influence the rate of AD due to their nano-sized structures, specific physicochemical properties, and the interactions between the substrate and microorganisms.

2.3.1 Zero-valent Metallic Nanomaterials

Zero-valent metallic nanomaterials (Fe, Ni, Cu, Co, Ag, Au, etc.) have unique and significant strong chemical reducibility, high efficiency, and large specific surface, and have thus been considered as one of the promising candidates for applications in environmental processes. As mentioned previously, AD is performed by a consortium of microorganisms. The flow of electrons within these communities governs the occurrence of reactions and their rates [26]. Interactions between acetogenesis and methanogenesis, mediated by interspecies electron transfer (DIET) between syntrophic bacteria and methanogenic archaea, plays a vital role in improving AD efficiency. Figure 2.3 shows the interactions of zero-valent metallic nanomaterials in AD. Table 2.1 summarizes the efforts of zero-valent metallic nanomaterials as reported in the literature to date.

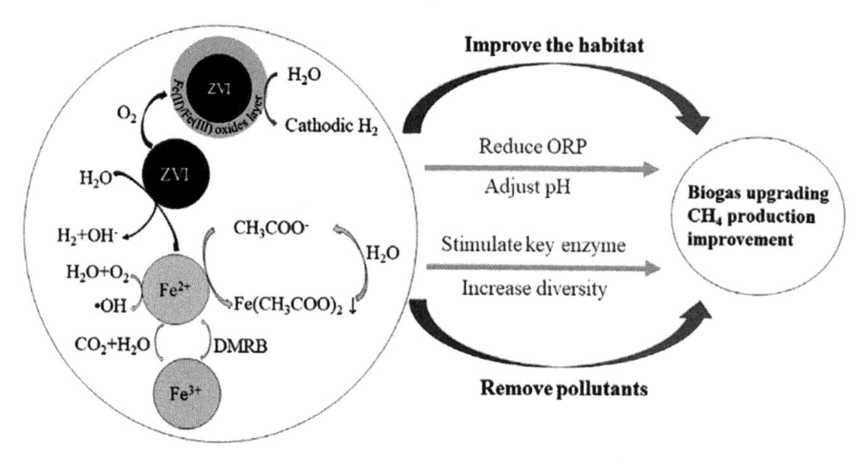

Figure 2.3 Effect of zero-valent metallic nanomaterials (ZVI) on AD. Obtained from Ye et al. [28].

Based on Table 2.1, one of the common trace elements investigated for their impact on AD is iron (Fe), which is essential for the AD process. The addition of Fe has been reported to result in the maximum biogas ever produced from the anaerobic incubation of cellulose through the application of bacterial inoculum and also metal oxides (Fe_3O_4) nanomaterials [39]. Furthermore, the addition of biosynthesized Fe nanomaterials collected from treated water sediment to anaerobic incubators improves biogas production from wastewater [40]. As reported in Table 2.1, a proper amount of zero-valent metallic nanomaterials

Table 2.1 Reported applications of zero-valent metal nanomaterials in AD processes.

Nanomaterial	Mean Size (nm)	Concentration	Substrate Type	Temperature of AD (°C)	Effects on the AD Process	Ref.
Ni	17	2 mg/L	manure slurry	37, batch	Increase in biogas yield by 101% Increase in methane yield by 74%	[29]
	100	5–10 mg/kg VS	sewage sludge	37, batch, CSTR	Increase in methane yield by 10%	[30]
Fe	9	5, 10, 20 mg/L	raw manure	37, batch	Increase in biogas yield by 45% Increase in methane yield by 59%	[31]
	20	10 g/L	sewage sludge	35, batch	Increase in biogas yield by 30.4% Increase in methane yield by 40.4%	[32]
	160	100 mg/L	sewage sludge	37, batch	Increase in methane yield by 25.2% Increase in COD removal by 10%	[33]
	45	1000 mg/L	domestic sludge	37, batch	Increase in biogas yield by 105.46%	[34]
Co	28	1 mg/L	manure slurry	37, batch	Increase in biogas yield by 64% Increase in methane yield by 86%	[29]
Cu	40–60	10–1500 mg/L	granular sludge	37, batch	Highly inhibitory towards aceloclastic and hydrogenotrophic methanogens	[35]
Ag, Fe, Mg	128	10 mg/g TSS	waste activated sludge	35, batch	Increase in methane yield by 120%	[36]
Ag	10–15	0.5, 1, 5, 100 mg/L	biosolids from wastewater treatment plants	37, batch	Up to complete inhibition of anaerobic biogas production	[37]
Au	20	75 mg/L	cellulose	37, 55, batch	Zero or slight toxicity on ordinary heterotrophic organisms, ammonia-oxidizing bacteria, and anaerobic bacteria	[38]

has a good effect on microbial activity, whereby the control of a certain dosage of zero-valent metallic nanomaterials has a positive effect, mainly manifested through a better long-term stability of the digester, such as greater substrate degradation, lower volatile fatty acids (VFAs) concentration and higher microbial abundance and methane production. However, due to the great variance of the AD operation and digestion substrates, there are still many challenges when designing an efficient and cost-effective AD system for practical solid waste recycling. Overall, the supplementation of zero valent metallic nanomaterials into the AD system presents notable influences on the performance of AD regarding process performance, increased biogas/methane production, as well as effluent quality.

2.3.2 Metal Oxide Nanomaterials

Metal oxide nanomaterials exhibit unique physical and chemical properties due to the presence of their active sites, small size, and high density of corner or edge surface sites [16]. There are numerous types of metal oxide nanomaterials, such as ZnO, CuO, TiO_2, MgO, NiO, Fe_2O_3 and CuO [27]. These metal oxide nanomaterials have shown outstanding achievements in transforming biomass into biogas through AD. Metal oxide ions are a prerequisite for enzymes and cofactors of chemical reactions occurring in the AD processes. More specifically, cobalt (Co) is vital for the degradation of methanol by methanogenic bacteria, whereas nickel (Ni) is a coenzyme in methanogenic archaea [41]. Zinc (Zn) and cobalt (Co) play an important role as structural ions in the transesterification factor of enzymes, while copper (Cu) is a fundamental factor for biological electron transport as a coenzyme [42].

Adding the above-mentioned metal nanomaterials and their oxides induces remarkable enhancements even at very low concentrations (approximately 10 mg/L), thereby stimulating the activities of microorganisms and key enzymes [7, 43]. This scenario would therefore result in more gas production and better effluent quality. Iron-based nanomaterials play an important role in increasing the concentration of acetate and butyrate—known as energy favourable VFAs for the biogas (methane) production phase [44]; this may consequently promote the AD process. Similarly, the addition of low concentrations (10 mg/g VS) of nano-ZnO positively influences biogas production [45]. The increases gas production may be attributed to the involvement of zinc-dependent enzyme ADH, which can catalyze the conversion between alcohols and aldehydes. Nevertheless, Mu et al. [46] reported the significant impact of nano-ZnO addition to anaerobic granular sludge (AGS), with >50 mg/g dosage of TSS. This scenario was likely the result of the nano-ZnO preventing the generation of all the functional groups in the extracellular polymeric substances (EPS), except polysaccharide contents. If the addition of nano-ZnO continues rising >100 mg/g TSS, the concentration of Zn^{2+} ions surpasses the chemical adsorption capacity of EPS (no longer trapping of the cations released from nano-ZnO), leading to further reduction in EPS production. As for the addition of nano-NiO at a low concentration of 5 mg/g VS, it showed great improvement in biogas production, by >15% [45,47]. Nano-NiO serves as a source of Ni^{2+} for several metal-enzymes, including [Fe-Ni] hydrogenase

Table 2.2 Metal oxide nanomaterials in AD processes.

Nanomaterial	Mean size (nm)	Concentration	Substrate Type	Effects on the AD Process	Ref.
Fe_3O_4	7	100 mg/L	crystalline cellulose	180% increase in biogas production and 8% increase in methane formation rate	[50]
	15-22	50-125 mg/L	municipal solid waste	117% increase in methane production	[51]
Fe_2O_3	128	100 mg/g TSS	waste activated sludge	117% increase in methane production	[36]
	N.A.	750 mg/L	granular sludge	38% increase in methane production	[52]
Al_2O_3	<50	1500 mg/L	granular sludge	No toxic effects on methanogenesis	[53]
MgO	154	500 mg/g TSS	waste activated sludge	99% decrease in methane production	[50]
CuO	30-50	11, 110, 330, 550, 1100 mg/L	municipal waste sludge	Up to 84% inhibitory effect on biogas production	[52]
ZnO	140	10, 300, 1500 mg/L	waste activated sludge	Up to 75.1% decrease in methane production	[54]
	145	5, 50, 100, 250, 500 mg/L	waste activated sludge	VFAs accumulation in the AD process; 25% reduction in biogas and 50% reduction in methane production	[55]
	120-140	42, 210, 1050 mg/L	mixed primary sludge	Decrease in the abundance of methanogenic archaea; inhibition of methane production	[56]
SiO_2	10-20	1500 mg/L	granular sludge	No toxic effects on methanogenesis	[57]
TiO_2	170	42, 210, 1050 mg/L	mixed primary sludge	No significant impact on methane production	[56]
	25	1500 mg/L	granular sludge	No toxic effects on methanogenesis	[58]

and acetyl-CoA synthetase (ACS) [48]. Hydrogenase catalyzes bio-hydrogen production and ACS assists the conversion of acetyl-CoA to acetate; thus, the addition of nano-NiO facilitates the conversion of acetaldehyde to acetyl-CoA [49]. In contrast to iron-based nanomaterials, the stimulated impacts of fermentative AD processes such as alterations in microbial activity and metabolic processes are significant, even at low concentrations, implying that the hydrogen-producing bacteria are more sensitive to released Ni^{2+} than Fe^{2+} cations.

Table 2.2 evaluates the impacts of metal oxide nanomaterials on the efficiency of AD processes. Moreover, the fate of metal oxide nanomaterials will vary with factors like type of biomass, size and concentration of metal oxide nanomaterials during the AD process.

2.3.3 Carbon-based Nanomaterials

Carbon-based nanomaterials like graphite, graphene, activated carbon (AC), biochar (BC), carbon cloth (CC), carbon nanotubes (CNTs), and their composites can enhance the efficiency of AD, given their capability for adsorbing chemicals onto their surfaces. This is because carbon serves as a habitation for microbial immobilization and as a provision for bioelectrical connections among cells, providing some essential elements for anaerobes [59]. According to a survey of 27 reactors [60], AD reactors supplemented with carbon-based nanomaterials (including BC, CC, CNT, graphene, and conductive polymers) reduced the lag times for methane formation by 10–75%, enhanced the methane production rates by 79–300%, and increased the methane production yields by 100–178%, compared to AD without supplementation.

Biochar (BC), a combustible carbon-rich solid material produced via gasification or pyrolysis of biomass wastes, has been touted as a promising additive with many desirable characteristics for enhancing the AD process [61–63]. Specifically, the porous structure of BC serves as a good immobilization matrix by mitigating ammonium (NH_4^+–N) inhibition, which significantly promotes the production of VFAs, along with enhancing the growth of archaea that favor the methanogenesis stage [64]. Owing to its high surface area and resultant high adsorption capability, BC is known to adsorb heavy metals, phosphates and other inhibitive organic compounds [65]. Moreover, BC improves digester buffering capacity and reduces acidification by affecting the alkalinity of the digester, and improves the granulation of the anaerobic sludge [62]. Additionally, it contributes to the utilization of fertilizer through the improved production of digestate, ensuring efficient nutrient uptake by crops and enhancing soil fertility [66]. As a biofertilizer, the addition of biochar to the digestate can contribute to nutrient retention, increase the carbon to nitrogen ratio and reduce nutrient leaching after application of the digestate mixture to the land [65]. Lee et al. [67] investigated the application of wood waste-derived biochar to food waste AD in order to evaluate the feasibility of its utilization to create a circular economy. This biochar was first applied for upgrading the biogas production from AD, before treating and recovering the nutrients in the solid portion of the digestate; this was finally employed as a biofertilizer for the organic cultivation of vegetables. It is worth mentioning that the amount of CO_2 absorbed by the biochar from the biogas was considered low (11.17 mg g^{-1}) and could potentially be increased through physical and chemical methods. Meanwhile, BC was able to remove around 31% of chemical oxygen demand (COD), 8% of the ammonia and almost 90% of the total suspended solids (TSS) from the digestate wastewater, which is better than dewatering via centrifugation. In another study, Song et al. [66] studied the agronomic performance of AD as a fertilizer, through vegetable cultivation experiments using four leafy vegetables (Chinese spinach, water spinach, Chinese cabbage and lettuce), by investigating the effect of dilution on AD performance. The results revealed that vegetable growth were favored when the original concentration of the digestate was diluted to 20–40% (v/v), where the shoot fresh weight of the vegetables grown was comparable to that of vegetables grown using a commercial fertilizer. Meanwhile, when applied in high concentrations, the high concentrations of AD were observed to induce ammonia and salt stress in vegetables. From microbial characteristics, AD application was also shown to introduce *Synergistetes* bacteria into the growing medium, but the overall bacterial diversity and composition were similar to those of the control treatment. Contingent upon the optimal upgrading of biogas, the concept of a circular economy based on biochar and anaerobic digestion appears to be feasible.

Similarly, granular activated carbon (GAC) and carbon cloth (CC) play a role akin to that of BC. Dang et al. [68] studied the growth of both bacterial and archaeal species capable of interspecies electron exchange, stimulated by GAC and CC supplementation to AD treating dog food (a substitute for the dry and organic part of municipal solid waste (OFMSW)). Methane production (772–1428 mmol vs. <80 mmol), volatile solids removal (78%–81% vs. 54%–64%) and COD removal efficiencies ($80% vs. 20%–30%) were all significantly higher in reactors amended with GAC or CC, compared to those without supplementation. OFMSW degradation was also significantly accelerated and VFA concentrations were substantially lower in AD supplementation with GAC or CC. In terms of metagenomics insights, 16S rDNA sequencing analysis revealed a shift in the archaeal community composition (methane formation). Specifically, *Methanosarcina* proportion decreased by 17%, while the *Methanosaeta* proportion increased by 5.6%. This outcome suggested that a carbon dioxide reduction pathway was favored by the GC or CC addition, rather than the common acetate decarboxylation pathway for methane formation. Overall, these results proved that both GAC and CC improve methane production performance by favorably altering the microbial community composition, particularly towards archaeal dominance. This modification enhances methane production efficiency and overall digester performance, while also influencing the functional genes associated with direct interspecies electron transfer [69].

Furthermore, graphene nanomaterial and activated charcoal have been assessed to enhance AD performance in ethanol. For instance, Lin et al. [16] observed that the addition of graphene significantly (1.0 g/L) produced the highest biomethane yield (695.0 ± 9.1 mL/g) and production rate (95.7 ± 7.6 mL/g/d), corresponding to an enhancement of 25.0% in biomethane yield and 19.5% in production rate, compared to those without the supplementation. The ethanol degradation constant was accordingly improved by 29.1% in the presence of graphene. Moreover, microbial analyses revealed that the electrogenic bacteria (*Geobacter* and *Pseudomonas*) along with archaea (*Methanobacterium* and *Methanospirillum*) participate in direct interspecies electron transfer (DIET) that favors the methanogenesis step. Besides, theoretical calculations provided evidence that graphene-based DIET can sustain a much higher electron transfer flux than conventional hydrogen transfer. Nevertheless, from a biological perspective, graphene is toxic to the microbial community and is detrimental for microorganisms capable of DIET [70].

Owing to the toxic nature of graphene, the combination of bioaugmentation and activated charcoal enables a better reactor performance from a greener perspective. Zhao et al. [71] investigated the addition of activated charcoal and found out that the degradation rate of VFAs was significantly increased, thus accelerating the biogas production rate (104 ~ 371%). Moreover, activated charcoal mainly contributes to the microorganism's immobilization to form bio-activated charcoal, given its high surface area and large number of pores that enables microorganisms adsorption. The biofilms on the activated charcoal decrease the distance between the microbes and delay the washout effect of microorganisms, thus enabling a longer-lasting effect of bioaugmentation. On top of that, 16S-rRNA analysis shows that the addition of activated charcoal leads to the enrichment of syntrophic microbial guilds (syntrophic VFAs oxidizing bacteria and hydrogenotrophic methanogens). The results again similarly suggest that the addition of activated charcoal promotes interspecies electron transfer.

Bose et al. [72] and Kang et al. [73] studied the effect of single-walled carbon nanotubes at ambient temperatures to promote DIET among microbes and methanogens in the AD processes. It was highlighted that the application of the conductive single-walled carbon nanotubes (SWCNTs) led to extended periods, up to 180 hours, of high biomass concentration at 100 mg/L, with no significant impact on the methane formation rate. Further research on AD by the addition of SWCNTs as nanomaterials on glucose can promote thermophilic AD process. Existing results show that the methane formation rate in cases with SWCNTs as an additive in the supplemented reactor is 100% higher than in controlled reactors. Simonin and Richaume [74] demonstrated that multi-walled carbon nanotubes (MWCNTs) had a significant effect on methane formation rate at ambient temperatures. They showed the active role of conductive nanomaterials in biogas production rate and biomass conversion at 1500 mg/L concentration. Studies suggest that MWCNTs can be a good and effective promoter of direct interspecies electron transfer to enhance the biogas production rate.

To summarize, the above-mentioned carbon-based nanomaterials favor the growth and development of microorganisms by providing genial adaptable environmental conditions in anaerobic conditions. The function of various carbon-based nanomaterials has been further illustrated in Figure 2.4. These carbon-based nanomaterials offer an excellent immobilization matrix for microorganisms, which increases activities of microbes, electron transfer between anaerobes, and biogas generation, which has been further detailed in Table 2.3.

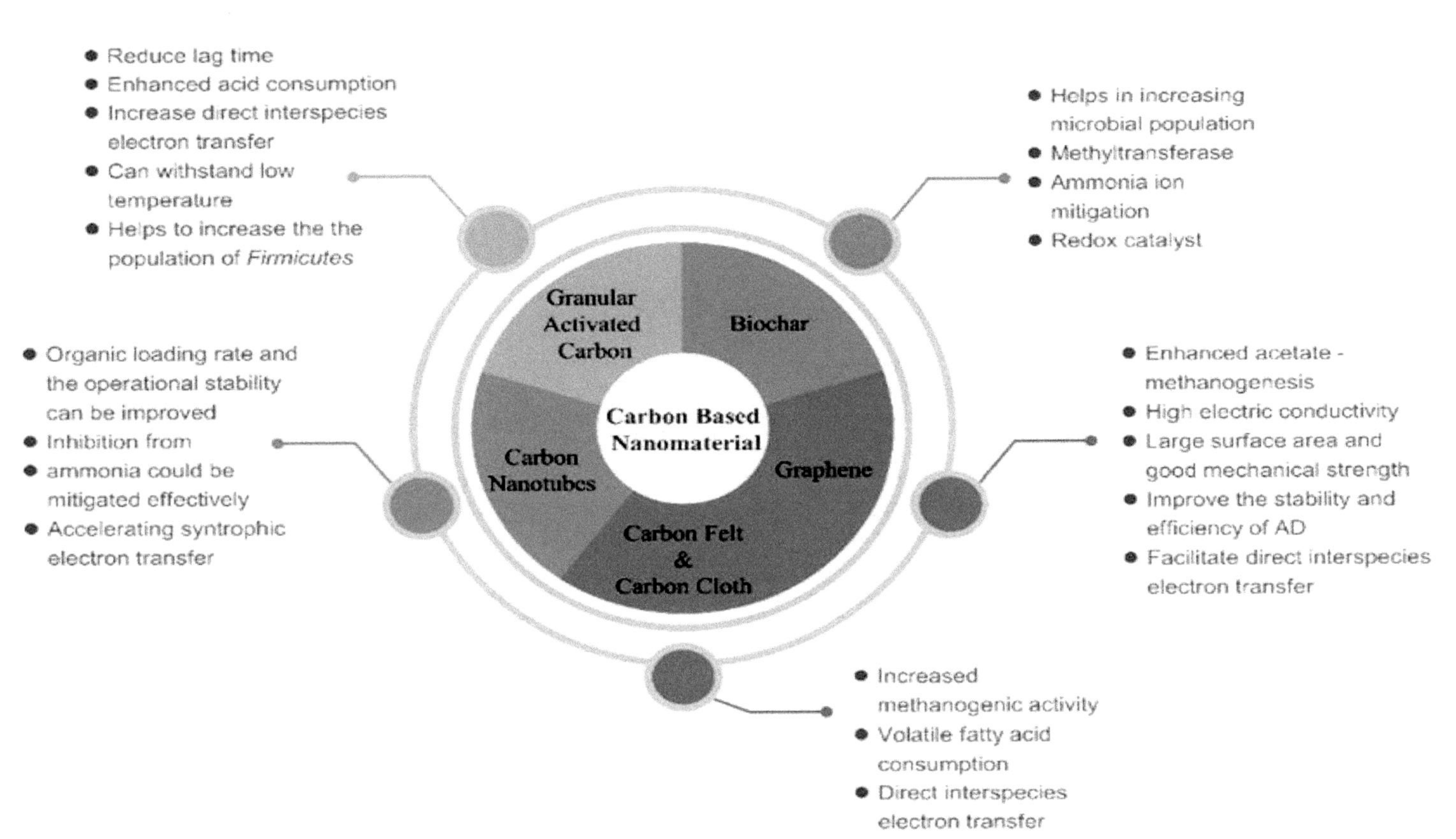

Figure 2.4 The function of various carbon-based nanomaterials in enhancing AD processes. Obtained from Jeyakumar and Vincent [75].

Table 2.3 List of the various applications of conductive carbon-based nanomaterials in the AD processes.

Nanomaterial	Mean Size (nm)	Concentration	Substrate Type	Effects on the AD Process	Ref.
Graphene	4–20	0.5–2 g/L	Ethanol	25% increase in methane yield and 19.5% increase in biogas production	[76]
	–	30–120	Glucose	Up to 51.4% increase in methane production rate	[77]
Fullerene (C_{60})	–	Up to 50,000 ng/Kg of biomass	Sucrose	No effect	[50]
Single-walled CNT	5–20 nm length 1–2 nm diameter	1000 mg/L	Sucrose	No impact on methanogenesis, 100% increase in methane production rate	[78]
	5–20 μm length 1–2 nm diameter	1000 mg/L	Glucose	92% increase in methane formation rate	[79]
Multi-walled CNT	–	1500 mg/L	Granular sludge	43% increase in methane production	[79]

2.4 CURRENT TRENDS AND PROSPECTS

Biogas (biomethanation) through AD is the most reliable waste-to-energy energy harvesting process. Microbial communities, including hydrolytic and fermentative bacteria, syntrophic bacteria, and methanogenic archaea, and their interspecies symbioses allow complex metabolisms for the volumetric reduction of organic waste in AD. Nevertheless, the biomass degradation time, hydrolysis step enhancement, and microbial activity remains to be the challenging parts in the AD process.

The catalytic role of nanomaterials is thus applied to enhance hydrolysis and methanogenesis in the overall biogas production through AD [80, 81]. Specifically, hydrolytic and fermentative bacteria, syntrophic bacteria, and methanogenic archaea associate via intricate symbiosis to allow the volumetric reduction and conversion of the organic matter of biomass into biogas (methane) during AD [81]. Organic waste, including polymers such as polysaccharides, proteins and lipids, and simple molecules including VFAs and ethanol from domestic, municipal, food processing, animal husbandry, and slaughterhouse units are cheap and abundant sources of substrates for AD [82–84]. In brief, the conversion efficiency of these organic resources into biomethane is relatively high (lipids: 94.8%; proteins: 71%; carbohydrates: 50.4%) compared to that of sewage sludge (40–50%) alone during AD, and contributes to about 69% of the atmospheric methane [85, 86]. As electron donors, organic waste matter facilitates DIET and maintains the electron pool during AD [87]. DIET can outcompete interspecies H_2/formate transfer in syntrophic methanogenic systems due to a higher electron transfer efficiency, via electrical conduits, thus minimizing the loss of intermediates. For instance, acetoclastic methanogens expresses syntrophy with acetogenic and exoelectrogenic syntrophs, signifying their dual capabilities for the oxidation of acetate and the reduction of carbon dioxide via DIET [88, 89]. Thus, the abundance of acetoclastic methanogens (>90%) outcompetes the population of hydrogenotrophs (<0.6%) in AD [90]. Insights into microbial communities and their surface area colonization, along with the stabilization of buffering mechanisms and the supply of sufficient nutrients, are crucial aspects enhanced by incorporating nanomaterials in the AD process. The impact on addressing substrate inhibition and optimizing organic matter degradation can be achieved through nanomaterials applications. Further investigation is warranted to understand the intricate interactions among microbes, nanomaterials, and the buffering capacity of AD, as well as the absorption impact of biomass content [91, 92].

Most of the work on AD process management has been narrowed and focused on pollution control and energy generation. The engineering feasibility, environmental sustainability and economic viability of nanomaterials make them the next technology for biogas production, i.e., researching towards achieving water recovery and energy production while minimizing solid discharge from the AD [10, 93]. At present, achieving economic and environmental stability is paramount in the development of AD technologies. This involves not only optimizing the efficiency and cost-effectiveness of AD processes but also ensuring minimal environmental impact, such as reducing GHGs and promoting resource recovery [94,95]. Reduction and recovery are the key to minimizing the environmental risk associated with nanomaterials. Most nanomaterials have recycling potential, decreasing the discharge of nanomaterials as such (in the form of effluents from industries) into the environment [96]. Hence, instead of using fresh nanomaterials for the AD process, used nanomaterials can be recovered and reused. The digestate, enriched with certain remediating nanomaterials like Zn and Cu, offers a unique opportunity for sustainable disposal practices. These nanomaterials can undergo remediation processes, such as Zn to ZnS conversion, rendering them suitable for reuse as agricultural fertilizers. In addition to this, production of nanomaterials using natural resources prevents the risk associated with effluent disposal [97].

In terms of the economic feasibility of using nanomaterials for biogas production, it mainly relies on the amount of biogas energy produced and revenue generated, along with the cost of nanomaterials utilized in the digestion process. For instance, Abdelwahab et al. [98] studied the effect of the usage of various nanomaterials (Ni, Fe and Fe_3O_4) in AD, and observed that Fe has a higher energy content of 403 kWh with a net profit of USD −676.5, but low net-energy (192.6 kWh) and net profit (7.2 USD) was achieved with the use of 5 mg/L of Fe nanomaterials. Additionally, Ni nanomaterials achieved a good net profit of USD 20.6 with 231 kWh energy at the dosage of 1 mg/L. Also, using 23.5 mg/L of NiO-TiO_2 nanomaterials produced 75.84 kWh of biogas. In this study, the cost of energy consumption was considered 0.0612 Euro/kWh, based on the price status of Denmark. Additionally, the consumable cost for TiO_2 nanomaterials was estimated as 2000 EUR/Ton. The net profit achieved was EUR 0.29, which was 7% higher than the control sample. In another study, Liu et al. [99] studied the usage of zero-valent metallic nanomaterials and Fe_3O_4 nanomaterials in AD. The economic evaluation confirmed that using Fe nanomaterials could save 272,400 USD/year and cut down carbon emissions by 1660 tCO_2/year, compared to the traditional AD process.

2.5 CONCLUSION

Further investigation on enhancing AD efficiency has been carried out by optimizing the parameters for enhancing microbial existence during the digestion process and improving the hydrolysis process. Nanomaterials in this regard would be a candidate to replace conventional materials or processes for more efficient sludge biodegradability. The usage of various nanomaterials in AD is witnessing a growing trend for sustainable and feasible applications in methane production. One of the major challenging facts here is the implementation of the process on an industrial pilot scale since degradation time, the catalytic role of nanomaterials, and microbial interactions with nanomaterials require further investigation. Combatting substrate inhibition, buffering stabilization, and microbial colonization area are highly focused during AD enhancement. Further research regarding the interactions between nanomaterials and microbes, buffering capacity, and the impact of biomass content is required. Hence, proper research is still necessary to estimate the syntrophic conversion of various substrates with nanomaterials in AD. A more thorough investigation is also important to validate the metal fractions and bioavailability in AD. Additional exploration of the synergistic effects of multi-nanomaterials on metal specifications, particularly the production of metal sulfides in sulfite-rich environments, is necessary. Impacts related to changes in particle properties of nano-additives with the same chemical components, including the depression degree and stability of nanomaterials in the liquid phase, remain as a gap in the literature. Overall, a series of studies are still to be performed in future to determine an optimal size range, optative morphologies, and the best dispersion solution, as well as appropriate pretreatment for each metallic nano-additive.

2.6 ACKNOWLEDGEMENTS

This work was supported by the Agency for Science, Technology and Research (A*STAR), Singapore, under the A*STAR Core – Central Funds (Project No: C2333017002, SIBER 2.0).

REFERENCES

[1] Tian, H., Yan, M., Zhou, J., Wu, Q., Tiong, Y.W., Lam, H.T., et al. 2023. A closed loop case study of decentralized food waste management: System performance and life cycle carbon emission assessment. Sci. Total Environ. 899.

[2] Wang, J. 2014. Decentralized biogas technology of anaerobic digestion and farm ecosystem: Opportunities and challenges. Front. Energy Res. 2: 1–12. https://doi.org/10.3389/fenrg.2014.00010.

[3] Markandan, K. and Chai, W.S. 2022. Perspectives on nanomaterials and nanotechnology for sustainable bioenergy generation. Materials. 15: 1–20. https://doi.org/10.3390/ma15217769.

[4] Tsapekos, P., Khoshnevisan, B., Alvarado-Morales, M., Zhu, X., Pan, J., Tian, H., et al. 2021.Upcycling the anaerobic digestion streams in a bioeconomy approach: A review. Renewable Sustainable Energy Rev. 151: 111635. https://doi.org/10.1016/j.rser.2021.111635.

[5] Yao, Y., Huang, G., An, C., Chen, X., Zhang, P., Xin, X., et al. 2020. Anaerobic digestion of livestock manure in cold regions: Technological advancements and global impacts. Renewable Sustainable Energy Rev. 119: 109494. https://doi.org/10.1016/j.rser.2019.109494.

[6] Parakh, S.K., Sharma, P., Tiong, Y.W. and Tong Y.W. 2023. Recent advances in anaerobic digestion of lignocellulosic resources toward enhancing biomethane production. Handbook of Biorefinery Research and Technology. V. Bisaria. 1–29.

[7] Zhu, X., Blanco, E., Bhatti, M. and Borrion, A. 2021. Impact of metallic nanoparticles on anaerobic digestion: A systematic review. Sci. Total Environ. 757. https://doi.org/10.1016/j.scitotenv.2020.143747.

[8] Rocha-Meneses, L., Hari, A., Inayat, A., Shanableh, A., Abdallah, M., Ghenai, C., et al. 2022. Application of nanomaterials in anaerobic digestion processes: A new strategy towards sustainable methane production. Biochem. Eng. J. 188: 1–10. https://doi.org/10.1016/j.bej.2022.108694.

[9] Sharma, P., Tiong, Y.W., Yan, M., Tian, H., Lam, H.T., Zhang, J., et al. 2023. Assessing *Stachytarpheta jamaicensis* (L.) Vahl growth response and rhizosphere microbial community structure after application of food waste anaerobic digestate as biofertilizer with renewable soil amendments. Biomass Bioenergy 178. https://doi.org/10.1016/j.biombioe.2023.106968.

[10] Yan, M., Tian, H., Song, S., Tan, H.T.W. and Lee, J.T.E. 2023. Effects of digestate-encapsulated biochar on plant growth, soil microbiome and nitrogen leaching. J. Environ. Manage. 334.

[11] Xu, S., Bu, J., Li, C., Tiong, Y.W., Sharma, P. and Liu, K., et al. 2024. Biochar enhanced methane yield on anaerobic digestion of shell waste and the synergistic effects of anaerobic co-digestion of shell and food waste. Fuel. 357.

[12] Wang, X., Song, X., Yuan, H., Li, X. and Zuo, X. 2021. Two-Step pretreatment of hydrothermal with ammonia for cow bedding: Pretreatment characteristics, anaerobic digestion performance and kinetic analysis. Waste Biomass Valorization. 12: 5675–5687. https://doi.org/10.1007/s12649-021-01395-0.

[13] Paudel, S.R., Banjara, S.P., Choi, O.K., Park, K.Y., Kim, Y.M. and Lee, J.W. 2017. Pretreatment of agricultural biomass for anaerobic digestion: Current state and challenges. Bioresour. Technol. 245: 1194–1205. https://doi.org/10.1016/j.biortech.2017.08.182.

[14] Aljbour, S.H., El-Hasan, T., Al-Hamiedeh, H., Hayek, B. and Abu-Samhadaneh K. 2021. Anaerobic co-digestion of domestic sewage sludge and food waste for biogas production: A decentralized integrated management of sludge in Jordan. J. Chem. Technol. Metall. 56: 1030–1038.

[15] Kesharwani, N. and Bajpai, S. 2021. Pilot scale anaerobic co-digestion at tropical ambient temperature of India: Digester performance and techno-economic assessment. Bioresour. Technol Reports. 15: 100715. https://doi.org/10.1016/j.biteb.2021.100715.

[16] Lin, R., Cheng, J., Zhang, J., Zhou, J., Cen, K. and Murphy, J.D. 2017. Boosting biomethane yield and production rate with graphene: The potential of direct interspecies electron transfer in anaerobic digestion. Bioresour. Technol. 239: 345–352. https://doi.org/10.1016/j.biortech.2017.05.017.

[17] Hassaneen, F.Y., Abdallah, M.S., Ahmed, N., Taha, M.M., Abd ElAziz, S.M.M., El-Mokhtar, M.A., et al. 2020. Innovative nanocomposite formulations for enhancing biogas and biofertilizers production from anaerobic digestion of organic waste. Bioresour. Technol. 309: 1–7. https://doi.org/10.1016/j.biortech.2020.123350.

[18] Dabirian, E., Hajipour, A., Mehrizi, A.A., Karaman, C., Karimi, F., Loke-Show, P., et al. 2023. Nanoparticles application on fuel production from biological resources: A review. Fuel. 331. https://doi.org/10.1016/j.fuel.2022.125682.

[19] Khan, A.U., Khan, M., Cho, M.H. and Khan, M.M. 2020. Selected nanotechnologies and nanostructures for drug delivery, nanomedicine and cure. Bioprocess. Biosyst. Eng. 43: 1339–1357. https://doi.org/10.1007/s00449-020-02330-8.

[20] Sadiku, M.N.O., Ashaolu, T.J. , Ajayi-Majebi, A. and Musa, S.M. 2021. Future of nanotechnology. Int. J. Sci. Adv. 2: 131–134. https://doi.org/10.51542/ijscia.v2i2.9.

[21] Choong, Y.Y., Norli, I., Abdullah, A.Z. and. Yhaya, M.F. 2016. Impacts of trace element supplementation on the performance of anaerobic digestion process: A critical review. Bioresour. Technol. 209: 369–379. https://doi.org/10.1016/j.biortech.2016.03.028.

[22] Zhao, W., Zhang, Y., Du, B., Wei, D., Wei, Q. and Zhao, Y. 2013. Enhancement effect of silver nanoparticles on fermentative biohydrogen production using mixed bacteria. Bioresour. Technol. 142: 240–245. https://doi.org/10.1016/j.biortech.2013.05.042.

[23] FitzGerald, J.A., Wall, D.M., Jackson, S.A., Murphy, J.D. and Dobson, A.D.W. 2019. Trace element supplementation is associated with increases in fermenting bacteria in biogas mono-digestion of grass silage. Renewable Energy. 138: 980–986. https://doi.org/10.1016/j.renene.2019.02.051.

[24] Li, H., Cui, F., Liu, Z. and Li, D. 2017. Transport, fate, and long-term impacts of metal oxide nanoparticles on the stability of an anaerobic methanogenic system with anaerobic granular sludge. Bioresour. Technol. 234: 448–455. https://doi.org/10.1016/j.biortech.2017.03.027.

[25] Yin, X., Wu, W., Maeke, M., Richter-Heitmann, T., Kulkarni, A.C., Oni, O.E., et al. 2019. CO_2 conversion to methane and biomass in obligate methylotrophic methanogens in marine sediments. ISME J. 13: 2107–2119. https://doi.org/10.1038/s41396-019-0425-9.

[26] Cheng, Q. and. Call, D.F 2016. Hardwiring microbes: Via direct interspecies electron transfer: Mechanisms and applications. Environ. Sci. Process. Impacts. 18: 968–980. https://doi.org/10.1039/c6em00219f.

[27] Jadhava, P., Muhammad, N., Bhuyar, P., Krishnan, S., Razak, A.S.A., Zularisam, A.W., et al. 2021. A review on the impact of conductive nanoparticles (CNPs) in anaerobic digestion: Applications and limitations. Environ. Technol. Innovation. 23. https://doi.org/10.1016/j.eti.2021.101526.

[28] Ye, W., Lu, J., Ye, J. and Zhou, Y. 2021. The effects and mechanisms of zero-valent iron on anaerobic digestion of solid waste: A mini-review. J. Cleaner Prod. 278: 123567. https://doi.org/10.1016/j.jclepro.2020.123567.

[29] Abdelsalam, E.M., Samer, M., Attia, Y.A., Abdel-Hadi, M.A., Hassan, H.E. and Badr, Y. 2017. Effects of Co and Ni nanoparticles on biogas and methane production from anaerobic digestion of slurry. Energy Convers. Manage. 141: 108–119. https://doi.org/10.1007/s12649-018-0374-y.

[30] Tsapekos, P., Alvarado-Morales, M., Tong, J. and Angelidaki, I. 2018. Nickel spiking to improve the methane yield of sewage sludge. Bioresour. Technol. 270: 732–737. https://doi.org/10.1016/j.biortech.2018.09.136.

[31] Abdelsalam, E., Samer, M., Attia, Y.A., Abdel-Hadi, M.A., Hassan, H.E. and Badr, Y. 2017. Influence of zero valent iron nanoparticles and magnetic iron oxide nanoparticles on biogas and methane production from anaerobic digestion of manure. Energy. 120: 842–853. https://doi.org/10.1016/j.energy.2016.11.137.

[32] Zhao, L., Ji, Y., Sun, P., Li, R., Xiang, F., Wang, H., et al. 2018. Effects of individual and complex ciprofloxacin, fullerene C60, and ZnO nanoparticles on sludge digestion: Methane production, metabolism, and microbial community. Bioresour. Technol. 267: 46–53. https://doi.org/10.1016/j.biortech.2018.07.024.

[33] Suanon, F., Sun, Q., Li, M., Cai, X., Zhang, Y., Yan, Y., et al. 2017. Application of nanoscale zero valent iron and iron powder during sludge anaerobic digestion: Impact on methane yield and pharmaceutical and personal care products degradation. J. Hazard. Mater. 321: 47–53. https://doi.org/10.1016/j.jhazmat.2016.08.076.

[34] Amen, T.W.M., Eljamal, O., Khalil, A.M.E. and Matsunaga, N. 2017. Biochemical methane potential enhancement of domestic sludge digestion by adding pristine iron nanoparticles and iron nanoparticles coated zeolite compositions. J. Environ. Chem. Eng. 5: 5002–5013. https://doi.org/10.1016/j.jece.2017.09.030.

[35] Gonzalez-Estrella, J., Sierra-Alvarez, R. and Field, J.A. 2013. Toxicity assessment of inorganic nanoparticles to acetoclastic and hydrogenotrophic methanogenic activity in anaerobic granular sludge. J. Hazard. Mater. 260: 278–285. https://doi.org/10.1016/j.jhazmat.2013.05.029.

[36] Wang, T., Zhang, D., Dai, L., Chen, Y. and Dai, X. 2016. Effects of metal nanoparticles on methane production from waste-activated sludge and microorganism community shift in anaerobic granular sludge. Sci. Rep. 6: 1–10. https://doi.org/10.1038/srep25857.

[37] Gitipour, A., Thiel, S.W., Scheckel, K.G. and Tolaymat, T. 2016. Anaerobic toxicity of cationic silver nanoparticles. Sci. Total Environ. 557–558: 363–368. https://doi.org/10.1016/j.scitotenv.2016.02.190.

[38] García, A., Delgado, L., Torà, J.A., Casals, E., González, E., Puntes, V., et al. 2012. Sánchez, effect of cerium dioxide, titanium dioxide, silver, and gold nanoparticles on the activity of microbial communities intended in wastewater treatment. J. Hazard. Mater. 199–200: 64–72. https://doi.org/10.1016/j.jhazmat.2011.10.057.

[39] Casals, E., Barrena, R., García, A., González, E., Delgado, L., Busquets-Fité, M., et al. 2014. Programmed iron oxide nanoparticles disintegration in anaerobic digesters boosts biogas production. Small. 10: 2801–2808. https://doi.org/10.1002/smll.201303703.

[40] Yazdani, M., Ebrahimi-Nik, M., Heidari, A. and Abbaspour-Fard, M.H. 2019. Improvement of biogas production from slaughterhouse wastewater using biosynthesized iron nanoparticles from water treatment sludge. Renewable Energy. 135: 496–501. https://doi.org/10.1016/j.renene.2018.12.019.

[41] Li, L., Lu, H., Tilley, D.R. and Qiu, G. 2015. Effect of time scale on accounting for renewable emergy in ecosystems located in humid and arid climates. Ecol. Modell. 315: 88–95. https://doi.org/10.1016/j.ecolmodel.2015.07.030.

[42] Zhong, D., Li, J., Ma, W. and Qian, F. 2020. Clarifying the synergetic effect of magnetite nanoparticles in the methane production process. Environ. Sci. Pollut. Res. 27: 17054–17062. https://doi.org/10.1007/s11356-020-07828-y.

[43] Lei, Y., Wei, L., Liu, T., Xiao, Y., Dang, Y., Sun, D., et al. 2018. Magnetite enhances anaerobic digestion and methanogenesis of fresh leachate from a municipal solid waste incineration plant. Chem. Eng. J. 348: 992–999. https://doi.org/10.1016/j.cej.2018.05.060.

[44] Wang, Y., Wang, D. and Fang, H. 2018. Comparison of enhancement of anaerobic digestion of waste activated sludge through adding nano-zero valent iron and zero valent iron. RSC Adv. 8: 27181–27190. https://doi.org/10.1039/c8ra05369c.

[45] Elreedy, A., Fujii, M., Koyama, M., Nakasaki, K. and Tawfik, A. 2019. Enhanced fermentative hydrogen production from industrial wastewater using mixed culture bacteria incorporated with iron, nickel, and zinc-based nanoparticles.Water Res. 151: 349–361. https://doi.org/10.1016/j.watres.2018.12.043.

[46] Mu, H., Zheng, X., Chen, Y., Chen, H. and Liu, K. 2012. Response of anaerobic granular sludge to a shock load of zinc oxide nanoparticles during biological wastewater treatment. Environ. Sci. Technol. 46: 5997–6003. https://doi.org/10.1021/es300616a.

[47] Engliman, N.S., Abdul, P.M., Wu, S.Y. and Jahim, J.M. 2017. Influence of iron (II) oxide nanoparticle on biohydrogen production in thermophilic mixed fermentation. Int. J. Hydrogen Energy. 42: 27482–27493. https://doi.org/10.1016/j.ijhydene.2017.05.224.

[48] Boer, J.L., Mulrooney, S.B. and Hausinger, R.P. 2014. Nickel-dependent metalloenzymes. Arch. Biochem. Biophys. 544: 142–152.

[49] Trifunović, D., Schuchmann, K. and Müller, V. 2016. Ethylene glycol metabolism in the acetogen Acetobacterium woodii. J. Bacteriol. 198: 1058–1065. https://doi.org/10.1128/JB.00942-15.

[50] Yang, Y., Zhang, Y., Li, Z., Zhao, Z., Quan, X. and Zhao, Z. 2017. Adding granular activated carbon into anaerobic sludge digestion to promote methane production and sludge decomposition. J. Cleaner Prod. 149: 1101–1108. https://doi.org/10.1016/j.jclepro.2017.02.156.

[51] Ali, A., Mahar, R.B., Soomro, R.A. and Sherazi, S.T.H. 2017. Fe_3O_4 nanoparticles facilitated anaerobic digestion of organic fraction of municipal solid waste for enhancement of methane production, Energy Sources, Part A: Recovery. Util. Environ. Eff. 39: 1815–1822. https://doi.org/10.1080/15567036.2017.1384866.

[52] Mustapha, N.A., Toya, S. and Maeda, T. 2020. Effect of Aso limonite on anaerobic digestion of waste sewage sludge. AMB Express. 10 (2020). https://doi.org/10.1186/s13568-020-01010-w.

[53] Kökdemir Ünşar, E. and Perendeci, N.A. 2018. What kind of effects do Fe_2O_3 and Al_2O_3 nanoparticles have on anaerobic digestion, inhibition or enhancement?. Chemosphere. 211: 726–735. https://doi.org/10.1016/j.chemosphere.2018.08.014.

[54] Olaya, W., Dilawar, H. and Eskicioglu, C. 2021. Comparative response of thermophilic and mesophilic sludge digesters to zinc oxide nanoparticles, Environ. Sci. Pollut. Res. 28: 24521–24534. https://doi.org/10.1007/s11356-020-09067-7.

[55] Zhang, J., Dong, Q., Liu, Y., Zhou, X. and Shi, H. 2016. Response to shock load of engineered nanoparticles in an activated sludge treatment system: Insight into microbial community succession. Chemosphere. 144: 1837–1844. https://doi.org/10.1016/j.chemosphere.2015.10.084.

[56] Zheng, X., Wu, L., Chen, Y., Su, Y., Wan, R., Liu, K., et al. 2015. Effects of titanium dioxide and zinc oxide nanoparticles on methane production from anaerobic co-digestion of primary and excess sludge. J. Environ. Sci. Health. Part A Toxic/Hazard. Subst. Environ. Eng. 50: 913–921. https://doi.org/10.1080/10934529.2015.1030279.

[57] Purnomo, D.M.J., Richter, F., Bonner, M., Vaidyanathan, R. and Rein, G. 2020. Role of optimisation method on kinetic inverse modelling of biomass pyrolysis at the microscale. Fuel. 262. https://doi.org/10.1016/j.fuel.2019.116251.

[58] Adekunle, K.F. and Okolie, J.A. 2015. A review of biochemical process of anaerobic digestion. Adv. Biosci. Biotechno. 06: 205–212. https://doi.org/10.4236/abb.2015.63020.

[59] Zhang, J. Zhao, W., Zhang, H., Wang, Z., Fan, C. and Zang, L. 2018. Recent achievements in enhancing anaerobic digestion with carbon- based functional materials. Bioresour. Technol. 266: 555–567. https://doi.org/10.1016/j.biortech.2018.07.076.

[60] Park, J.H., Kang, H.J., Park, K.H. and Park, H.D. 2018. Direct interspecies electron transfer via conductive materials: A perspective for anaerobic digestion applications. Bioresour. Technol. 254: 300–311. https://doi.org/10.1016/j.biortech.2018.01.095.

[61] Zhang, J., Cui, Y., Zhang, T., Hu, Q., Wah Tong, Y., He, Y., et al. 2021. Food waste treating by biochar-assisted high-solid anaerobic digestion coupled with steam gasification: Enhanced bioenergy generation and porous biochar production. Bioresour. Technol. 331: 125051. https://doi.org/10.1016/j.biortech.2021.125051.

[62] Zhang, L., Lim, E.Y., Loh, K.C., Ok, Y.S., Lee, J.T.E., Shen, Y., et al. 2020. Biochar enhanced thermophilic anaerobic digestion of food waste: Focusing on biochar particle size, microbial community analysis and pilot-scale application. Energy Convers. Manage. 209: 112654. https://doi.org/10.1016/j.enconman.2020.112654.

[63] Zhou, H., Brown, R.C. and Wen, Z. 2020. Biochar as an additive in anaerobic digestion of municipal sludge: Biochar properties and their effects on the digestion performance. ACS Sustainable Chem. Eng. 8: 6391–6401. https://doi.org/10.1021/acssuschemeng.0c00571.

[64] Lü, F., Luo, C., Shao, L. and He, P. 2016. Biochar alleviates combined stress of ammonium and acids by firstly enriching Methanosaeta and then methanosarcina. Water Res. 90: 34–43. https://doi.org/10.1016/j.watres.2015.12.029.

[65] Fagbohungbe, M.O., Herbert, B.M.J., Hurst, L., Ibeto, C.N., Li, H., Usmani, S.Q., et al. 2017. The challenges of anaerobic digestion and the role of biochar in optimizing anaerobic digestion. Waste Manage. 61: 236–249. https://doi.org/10.1016/j.wasman.2016.11.028.

[66] Song, S., Lim, J.W., Lee, J.T.E., Cheong, J.C., Hoy, S.H., Hu, Q., et al. 2021. Food-waste anaerobic digestate as a fertilizer: The agronomic properties of untreated digestate and biochar-filtered digestate residue. Waste Manage. 136: 143–152. https://doi.org/10.1016/j.wasman.2021.10.011.

[67] Lee, J.T.E., Ok, Y.S., Song, S., Dissanayake, P.D., Tian, H., Tio, Z.K., et al. 2021. Biochar utilisation in the anaerobic digestion of food waste for the creation of a circular economy via biogas upgrading and digestate treatment. Bioresour. Technol. 333 (2021) 125190. https://doi.org/10.1016/j.biortech.2021.125190.

[68] Dang, Y., Sun, D., Woodard, T.L., Wang, L.Y., Nevin, K.P. and Holmes, D.E. 2017. Stimulation of the anaerobic digestion of the dry organic fraction of municipal solid waste (OFMSW) with carbon-based conductive materials. Bioresour. Technol. 238: 30–38. https://doi.org/10.1016/j.biortech.2017.04.021.

[69] Park, J.H., Park, J.H., Je Seong, H., Sul, W.J., Jin, K.H. and Park, H.D. 2018. Metagenomic insight into methanogenic reactors promoting direct interspecies electron transfer via granular activated carbon. Bioresour. Technol. 259: 414–422. https://doi.org/10.1016/j.biortech.2018.03.050.

[70] Qi, Q., Sun, C., Zhang, J., He, Y. and Wah Tong, Y. 2021. Internal enhancement mechanism of biochar with graphene structure in anaerobic digestion: The bioavailability of trace elements and potential direct interspecies electron transfer. Chem. Eng. J. 406: 126833. https://doi.org/10.1016/j.cej.2020.126833.

[71] Zhao, J., Li, Y. and Euverink, G.J.W. 2022. Effect of bioaugmentation combined with activated charcoal on the mitigation of volatile fatty acids inhibition during anaerobic digestion. Chem. Eng. J. 428. https://doi.org/10.1016/j.cej.2021.131015.

[72] Bose, S., Kuila, T., Mishra, A.K., Rajasekar, R., Kim, N.H. and Lee, J.H. 2012. Carbon-based nanostructured materials and their composites as supercapacitor electrodes. J. Mater. Chem. 22: 767–784. https://doi.org/10.1039/c1jm14468e.

[73] Kang, S., Pinault, M., Pfefferle, L.D. amd Elimelech, M. 2007. Single-walled carbon nanotubes exhibit strong antimicrobial activity. Langmuir. 23: 8670–8673. https://doi.org/10.1021/la701067r.

[74] Simonin, M. and Richaume, A. 2015. Impact of engineered nanoparticles on the activity, abundance, and diversity of soil microbial communities: A review. Environ. Sci. Pollut. Res. 22: 13710–13723. https://doi.org/10.1007/s11356-015-4171-x.

[75] Jeyakumar, R.B. and Vincent, G.S. 2022. Recent advances and perspectives of nanotechnology in anaerobic digestion: A new paradigm towards sludge biodegradability. Sustainability (Switzerland). 14. https://doi.org/10.3390/su14127191.

[76] Li, L., Geng, S., Li, Z. and Song, K. 2020. Effect of microplastic on anaerobic digestion of wasted activated sludge. Chemosphere. 247. https://doi.org/10.1016/j.chemosphere.2020.125874.

[77] Liu, Y., Li, Y., Gan, R., Jia, H., Yong, X., Yong, Y.C., et al. 2020. Enhanced biogas production from swine manure anaerobic digestion via in-situ formed graphene in electromethanogenesis system. Chem. Eng. J. 389. https://doi.org/10.1016/j.cej.2020.124510.

[78] Ambuchi, J.J., Zhang, Z., Shan, L., Liang, D., Zhang, P. and Feng, Y. 2017. Response of anaerobic granular sludge to iron oxide nanoparticles and multi-wall carbon nanotubes during beet sugar industrial wastewater treatment. Water Res. 117: 87–94. https://doi.org/10.1016/j.watres.2017.03.050.

[79] Gil, A., Siles, J.A., Serrano, A., Chica, A.F. and Martin, M.A. 2019. Effect of variation in the C/[N1P] ratio on anaerobic digestion. Environ. Prog. Sustainable Energy 38: 228–236. https://doi.org/10.1002/ep.

[80] Lovley, D.R. 2017. Happy together: Microbial communities that hook up to swap electrons. ISME J. 11: 327–336. https://doi.org/10.1038/ismej.2016.136.

[81] Saha, S., Basak, B., Hwang, J.H., Salama, E.S., Chatterjee, P.K. and Jeon, B.H. 2020. Microbial symbiosis: A network towards biomethanation.Trends Microbiol. 28: 968–984. https://doi.org/10.1016/j.tim.2020.03.012.

[82] Mendoza, Á., Morales, V., Sánchez-Bayo, A., Rodríguez-Escudero, R., González-Fernández, C., Bautista, L.F., et al. 2020. The effect of the lipid extraction method used in biodiesel production on the integrated recovery of biodiesel and biogas from Nannochloropsis gaditana, Isochrysis galbana and Arthrospira platensis. Biochem. Eng. J. 154: 107428. https://doi.org/10.1016/j.bej.2019.107428.

[83] Lee, J.T.E., Khan, M.U., Dai, Y., Tong, Y.W. and Ahring, B.K. 2021. Influence of wet oxidation pretreatment with hydrogen peroxide and addition of clarified manure on anaerobic digestion of oil palm empty fruit bunches. Bioresour. Technol. 332: 125033. https://doi.org/10.1016/j.biortech.2021.125033.

[84] Song, Y., Liu, J., Chen, M., Zheng, J., Gui, S. and Wei, Y. 2021. Application of mixture design to optimize organic composition of carbohydrate, protein, and lipid on dry anaerobic digestion of OFMSW: Aiming stability and efficiency. Biochem. Eng. J. 172: 108037. https://doi.org/10.1016/j.bej.2021.108037.

[85] Alibardi, L. and Cossu, R. 2016. Effects of carbohydrate, protein and lipid content of organic waste on hydrogen production and fermentation products. Waste Manage. 47: 69–77. https://doi.org/10.1016/j.wasman.2015.07.049.

[86] Davidson, T.A., Audet, J., Jeppesen, E., Landkildehus, F., Lauridsen, T.L., Søndergaard, M., et al. Synergy between nutrients and warming enhances methane ebullition from experimental lakes. Nat. Clim. Change. 8 (2018) 156–160. https://doi.org/10.1038/s41558-017-0063-z.

[87] Kucek, L.A., Xu, J., Nguyen, M. and Angenent, L.T. Waste conversion into n-caprylate and n-caproate: Resource recovery from wine lees using anaerobic reactor microbiomes and in-line extraction. Front. Microbiol. 7: 1–14. https://doi.org/10.3389/fmicb.2016.01892.

[88] Rotaru, A.E., Shrestha, P.M., Liu, F., Shrestha, M., Shrestha, D., Embree, M. 2014. A new model for electron flow during anaerobic digestion: Direct interspecies electron transfer to methanosaeta for the reduction of carbon dioxide to methane. Energy Environ. Sci. 7: 408–415. https://doi.org/10.1039/c3ee42189a.

[89] Saha, S., Jeon, B.H., Kurade, M.B., Govindwar, S.P., Chatterjee, P.K., Oh, S.E., et al. 2019. Interspecies microbial nexus facilitated methanation of polysaccharidic wastes. Bioresour. Technol. 289. https://doi.org/10.1016/j.biortech.2019.121638.

[90] Zhao, Z., Zhang, Y., Wang, L. and Quan, X. 2015. Potential for direct interspecies electron transfer in an electric-anaerobic system to increase methane production from sludge digestion. Sci. Rep. 5: 1–12. https://doi.org/10.1038/srep11094.

[91] Batstone, D.J. and Virdis, B. 2014. The role of anaerobic digestion in the emerging energy economy. Curr. Opin. Biotechnol. 27: 142–149. https://doi.org/10.1016/j.copbio.2014.01.013.

[92] Uhlenhut, F., Schlüter, K. and Gallert, C. 2018. Wet biowaste digestion: ADM1 model improvement by implementation of known genera and activity of propionate oxidizing bacteria. Water Res. 129: 384–393. https://doi.org/10.1016/j.watres.2017.11.012.

[93] Ma, Y. and Liu, Y. 2019. Turning food waste to energy and resources towards a great environmental and economic sustainability: An innovative integrated biological approach. Biotechnol. Adv. 37: 107414. https://doi.org/10.1016/j.biotechadv.2019.06.013.

[94] Huang, J., Xiao, J., Chen, M., Cao, C., Yan, C., Ma, Y., et al. 2019. Fate of silver nanoparticles in constructed wetlands and its influence on performance and microbiome in the ecosystems after a 450-day exposure. Bioresour. Technol. 281: 107–117. https://doi.org/10.1016/j.biortech.2019.02.013.

[95] Tiong, Y.W., Sharma, P., Tian, H., Tsui, T.-H., Lam, H.T. and Tong, Y.W. 2023. Startup performance and microbial communities of a decentralized anaerobic digestion of food waste. Chemosphere. 318. https://doi.org/https://doi.org/10.1016/j.chemosphere.2023.137937.

[96] Gupta, R. and Xie, H. 2018. Nanoparticles in daily life: Applications, toxicity and regulations. J. Environ. Pathol Toxicol. Oncol. 37: 209–230. https://doi.org/10.1615/JEnvironPatholToxicolOncol.2018026009.Nanoparticles.

[97] Tetteh, E.K., Amo-Duodu, G. and Rathilal, S. 2021. Synergistic effects of magnetic nanomaterials on post-digestate for biogas production. Molecules. 26. https://doi.org/10.3390/molecules26216434.

[98] Abdelwahab, T.A.M., Mohanty, M.K., Sahoo, P.K. and Behera, D. 2020. Impact of iron nanoparticles on biogas production and effluent chemical composition from anaerobic digestion of cattle manure. Biomass Convers. Biorefin. 5583–5595. https://doi.org/10.1007/s13399-020-00985-7.

[99] Liu, Y., Zhang, Y. and Ni, B.J. 2015. Zero valent iron simultaneously enhances methane production and sulfate reduction in anaerobic granular sludge reactors. Water Res. 75: 292–300. https://doi.org/10.1016/j.watres.2015.02.056.

Chapter **3**

Biomass-based Composite Materials: Processing, Properties and Applications

Vishnuvarthanan Mayakrishnan*[1], Ramji Vaidhyanathan[2] and Ragavanantham Shanmugam[3]

[1]Department of Mechanical Engineering, Kalasalingam Academy of Research and Education, Krishnankovil, Srivilliputtur 626126, Tamil Nadu, India

[2]Department of Civil Engineering, College of Engineering, Guindy, Anna University, Chennai, Tamil Nadu, India

[3]Department of Engineering Technology, College of Science and Technology, Fairmont State University, Fairmont, WV-26554, USA

3.1 INTRODUCTION

In the consumer world today, packaging plays a very significant and vital role in protecting and improving the shelf life of products. Without packaging, material handling would be very tough and ineffective. It can be considered as a coordinated system for the product, designed for efficient delivery, maintaining high quality from the producer to the end user, through goods management, transportation, etc. In the modern world, the growth of the consumer market is based on the innovations and the efficiency of the packaging sector. In the current industrialized societies, packaging plays an important role in various fields, particularly automotive, medical, food, electronics, etc.

*For Correspondence: vishnuvarthanan.india@gmail.com

Among the above mentioned domains, food industry is the largest user of packaging materials. The main purpose of food packaging is to protect the food in an effective manner that satisfies both the industries and the consumer's requirements, while also maintaining the safety of the food and minimizing the negative impact on the environment. There are many types of packaging materials which are used in the food packaging sector. In general, the common materials used for food packaging are paper, metal and glass. However, in the modern era, plastics have become the material of choice for food packaging applications because of their low cost and functional advantages over traditional materials.

The main aim of food packaging materials is to protect the food from contamination and improve the quality and safety of food during storage, transportation and distribution [1]. Nowadays, most food packaging materials are made up of conventional synthetic non-biodegradable petroleum-based plastic materials because of their availability, cost effectiveness, light weight and easy formability [2]. These food packaging materials are non-recyclable and toxic, with high environmental and human health risks [3]. Hence, it is a need of the hour to prepare active and sustainable food packaging materials as innovative replacements with eco-friendly and enhanced functional properties [4]. At present, the renewable and eco-friendly materials which likely satisfy such requirements are biomass-based materials and composites [5], synthetic bio-based polymers and composites [6] and they have been used as materials for food packaging applications. Biomass materials such as polysaccharides and proteins, and synthetic bio-based polymers such as polylactic acid (PLA), polyhydroxybutyrate (PHB) and polyhyroxyalkonates (PHA) have been widely used. The advantages of using these types of materials in food packaging applications are their environmentally friendliness, availability, renewability, biodegradability and elimination of human health risks [7].

We need food packaging materials with multifunctional characteristics such as mechanical, barrier, optical and antibacterial activities, oxygen scavenging, observing of spoilage and freshness [8]. In this respect, biomass-based films have attracted interest for use as food packaging materials and have been used for eco-friendly food packaging. However, their use has been restricted because of their inadequate/limited properties compared to conventional petroleum-based plastic materials [9]. Their properties can be improved by adding reinforcing compounds and forming the composites. But most of the reinforced materials yield poor polymer matrix-filler interactions which tend to improve the properties with reduced filler dimensions (nanomaterials). Various nanomaterials such as nanoclays, metal oxide nanoparticles and metallic particles have also been explored. The use of fillers with at least one dimension in the nanoscale produces nanocomposites. A proper dispersion of nanoparticles in the polymer matrix produces a very good matrix and filler interfacial area, which changes the mobility of the molecule, the relaxation behaviour and the thermal, barrier and mechanical properties of the material. Fillers with high aspect ratios are particularly interesting because of their high surface area, and provide better reinforcing effects.

The incorporation of antimicrobial materials into food packaging materials has received significant attention recently [10]. Films with antimicrobial properties can help restrict the growth of pathogenic and spoilage microorganisms. The antimicrobial properties contributed by antimicrobial agents impregnating nanocomposites and films are desirable because of the suitable structural integrity and the barrier and mechanical properties imparted by the nanocomposite matrix. Materials in the nanoscale have a higher surface to volume ratio, compared to their microscale counterparts. This allows nanomaterials to bind to more copies of biological molecules, resulting in greater efficiencies. Nanomaterials have been studied for antimicrobial activity and can be used as killing agents, growth inhibitors or antibiotic carriers. The combination of nanomaterial with the antimicrobial effect of materials has given rise to a new generation of materials.

This chapter presents the comprehensive summary of the preparation and properties of multifunctional biomass-based bio nanocomposite films for active food packaging applications. The case study of Carrageenan/MMTK10/AgNPs bio nanocomposite films and their properties have also been discussed herein.

3.2 BIOMASS-BASED BIO NANOCOMPOSITE FILMS

3.2.1 Carrageenan-based Bio Nanocomposite Films

Carrageenan is a naturally occurring anionic sulfated linear polysaccharide and is extracted from the red seaweed which belongs to the family of Rhodophyceae [11]. The word carrageenan originated from the inhabitants of Caraghen on the coast of South Ireland, where the red algae extracts were used in foods and medicines. The major constituents of carrageenan are co-polysaccharides with β-d-galactose and 3, 6-anhydro-α-galactose, with density variation in the sulfated group. To extract carrageenan, the raw materials used are *Gigartina stellata, Iridaea spp, Chondrus crispus, Kappaphycus spp* and *Eucheuma spp.* The carbohydrate residues in carrageenan commonly exist as uronic acids, glucose and xylose. The disaccharides vary by sulfated units; in commercial carrageenan, the sulfate content is 22–28% by weight. Other cations such as calcium, potassium, magnesium, ammonium and sodium are also present in the form of galactose esters, which are also soluble fibers. Several types of carrageenan exist, with slight variation in properties and chemical structures. The most common carrageenan of the highest commercial interest are kappa, iota and lambda. In the FDA, carrageenan falls under the list of materials for consumption and topical applications and "Generally Recognized as Safe" (GRAS). Carrageenan can be classified based on the position and the amount of sulphate groups. food sectors, kappa carrageenan is normally used because of its excellent physical and functional properties such as gelling, stabilizing ability, thickening and emulsifying. It has also been used to enhance the quality of cheese, puddings and dairy sweets and can also be used as binders and stabilizers in the meat manufacturing sectors and for the production of sausages, patties and low-calorie sandwiches. In addition, it may be used as an oxygen barrier to resist lipid oxidation in meat products, pet foods, water-based foods, baby foods and supplement drinks.

3.2.2 Pectin-based Bio Nanocomposite Films

Pectin is a family of oligosaccharides and polysaccharides with the most complex macromolecule structure in nature [12]. The major sources for pectin are citrus peels and apple pomace. In pectin, galacturonic acid is present in very high amounts; therefore, the FAO and the EU have fixed that pectin must contain at least 65% of galacturonic acid for use in various applications. Pectin may form a gel structure when galacturonan is cross linked to form a three-dimensional crystalline network in which solutes and water are trapped. The gelling properties of pectin depends on several factors such as the pectin type, degree of methyl esterification, degree of acetylation, pH, sugar, calcium and temperature. Pectin is an important ingredient used without any limitation in food industries and is "Generally Recognized as Safe" (GRAS) by the FDA. It has been used in the food industry mainly as a stabilizing, gelling and thickening agent in many products like yoghurt drinks, jams, ice creams and fruity milk drinks. One of the important features of pectin films is their ability to carry and release a mixture of active compounds like flavourings, antibrowning, antioxidants and antimicrobial compounds. There are many antimicrobial substances which have been incorporated with pectin in order to obtain antimicrobial active packaging that extends the shelf life of the product and reduces the growth of microorganisms.

3.2.3 Chitosan-based Bio Nanocomposite Films

Chitosan is a polysaccharide composed mainly of β-(1,4)-linked 2-deoxy-2-amino-D-glucopyranose units and is the deacetylated product of chitin [13]. The good structure of chitosan is distinct due to the overall or bulk content of D-hexosamine residues and also their distribution along the biopolymer chain. Chitosan and chitin are the second most abundant natural biopolymers, next to cellulose. Chitosan is well known for its biocompatible, non- toxic and good biodegradable properties. Similarly, its inherent antimicrobial properties have made it a viable material for food packaging. However, the use of chitosan in food packaging applications has been limited due to its low strength, moisture and gas barrier properties that are not enough to make it congregate. Moreover, the limited performance and high cost of biopolymers restrict their competitiveness compared to conventional plastic materials. In order to enhance the properties of biopolymers and to extend their possible applications, composites are produced. For this, chitosan is blended with gelatin and silver nanoparticles and the properties have been studied with respect to other food packaging applications [14]. The nanocomposite film prepared by blending chitosan with titanium dioxide is used as an active food packaging material [15].

3.2.4 Cellulose-based Bio Nanocomposite Films

Cellulose-based packaging materials are widely used because of their abundance, renewable character, biocompatibility, biodegradability and other specific properties [16]. In particular, nanocellulose has gained much attention in the past few years because of its commendable capabilities in various applications such as food packaging, biomedical, fibers, adhesives and automotive applications. Nanocellulose is a biodegradable, renewable and abundant nanomaterial. It has various outstanding properties such as specific strength, thermal stability and effortless chemical functionalization. Nanocellulose is currently being used as a coating or film in bio-based food packaging applications because of its low cost, non-toxicity, biodegradability, renewability and high barrier property against gases. Nanocellulose film can be a good alternative packaging material, albeit with some limitations in mechanical, water vapour barrier, wettability and antibacterial properties. Several efforts have been made to resolve these limitations by blending in cellulose or coating them on cellulose films to increase the usage of nanocellulose in food packaging applications. Various research works like those that study the effect of dextran-coated silver nanoparticles in nanocellulose have investigated their barrier and antimicrobial properties for food packaging applications [17] and extended the ground meat shelf life through chitosan and nanocellulose biocomposites [18].

3.3 CASE STUDY

In this study, we discuss about the development of montmorillonite (MMTK10) and silver nanoparticles (AgNPs) loaded carrageenan nanocomposite coatings coated on oxygen plasma surface modified polypropylene (PP) films, and the effect of nanocomposite coatings on surface, mechanical, barrier and optical properties. The detailed procedure and the results of the oxygen plasma treatment of PP films are also discussed [19]. The AgNPs were green synthesized by *Digitalis purpurea* plant and the characterization and properties were investigated [20]. The carrageenan and MMTK10 used in the study were pre-dried in a hot air oven for about 6 hrs. The MMTK10 with different wt% 1, 3, 5 and 7 was dispersed separately in 100 mL of double distilled water. The solutions were stirred using magnetic stirrers for 12 hours and were subsequently homogenized for 15 minutes at 5000 rpm to get a completely swollen clay mineral. The amount of carrageenan used for the preparation of nanocomposites was optimized. 3 grams of carrageenan powder with 0.5 mL of glycerol and the prepared clay solution with different wt% was mixed separately with 20μg of AgNPs solution and heated for about 30 minutes at 60°C, vigorously mixed using a magnetic stirrer. Then, the solutions were sonicated for 45 minutes to avoid agglomeration. The prepared carrageenan nanocomposites with different wt% 0, 1, 3, 5 and 7 MMTK10 nanoclay were designated as CK0, CK1, CK3, CK5 and CK7 respectively. The prepared nanocomposites were coated on the surface modified PP films by K coater. The carrageenan-based nanocomposite solutions were taken in a syringe and poured on the oxygen plasma treated polypropylene of dimensions 15 cm × 10 cm. The coated PP films were dried at room temperature for 24 hours. The average coating thickness obtained was about 24 μm, depending on the specification of the selected rod for coating.

The contact angle results are shown in Figure 3.1. The contact angle of the surface modified PP film is 60°. After the carrageenan coating, it reduced to 48.1°, due to the hydrophilic nature of carrageenan. It may be mainly attributed to the functional hydrophilic carboxyl and hydroxyl groups [21]. After the addition of 1 wt% MMTK10, the contact angle value reduced to 46.3°. After the incorporation of AgNPs, the values increased to 47.2° for K10. After the addition of 3 wt% of MMTK10, the value further reduced to 45.8°. There was a minimal reduction of contact angle by the addition of 5 wt% nanoclay.

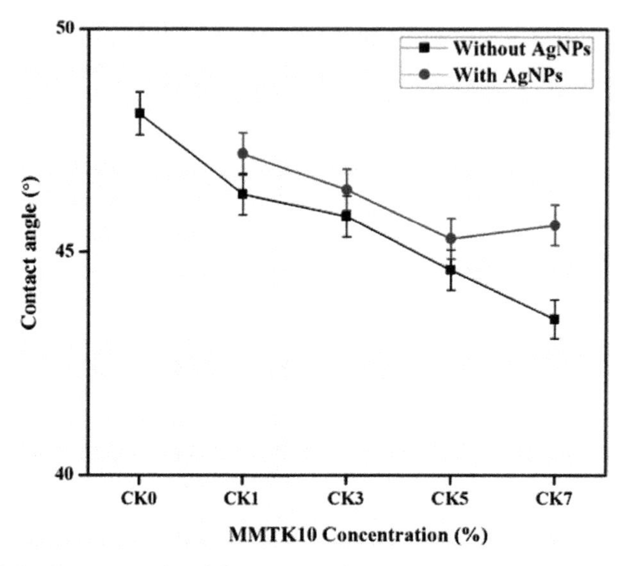

Figure 3.1 Contact angle of Carrageenan/MMTK10/AgNPs bio nanocomposite coated films.

In both 3 wt% and 5 wt%, the value increased after the addition of AgNPs. There was a decrease in the contact angle value by increasing the loading of MMTK10 nanoclay content in carrageenan. The decrease in the contact angle value was due to the hydrophilic nature of MMTK10. Thereafter, the values increased after the addition of AgNPs with the nanoclay. This result indicates the influence of AgNPs on the hydrophobicity of the coated films. It can be further inferred that, the carrageenan nanocomposite coating with AgNPs have a relatively higher contact angle compared to the coated film without AgNPs [22].

The tensile strength of the oxygen plasma surface modified PP coated with carrageenan-based nanocomposites is shown in Figure 3.2. The mechanical properties of biopolymer coatings tend to rely on the synthetic substrate [23]. The tensile strength of the uncoated PP film is 4.680 MPa. After the carrageenan coating, the value increased to 5.2 MPa. The incorporation of MMTK10 at 1 wt%, the tensile strength was found to be 7.5 MPa. There was a significant increase in tensile strength after the addition of AgNPs and the value was found to be 10.9 MPa. In the case of 3 wt% MMTK10, the tensile strength was about 7.9 MPa. After mixing AgNPs, the value was found to be 11.4 MPa. The maximum tensile strength value was observed with the incorporation of AgNPs with 5% of MMTK10.

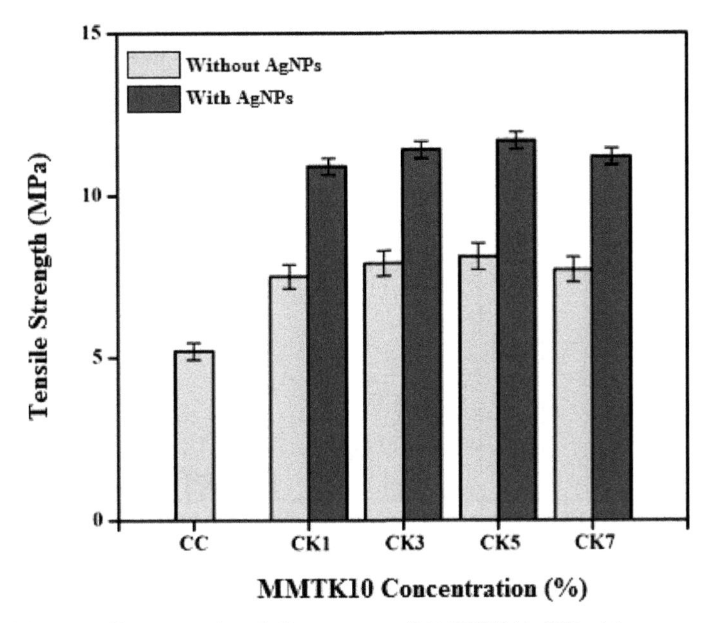

Figure 3.2 Tensile strength of Carrageenan/MMTK10/AgNPs bio nanocomposite coated films.

The highest tensile strength was noticed in the case of MMTK10 because of the large surface area. These results indicate that, with an increase in the concentration of nanoclay, the tensile strength also increases. With further increase in the nanoclay concentration, the tensile strength decreases. This is because of the agglomeration of the nanoclay in the polymer matrix, which leads to the reduction in the reinforcement action of nanofillers [24]. Generally, the increase in tensile strength is due to the exfoliated structure formation and a possible strain-induced alignment of nanoparticles in the polymer matrix. The increase in the tensile strength of the coated composite films is mainly because of the physical attraction between the components.

The packaging may undergo physical, chemical and biological deterioration once it interacts with oxygen and water. The quality loss of the packaged food is due to the passage of gaseous molecules. To maintain the quality of the packaged food, the oxygen barrier property of the coated film is very crucial in determining suitability for various applications [25]. The oxygen transmission rate (OTR) of the uncoated oxygen plasma treated polypropylene film and the carrageenan nanocomposite coated films are shown in Figure 3.3. The oxygen transmission rate of the uncoated oxygen plasma treated film was 1853.32 cc/m^2.day.atm. After the carrageenan coating, the OTR significantly decreased to 1499.81 cc/m^2. day.atm. By the addition of 1 wt% of MMTK10 with carrageenan, the OTR decreased to 1034.88 cc/m^2.day.atm. After the incorporation of AgNPs with 1 wt% MMTK10, the OTR was 1002.83 cc/m^2.day.atm. The OTR decreased to 592.96 cc/m^2.day atm for 3 wt% MMTK10 clay and it further decreased to 574.71 cc/m^2.day.atm after the incorporation of AgNPs.

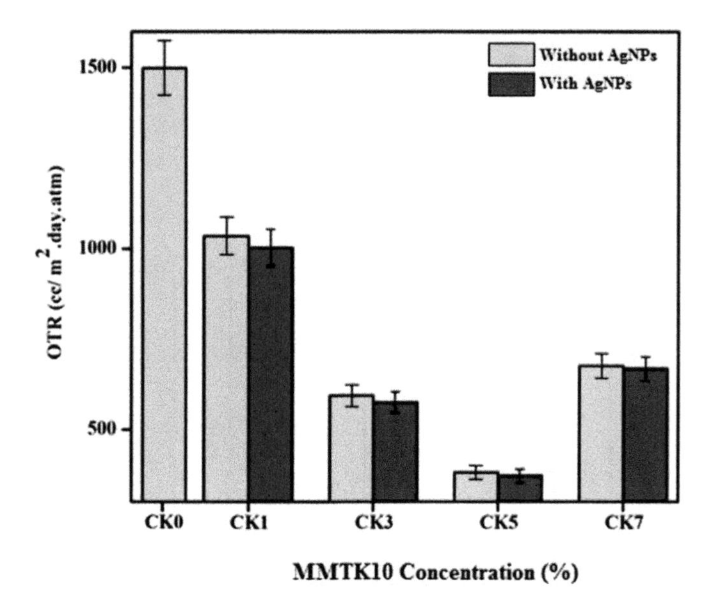

Figure 3.3 OTR of Carrageenan/MMTK10/AgNPs bio nanocomposite coated films.

It has been reported that the incorporation of nanoclay or metal nanoparticles in the polymeric matrices reduces gas permeability [26]. These materials have the ability to act as a barrier for oxygen and control the transport of oxygen to extend the shelf life of food. The OTR continuously decreased after the addition of nanoclay and AgNPs to carrageenan. There was a large reduction in OTR after the addition of MMTK10 to carrageenan, possibly due to the existence of a more tortuous path in the diffusing molecules, that can go around the impenetrable platelets. The formation of a tortuous path in carrageenan was due to the presence of MMTK10, which decreases the permeation of oxygen. After the incorporation of AgNPs, further decrease in OTR values compared to MMTK10-incorporated carrageenan was noticed. It was because of the shape and the structure of the nanofillers – layered plate- like structured clay minerals and the spherical shape of the AgNPs. The dispersion of clay mineral in the polymer matrix with AgNPs may also develop further efficient tortuous paths for oxygen transmission. In comparison, the highest reduction of OTR was observed at 5 wt% of MMTK10 and the value was 380.03 cc/m^2.day.atm. The measured value of OTR for 7 wt% of MMTK10 was about 674.59 cc/m^2.day.atm and the increase in value may be due to the agglomeration of MMTK10 in the polymer matrix. As was noticed, the barrier properties are high for MMT (~100) because of a high aspect ratio [27]. At lower % levels, the MMTK10 was well dispersed in the carrageenan matrix and can increase the tortuosity of the diffusion path, leading to a significant improvement in the barrier properties. The aspect ratio of layered silicates is very high, compared to the other fillers used for the preparation of nanocomposites. The effective aspect ratio of the layered silicate is reduced because of agglomeration and decrease in the length of the diffusion path.

To increase the shelf life of food, one of the main functions of a packaging material is to decrease the transfer of moisture between the food and the surrounding atmosphere, the water vapour transmission rate should be as low as possible [28]. Water vapour transmission rate (WVTR) of the uncoated oxygen plasma treated PP and the carrageenan nanocomposites coated PP films are shown in Figure 3.4. The WVTR of the uncoated PP film was 9.62 g/m^2/day; after the coating of carrageenan, decreased to 7.65 g/m^2/day. With the addition of 1 wt% of MMTK10 in carrageenan, the value of WVTR was 5.22 g/m^2/day. After the incorporation of AgNPs, the values of WVTR further decreased to 5.19 g/m^2 day for MMTK10. The decrease in WVTR was due to the effect of nanoclay and nanosilver. In the case of 3 wt% MMTK10, the WVTR values further decreased to 2.98 g/m^2/day. After the incorporation of AgNPs, it was 2.90 g/m^2/day. The decrease in the permeability of the coated films was due to the presence of nanoparticles on the surface of the film in uniform dispersion. MMT and other layered silicates have the ability to improve the water vapour barrier for synthetic as well as biopolymer-based nanocomposites.

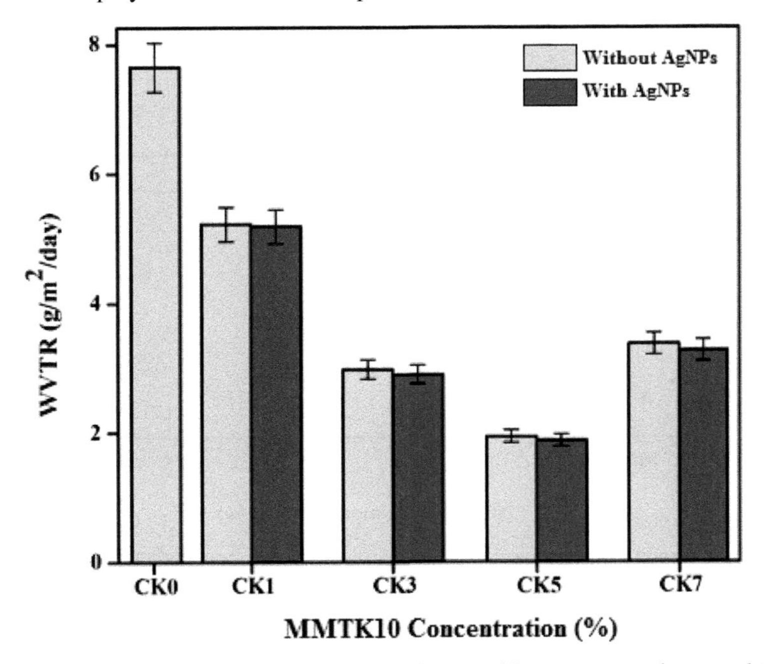

Figure 3.4 WVTR of Carrageenan/MMTK10/AgNPs bio nanocomposite coated films.

An efficient reduction of water vapour values was obtained for the carrageenan with 5 wt% MMTK10 and the value was 1.94 g/m^2/day. After the addition of AgNPs, the value of WVTR decreased and it was 1.88 g/m^2/day. It was found that there was a significant increase in WVTR with increase in clay content for 5 wt% of MMTK10. It was inferred that, similar to the OTR results, the increase in WVTR could be attributed to the aggregated structure of carrageenan nanocomposites rather than the exfoliated structure. Here, the materials acted as a discontinuous barrier for the diffusion of water vapour in the film matrix and thus the path length of the diffusion decreased. This results in an increase in the water vapour transmission rate. This in turn causes agglomeration of clays and nanoparticles and may present channels for water vapour to pass more readily [29].

For packaging applications, opacity is one of the most important parameters since it affects product appearance. Opacity is an indication of the amount of light not allowed to travel through the material and low transparency indicates high opacity. The opacity values of oxygen plasma treated PP and the carrageenan nanocomposites coated PP films are shown in Figure 3.5. The opacity of the uncoated PP is 9.44%. But in the case of carrageenan nanocomposite coated PP, the value of opacity was higher because of the presence of light blockage particles.

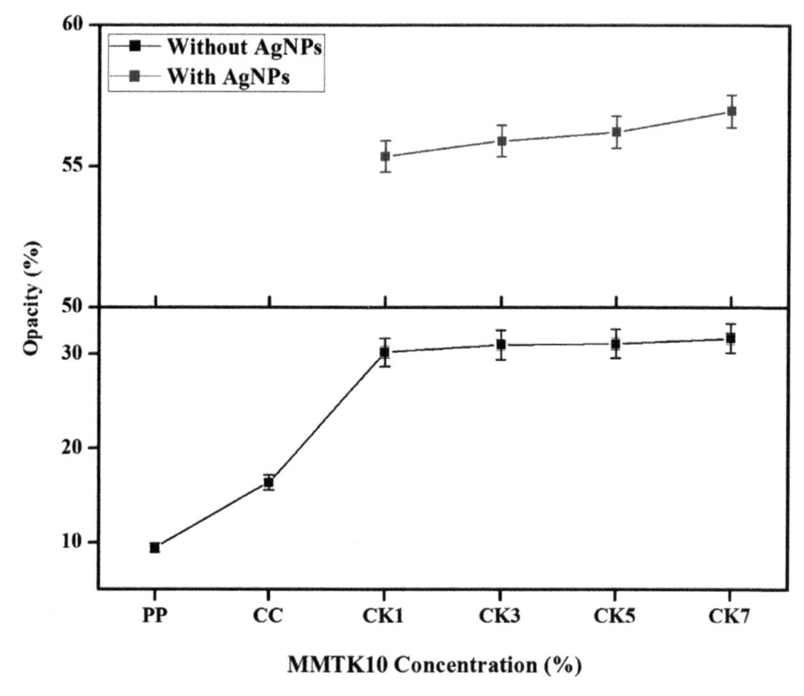

Figure 3.5 Opacity of Carrageenan/MMTK10/AgNPs bio nanocomposite coated films.

After the carrageenan coating, the opacity value increased to 16.34%. The incorporation of MMTK10 in carrageenan leads to an increase in the value of opacity. The opacity value for 1 wt% K10 was 30.21%. It was noted that, the value increased to 30.98% for 3 wt% K10 and the maximum value of 31.68% was observed at 7 wt%. This shows that the values of opacity of the coated films are directly proportional to the concentration of K10. The opacity gradually increased after the increase of MMTK10 content. The increase of clay content in the carrageenan coating showed the higher value of opacity; it was due to the strong scattering of light. After the incorporation of AgNPs with MMTK10 in carrageenan, the value of opacity increased to 55.34%. It was noticed that the value of opacity observed was higher after the loading of AgNPs in the carrageenan-based coating, compared to the clay- loaded carrageenan. Increase in the value of opacity was dependent on the type of nanofillers used. The AgNPs-containing carrageenan coating showed a very high value of opacity. This was because of the prevention of light transmission in the polymer matrix by the dispersion of nanoparticles [30]. It may be well understood that the well dispersed clay platelets in the polymer matrix of about 1nm thickness do not obstruct the passage of light because of the size of clay platelets being smaller than the wavelength of visible light. Based on these results, the film is desirable for use in food packaging, due to the prevention of light penetration, which averts light-driven reactions such as discolouration, lipid oxidation, off flavour formation and nutrient loss [31].

3.4 CONCLUSION

Renewable biomass-based bio nanocomposite films can be extensively used in active food packaging applications. They can overcome the environmental problems caused by synthetic plastic materials and provide an alternative to conventional food packaging materials. Particularly, the addition of functional nanomaterials with biomass-based materials can enhance their important characteristics such as surface, optical, barrier and mechanical properties. It can extend the shelf life of food and ensure food quality. Hence, biomass-based bio nanocomposite films can be proposed to contribute to the improvement of efficiency of environmentally sustainable food packaging films with high multifunctional properties.

REFERENCES

[1] Han, J.W., Ruiz-Garcia, L., Qian, J.P. and Yang, X.T. 2018. Food packaging: A comprehensive review and future trends. Compr. Rev. Food Sci. Food Saf. 17(4): 860–877.

[2] Zhao, X., Katrina, C. and Yael V. 2020. Narrowing the gap for bioplastic use in food packaging: an update. Enviro. Sci. Technol. 54(8): 4712–4732.

[3] Rochman, C.M., Mark, A.B., Benjamin, S.H., Brian, T.H., Eunha, H., Hrissi, K.K., et al. 2013. Classify plastic waste as hazardous. Nature. 494(7436): 169–171.

[4] Wang, H., Jun, Q. and Fuyuan D. 2018. Emerging chitosan-based films for food packaging applications. J. Agric. Food. Chem. 66(2): 395–413.

[5] Deng, J., En-Qing, Z., Gao-Feng, X., Nithesh, N., Vignesh, M., Ming-Guo, M., et al. 2022. Overview of renewable polysaccharide-based composites for biodegradable food packaging applications. Green Chem. 24(2): 480–492.

[6] Asgher, M., Sarmad, A.Q., Muhammad, B. and Hafiz, M.N. Iqbal. 2020. Bio-based active food packaging materials: Sustainable alternative to conventional petrochemical-based packaging materials. Food Res. Int. 137: 109625.

[7] Atta, O.M., Sehrish, M., Ajmal, S., Mazhar, Ul-I., Muhammad, W.U. and Guang Y. 2022. Biobased materials for active food packaging: A review. Food Hydrocolloids. 125: 107419.

[8] Chen, W., Shaobo, M., Qiankun, W., David, J.M., Xuebo, L., To N., et al. 2022. Fortification of edible films with bioactive agents: A review of their formation, properties, and application in food preservation. Crit. Rev. Food Sci. Nutr. 62(18): 5029–5055.

[9] Kadzińska, J., Monika, J., Stanisław, K., Joanna, B. and Andzej L. 2019. An overview of fruit and vegetable edible packaging materials. Packag. Technol. Sci. 32(10): 483–495.

[10] Ramji, V. and Vishnuvarthanan, M. 2022a. Influence of NiO supported silica nanoparticles on mechanical, barrier, optical and antibacterial properties of polylactic acid (PLA) bio nanocomposite films for food packaging applications. Silicon. 14(2): 531–538.

[11] Ramji, V. and Vishnuvarthanan, M. 2022b. Chitosan ternary bio nanocomposite films incorporated with MMT K10 nanoclay and spirulina. Silicon. 14(3): 1209–1220.

[12] Vishnuvarthanan, M. and Rajeswari, N. 2019a. Preparation and characterization of carrageenan/silver nanoparticles/Laponite nanocomposite coating on oxygen plasma surface modified polypropylene for food packaging. J. Food Sci. Technol. 56: 2545–2552.

[13] Vishnuvarthanan, M. and Rajeswari, N. 2019b. Food packaging: pectin–laponite–Ag nanoparticle bionanocomposite coated on polypropylene shows low O_2 transmission, low Ag migration and high antimicrobial activity. Environ. Chem. Lett. 17: 439–445.

[14] Kumar, S., Shukla, A., Baul, P.P., Mitra, A. and Halder, D. 2018. Biodegradable hybrid nanocomposites of chitosan/gelatin and silver nanoparticles for active food packaging applications. Food Packag. Shelf Life. 16: 178–184.

[15] Siripatrawan, U. and Kaewklin, P. 2018. Fabrication and characterization of chitosan-titanium dioxide nanocomposite film as ethylene scavenging and antimicrobial active food packaging. Food Hydrocolloids. 84: 125–134.

[16] Shanmugam, R., Mayakrishnan, V., Kesavan, R., Shanmugam, K., Veeramani, S. and Ilangovan, R. 2022. Mechanical, barrier, adhesion and antibacterial properties of pullulan/graphene bio nanocomposite coating on spray coated nanocellulose film for food packaging applications. J. Polym. Environ. 1–9.

[17] Lazić, V., Vivod, V., Peršin, Z., Stoiljković, M., Ratnayake, I.S., Ahrenkiel, P.S., et al. 2020. Dextran-coated silver nanoparticles for improved barrier and controlled antimicrobial properties of nanocellulose films used in food packaging. Food Packag. Shelf Life. 26: 100575.

[18] Dehnad, D., Mirzaei, H., Emam-Djomeh, Z., Jafari, S.M. and Dadashi, S. 2014. Thermal and antimicrobial properties of chitosan–nanocellulose films for extending shelf life of ground meat. Carbohydr. Polym. 109: 148–154.

[19] Vishnuvarthanan, M. and Rajeswari, N. 2015. Effect of mechanical, barrier and adhesion properties on oxygen plasma surface modified PP. Innovative Food Sci. Emerg. Technol. 30: 119–126.

[20] Vishnuvarthanan, M. and Rajeswari, N. 2017. Plant mediated greener approach for synthesis of silver nanoparticles from Digitalis purpurea plant and its antibacterial activity. Int. J. Nanopart. 9(3): 166–179.

[21] Shankar, S. and Rhim, J.W. 2017. Preparation and characterization of agar/lignin/silver nanoparticles composite films with ultraviolet light barrier and antibacterial properties. Food Hydrocolloids. 71: 76–84.

[22] Rhim, J.W. and Wang, L.F. 2014. Preparation and characterization of carrageenan-based nanocomposite films reinforced with clay mineral and silver nanoparticles. Appl. Clay Sci. 97: 174–181.

[23] Khwaldia, K., Arab-Tehrany, E. and Desobry, S. 2010. Biopolymer coatings on paper packaging materials. Compr. Rev. Food Sci. Food Saf. 9(1): 82–91

[24] Swain, S., Sharma, R.A., Bhattacharya, S. and Chaudhary, L. 2013. Effects of nano-silica/nano-alumina on mechanical and physical properties of polyurethane composites and coatings. Trans. Electr. Electron. Mater. 14(1): 1–8.

[25] Vishnuvarthanan, M., Dharunya, R., Jayashree, S., Karpagam, B. and Sowndharya, R. 2019. Environment-friendly packaging material: banana fiber/cowdung composite paperboard. Environ. Chem. Lett. 17: 1429–1434.

[26] Casariego, A.B.W.S., Souza, B.W.S., Cerqueira, M.A., Teixeira, J.A., Cruz, L., Díaz, R., et al. 2009. Chitosan/clay films' properties as affected by biopolymer and clay micro/nanoparticles' concentrations. Food Hydrocolloids. 23(7): 1895–1902.

[27] Tang, X. and Alavi, S. 2012. Structure and physical properties of starch/poly vinyl alcohol/laponite RD nanocomposite films. J. Agric. Food. Chem. 60(8): 1954–1962.

[28] Alves, V.D., Mali, S., Beléia, A. and Grossmann, M.V.E. 2007. Effect of glycerol and amylose enrichment on cassava starch film properties. J. Food Eng. 78(3): 941–946.

[29] Johansson, C. and Clegg, F. 2015. Effect of clay type on dispersion and barrier properties of hydrophobically modified poly (vinyl alcohol)–bentonite nanocomposites. J. Appl. Polym. Sci. 132(28).

[30] Kanmani, P. and Rhim, J.W. 2014. Physical, mechanical and antimicrobial properties of gelatin based active nanocomposite films containing AgNPs and nanoclay. Food Hydrocolloids. 35: 644–652.

[31] Ramos, Ó.L., Reinas, I., Silva, S.I., Fernandes, J.C., Cerqueira, M.A., Pereira, R.N., et al. 2013. Effect of whey protein purity and glycerol content upon physical properties of edible films manufactured therefrom. Food hydrocolloids. 30(1): 110–122.

Chapter **4**

Nanomaterials in Microalgae for the Sustainable Production of Bioactive Compounds

Angela Paul Peter
School of Engineering, Faculty of Innovation and Technology,
Taylor's University Lakeside Campus
Email: angela.peter@taylors.edu.my

4.1 INTRODUCTION

The use of nanotechnology is widespread globally in the twenty-first century. Numerous industries, including health care, biotechnology, national security, information technology, and nanoelectronics, have used nanotechnology for a broad range of purposes. For instance, nano-sorbents have a wide range of features, such as a considerable capacity for sorption, that make them effective and powerful for the bioremediation industry's need to clean wastewater. Numerous research projects have been engaged to discover the optimum bio-sorbent capable of effectively removing heavy metals [1]. Moreover, wastewater treatment has made use of nano-silver (Ag) particles that naturally have an antimicrobial impact. It might be possible to clean flowing water of germs like *Escherichia coli* and *Enterococcus faecalis* by attaching nano-Ag particles to a specially developed blotting paper [2]. Similarly, the highly reactive nano zero-valent iron (nZVI) has been helping clean wastewater. The extremely small nZVI with a large specific surface area has significant impact such as the precipitation of contaminants, enhanced absorption that promotes the removal of a wide spectrum of pollutants from wastewater, via adsorption, reduction, precipitation, and oxidation in the presence of dissolved oxygen [2]. There are two techniques of synthesising nanoparticles: top-down approaches and bottom-up approaches that combine biological, chemical, and physical processes as shown in Figure 4.1 [3].

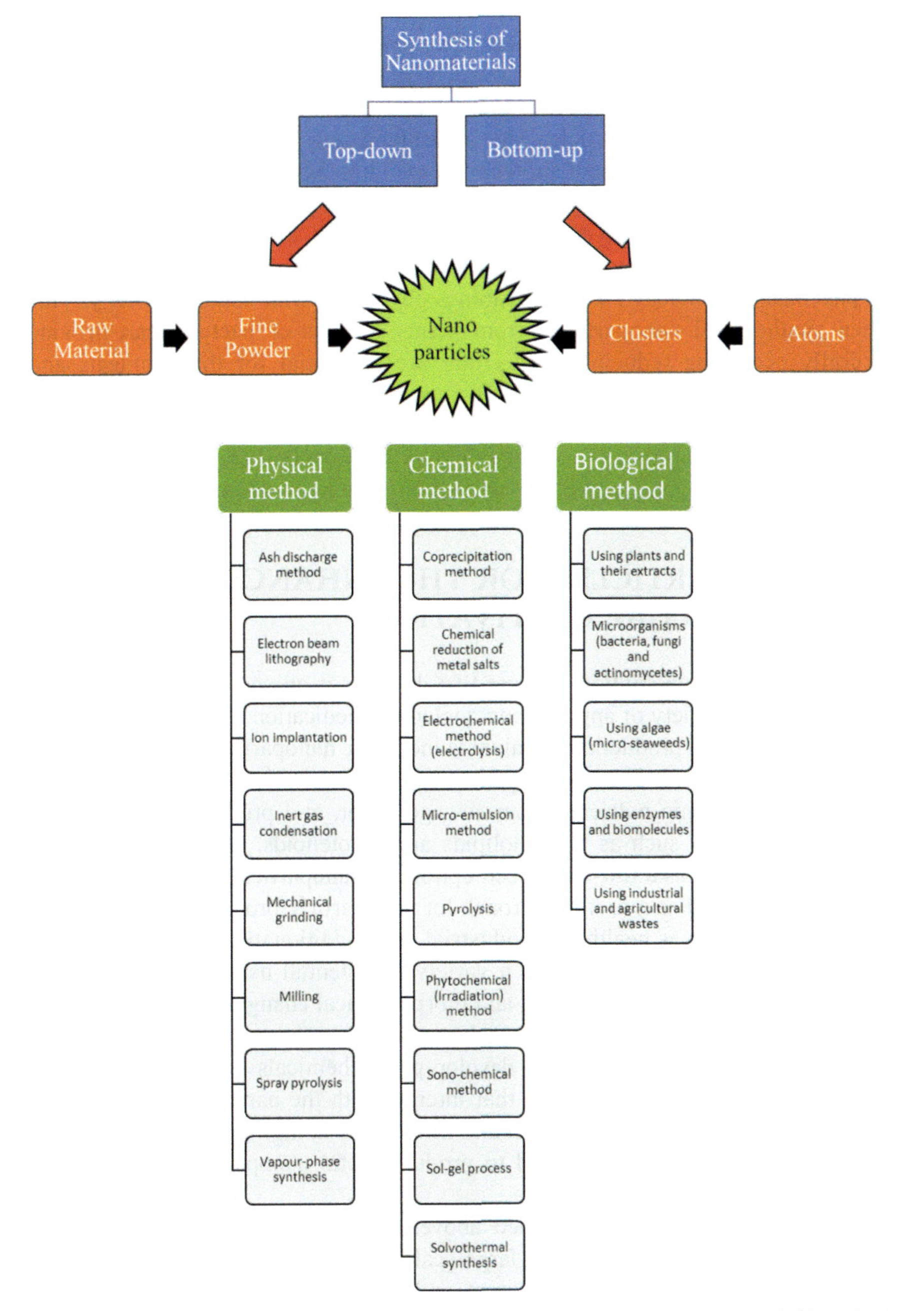

Figure 4.1 Nanoparticle synthesis approaches for physical, chemical, and biological synthesis methods.

Nanoparticle synthesis processes are constantly met with several drawbacks due to their harmful impacts on the environment, lengthy manufacturing processes, prohibitive costs, and the chemical and physical approaches employed [4, 5]. The high surface energy and reactive characteristics of conventionally manufactured nanoparticles are generally unstable in their natural state. On the other hand, a more cost-effective and ecologically friendly method of addressing a variety of environmental contamination issues is through the green synthesis of nanoparticles. As a result, a new method for synthesising nanoparticles, named 'phyconanotechnology', has been developed. The term 'Phytonanotechnology' refers to the production of nanomaterials using plant biomolecules, whereas 'myconanotechnology' refers to the mycosynthesis of inorganic nanoparticles by fungus. Parallel to this, there is the developing field of 'phyconanotechnology', which combines phycology with nanotechnology. This green synthesis of nanoparticles from plants, fungus or bacteria can provide naturally occurring reducing agents that are beneficial for the synthesis and amalgamation of nanomaterials [6]. By combining phycochemical components with contemporary technology, it is possible to generate nanoscale particles and materials. Combining these two methods may improve conversion and minimize the toxic effects of generated nanoparticles on the environment.

4.2 NANOPARTICLES FOR THE ENHANCEMENT OF MICROALGAE CULTIVATION

Microalgae have been studied for over two decades as an ecologically acceptable alternative for a variety of applications, including medication delivery, wastewater purification, and the production of valuable inorganic nanoparticles [7]. Microalgae provide a cost-effective solution for reducing carbon dioxide emissions, removing inorganic and organic nutrients from sewage water, and producing value-added bioactive chemicals such as phospholipids and carotenoids. Microalgae has also been recommended as a low-cost green option for nanoparticle manufacturing [8]. Algae are suggested as a useful approach for nanoparticle production for a variety of applications such as healthcare, industrial, etc. Additionally, it is proposed that various species of microalgae have a variety of potential uses in the creation of metal nanoparticles [9]. The growth and morphological changes of microalgae can be impacted and interfered with by a large concentration of heavy metals [10]. Microalgae, on the other hand, can develop phycochemicals such metal chelating agents and reactive oxygen species that interact with the nanosized metal nuclei. The interaction between metal nuclei and microalgae at high metal concentrations have proved to be a useful method to produce metal nanoparticles for use in a variety of applications.

Besides the alternative described above, significant research is required to enhance microalgae culture for the large-scale commercial generation of biofuels by raising their productivity, lipid content, and consumption efficiencies of CO_2 and light [11, 12]. To promote photosynthesis and algae development, there is growing interest in using functional nanomaterials to enhance CO_2 adsorption and light conversion efficiencies in algal culture. Nevertheless, the effect of nanomaterials on the growth of algae are dependent on the quantities and distinctive characteristics of the nanomaterial (such as size, crystal structure, and oxidation state), culture medium, and algae species. The effects of nanoparticles on microalgae, at low and high concentrations, are shown in Figure 4.2.

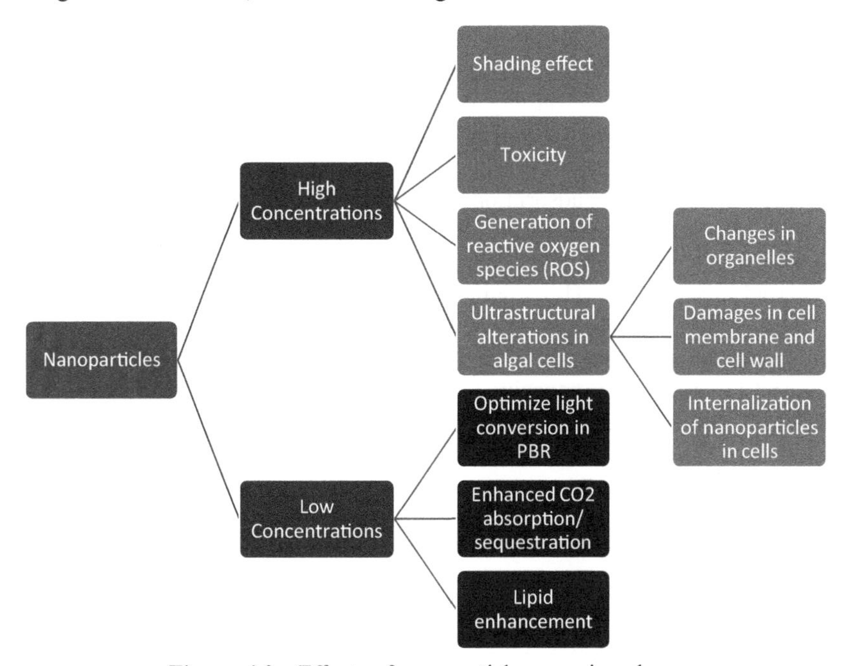

Figure 4.2 Effects of nanoparticles on microalgae.

4.2.1 Metallic Nanoparticles as Micronutrients

As micronutrients, trace metals are essential for the development of microalgae. Their effectiveness is determined by their concentrations in the culture medium and their beneficial or detrimental influence on other environmental conditions [13]. According to Pádrová et al., adding a trace quantity of nZVI (1.7 to 5.1 mg L1) can boost the growth of green algae (*Desmodesmus subspicatus, Dunaliella salina, Parachlorella kessleri*, and *Raphidocelis subcapitata*) and *eustigmatophycean* algae (*Nannochloropsis limnetica* and *Trachydiscus minutus*) [13]. Moreover, Kadar et al. demonstrated that while exposed to uncoated and coated nZVI powder, *Tetraselims suecica* and *Pavlova lutheri's* lipid levels increased by 41.9% and 46.34%, respectively [14]. Cao et al. discovered that the optimal $FeCl_{36}H_2O$ concentration for the greatest lipid content and growth rate of *Chlorella minutissima* was between 0.05 and 0.1 mM [15]. Most studies indicate that increasing iron concentrations can enhance algal growth rate and lipid content, although concentrations over 0.002 g/L and 0.001 g/L have negative effects on biomass and lipid synthesis, respectively, due to inhibitory effects [13].

4.2.2 CO_2 Supply from Nanoparticles

Microalgae can absorb and convert atmospheric CO_2 into oxygen and biomass, during photosynthesis. While considering the importance of CO_2 as a carbon source in green algae culture, improving CO_2 bio-fixation can boost algal productivity while reducing atmospheric CO_2 [16]. To retain the CO_2 gas in the culture, nanostructured adsorbents can be used to adsorb CO_2 at an acidic pH value and increase algal development by managing the availability of CO_2 in the culture medium via an adsorption/desorption cycle. Moreover, the CO_2 adsorption capacity and selectivity of adsorbents may be improved by adding certain heteroatoms such as nitrogen, oxygen and sulphur to the nanostructured surface to provide basic sites [17]. It has been observed that amine impregnation on zeolite enhances CO_2 adsorption by up to 4.44 mmol/g [18]. According to recent research, nanostructured adsorbents have more CO_2 collecting capability and reusability through several adsorption/desorption cycles, compared to other common adsorbents [19]. This might be due to their large specific surface area and functionality, which can provide more accessible CO_2 adsorption sites. Carbonization of pinecone shells at 650°C and subsequent activation by KOH increases CO_2 adsorption capacity to 7.63 mmol/g and 2.35 mmol/g at 0°C at 1 bar and 0.15 bar pressure, respectively, due to increased specific surface area and porosity [20].

4.2.3 Nanoparticles Usage for During Algal Photosynthesis

The removal of light in undesirable wavelength ranges can reduce inadequate and wasteful lighting, hence increasing the chlorophyll content of green algae. A study found that replacing white light with blue light increases chlorophyll production in algae [21]. Certain light, at specified wavelengths and intensities, can be utilised by photoactive pigments in algae, while others act as photo-inhibitors. As a result, filtering of light by enhancing beneficial light with specified wavelengths and eliminating photo-inhibitor light can considerably enhance photosynthetic efficiency and algal biomass output. Torkamani et al. discovered that by adding silver nanoparticles to filter the absorbed light under localizing surface plasmon resonance (LSPR), the photoactivity and growth of *Chlamydomonas reinhardtii* (green alga) and *Cyanothece 51142* (green-blue alga) were increased [21]. Adding nanoparticles to the algal growth medium has the advantage of ensuring that light is efficiently distributed to the algae cells. Adding silica nanoparticles increases the chlorophyll concentration and subsequent development of *Chlamydomonas reinhardtii* by uniformly illuminating the algae cells [22]. Nanoparticles can be employed to disperse local strong incident light in a culture because of their fluid nature in the culture, facilitating flexible and inadequate backscattering [11].

4.3 NANOTECHNOLOGY APPLICATIONS IN MICROALGAE HARVESTING

For the commercial production of microalgae, harvesting technology is essential. The physiognomic structure, cell density, algal size, ultimate moisture content, and reusability of the growth medium plays a significant role in determining the best method for harvesting algae [23]. Centrifugation, flocculation with coagulants, precipitation with a pH increase, filtering, and flotation are all common harvesting techniques [24]. Magnetophoretic harvesting, which involves tagging algal cells with magnetic particles and then separating them from the culture medium, using an external magnetic field, has emerged as an energy-efficient and time saving microalgal harvesting approach [25]. The large specific surface area, superparamagnetic nature, and biocompatibility of Fe_3O_4 NPs make them suitable for such applications. Nevertheless, attaching magnetic Fe_3O_4 NPs to negatively charged algal cells necessitates a certain pH range in which the zeta potential of magnetic nanoparticles exhibits a positively charged surface [26]. As a result, the coating of cationic compounds onto Fe_3O_4 NPs is required. These cationic compounds are polymers like polyethylenimine and polyamidoamine (PAMAM). Hu et al. found that 20 mg/L of a Fe_3O_4-polyethylenimine nanocomposite increases the harvesting effectiveness and adsorption capacity of *Chlorella eppipsoidea* by 97% and 93.46 g dry microalgal cell weight/g nanocomposite, respectively [27]. It was discovered that Fe_3O_4 @ arginine nanoparticles might increase the ability of *Chlorella sp.* cells to be harvested by 95% at a dose of 200 mg/L. The amount of amine groups in amino acid molecules, as well as the amino acid concentration of nanoparticles, have a large impact on the harvesting performance [28].

4.4 FUTURE PROSPECTS IN THE DEVELOPMENT OF MICROALGAE VIA NANOMATERIALS

The current method of producing energy and fuels from petroleum is unsustainable. Lately, the potential of microalgae to accumulate a variety of high-value chemicals and produce biofuel has captured the attention of those involved in the development of microalgae. Hence, commercial, and economic potential, high production, sustainability, and environmental friendliness of microalgae feedstocks and processes must be improved rapidly. This need makes it difficult to increase the output and quality of microalgae biomass production through culture. Moreover, compound recovery, energy consumption, selectivity, and scalability of the extraction method remain to be challenging. Microalgae cultivation and application must comply with environmental, sanitary and safety requirements, especially when intended for human consumption. Improved production efficiency of biofuels and numerous important products utilizing microalgae biomass as feedstock is still required in a few biorefinery processes, which have yet to be optimised. The relevant processes have certain limitations, despite ongoing efforts to upgrade them for better efficiency and more effective use of nanotechnology.

The long-term stability, marketability, and culture requirements of microalgae-based products still need to be thoroughly established. Specifically, additional studies on the sustainability, economic and environmental aspects of microalgae production are required. Additionally, with regards to nanotechnology and its applications, the impacts of problems such as energy consumption, manufacturing prices, yields, loss, and performance on entire microalgae biorefinery process, as well as environmental friendliness also need to be addressed.

4.5 CONCLUSION

The potential of microalgae as a tool in bio-nanotechnology is still being explored. Due to their photosynthetic capabilities and effective water utilisation, microalgae are low-cost to develop and maintain. Algae produce significantly more biomass per acre of land than terrestrial plants and are extremely efficient at producing large amounts of goods. Algae can be genetically modified, chosen and optimised to produce goods of importance since they are single-celled microorganisms, like other microbes, with the added advantage of requiring less energy input due to photosynthesis. Natural metal detoxifying pathways in microalgae convert them into very effective biomanufacturers for metallic nanoparticle production.

REFERENCES

[1] Chai, W.S. 2021. A review on conventional and novel materials towards heavy metal adsorption in waterwatre treatment application. 296.

[2] Abdelbasir, S.M. and Shalan, A.E. 2019. An overview of nanomaterials for industrial wastewater treatment. Korean J. Chem. Eng. 36: 1209–1225. https://doi.org/10.1007/s11814-019-0306-y.

[3] Patra, J.K. and Baek, K.H. 2014. Green nanobiotechnology: Factors affecting synthesis and characterization techniques. J. Nanomater. 2014: 417305. https://doi.org/10.1155/2014/417305.

[4] Agarwal, P., Gupta, R. and Agarwal, N. 2019. Advances in synthesis and applications of microalgal nanoparticles for wastewater treatment. J. Nanotechnol. 2019: 7392713. https://doi.org/10.1155/2019/7392713.

[5] Markandan, K. and Chai, W.S. 2022. Perspectives on nanomaterials and nanotechnology for sustainable bioenergy generation. Materials (Basel). 15: 1–20. https://doi.org/10.3390/ma15217769.

[6] Katas, H., Moden, N.Z., Lim, C.S., Celesistinus, T., Chan, J.Y., Ganasan, P., et al. 2018. Biosynthesis and potential applications of silver and gold nanoparticles and their chitosan-based nanocomposites in nanomedicine. J. Nanotechnol. 2018: 4290705. https://doi.org/10.1155/2018/4290705.

[7] Peter, A.P., Koyande, A.K., Chew, K.W., Ho, S.-H., Chen, W.-H., Chang, J.-S., et al. 2022. Continuous cultivation of microalgae in photobioreactors as a source of renewable energy: Current status and future challenges. Renew. Sustain. Energy Rev. 154: 111852. https://doi.org/10.1016/j.rser.2021.111852.

[8] Jacob, J.M., Ravindran, R., Narayanan, M., Samuel, S.M., Pugazhendhi, A. and Kumar, G. 2021. Microalgae: A prospective low cost green alternative for nanoparticle synthesis, Curr. Opin. Environ. Sci. Heal. 20: 100163. https://doi.org/10.1016/j.coesh.2019.12.005.

[9] Negi, S. and Singh, V. 2018. Algae: A potential source for nanoparticle synthesis. J. Appl. Nat. Sci. 10: 1134–1140. https://doi.org/10.31018/jans.v10i4.1878.

[10] Bao, Z. and Lan, C.Q. 2018. Mechanism of light-dependent biosynthesis of silver nanoparticles mediated by cell extract of Neochloris oleoabundans. Colloids Surfaces B Biointerfaces. 170: 251–257. https://doi.org/10.1016/j.colsurfb.2018.06.001.

[11] Vargas-Estrada, L., Torres-Arellano, S., Longoria, A., Arias, D.M., Okoye, P.U. and Sebastian, P.J. 2020. Role of nanoparticles on microalgal cultivation: A review. Fuel. 280: 118598. https://doi.org/10.1016/j.fuel.2020.118598.

[12] Markandan, K., Sankaran, R., Tiong, Y.W., Siddiqui, H. and Khalid, M. 2023. A review on the progress in chemo-enzymatic processes for CO_2 conversion and upcycling. Catalysts. 13(3): 1–26.

[13] Sajjadi, B., Chen, W.Y., Raman, A.A.A. and Ibrahim, S. 2018. Microalgae lipid and biomass for biofuel production: A comprehensive review on lipid enhancement strategies and their effects on fatty acid composition. Renew. Sustain. Energy Rev. 97 200–232. https://doi.org/10.1016/j.rser.2018.07.050.

[14] Kadar, E., Rooks, P., Lakey, C. and White, D.A. 2012. The effect of engineered iron nanoparticles on growth and metabolic status of marine microalgae cultures. Sci. Total Environ. 439: 8–17. https://doi.org/10.1016/j.scitotenv.2012.09.010.

[15] Cao, J., Yuan, H.L,. Li, B.Z. and Yang, J.S. 2014. Significance evaluation of the effects of environmental factors on the lipid accumulation of Chlorella minutissima UTEX 2341 under low-nutrition heterotrophic condition. Bioresour. Technol. 152: 177–184. https://doi.org/10.1016/j.biortech.2013.10.084.

[16] Ren, H.Y., Dai, Y.Q., Kong, F., Xing, D., Zhao, L., Ren, N.Q. et al. 2020. Enhanced microalgal growth and lipid accumulation by addition of different nanoparticles under xenon lamp illumination, Bioresour. Technol. 297: 122409. https://doi.org/10.1016/j.biortech.2019.122409.

[17] Lu, C., Shi, X., Liu, Y., Xiao, H., Li, J. and Chen, X. 2021. Nanomaterials for adsorption and conversion of CO_2 under gentle conditions. Mater. Today. 50: 385–399. https://doi.org/10.1016/j.mattod.2021.03.016.

[18] H. Cheng, H. Song, S. Toan, B. Wang, K.A.M. Gasem, M. Fan, F. Cheng. 2021. Experimental investigation of CO_2 adsorption and desorption on multi-type amines loaded HZSM-5 zeolites. Chem. Eng. J. 406: 126882. https://doi.org/10.1016/j.cej.2020.126882.

[19] Morais, M., Vargas, B., Vaz, B., Cardias, B. and Costa, J. 2021. Nanobiotechnology: advances in the use of nanomaterials to increase CO_2 biofixation by microalgae. Res. Sq. 1–22.

[20] Li, K., Tian, S., Jiang, J., Wang, J., Chen, X. and Yan, F. 2016. Pine cone shell-based activated carbon used for CO_2 adsorption, J. Mater. Chem. A. 4: 5223–5234. https://doi.org/10.1039/c5ta09908k.

[21] Eroglu, E., Eggers, P.K., Winslade, M., Smith, S.M. and Raston, C.L. 2013. Enhanced accumulation of microalgal pigments using metal nanoparticle solutions as light filtering devices, Green Chem. 15: 3155–3159. https://doi.org/10.1039/c3gc41291a.

[22] Giannelli, L. and Torzillo, G. 2012. Hydrogen production with the microalga Chlamydomonas reinhardtii grown in a compact tubular photobioreactor immersed in

a scattering light nanoparticle suspension. Int. J. Hydrogen Energy. 37: 16951–16961. https://doi.org/10.1016/j.ijhydene.2012.08.103.

[23] Singh, G. and Patidar, S.K. 2018. Microalgae harvesting techniques: A review. J. Environ. Manage. 217: 499–508. https://doi.org/10.1016/j.jenvman.2018.04.010.

[24] Kim, J., Yoo, G., Lee, H., Lim, J., Kim, K., Kim, C.W. et al. 2013. Methods of downstream processing for the production of biodiesel from microalgae. Biotechnol. Adv. 31: 862–876. https://doi.org/10.1016/j.biotechadv.2013.04.006.

[25] Prochazkova, G., Safarik, I. and Branyik, T. 2013. Harvesting microalgae with microwave synthesized magnetic microparticles. Bioresour. Technol. 130: 472–477. https://doi.org/10.1016/j.biortech.2012.12.060.

[26] Ge, S., Agbakpe, M., Wu, Z., Kuang, L., Zhang, W. and Wang, X. 2015. Influences of surface coating, UV irradiation and magnetic field on the algae removal using magnetite nanoparticles, Environ. Sci. Technol. 49: 1190–1196. https://doi.org/10.1021/es5049573.

[27] Hu, Y.R., Guo, C., Wang, F., Wang, S.K., Pan, F. and Liu, C.Z. 2014. Improvement of microalgae harvesting by magnetic nanocomposites coated with polyethylenimine, Chem. Eng. J. 242: 341–347. https://doi.org/10.1016/j.cej.2013.12.066.

[28] Liu, P., Wang, T., Yang, Z., Hong, Y., Xie, X. and Hou, Y. 2020. Effects of Fe_3O_4 nanoparticle fabrication and surface modification on Chlorella sp. harvesting efficiency, Sci. Total Environ. 704: 135286. https://doi.org/10.1016/j.scitotenv.2019.135286.

Chapter **5**

Sustainable Nanocomposites in Membrane Technology: An Overview in Water Treatment

Revathy Sankaran
Faculty of Medicine and Health Sciences,
UCSI University, Kuala Lumpur, Malaysia
Email: revathy@ucsiuniversity.edu.my

5.1 INTRODUCTION

The composite materials known as 'nanocomposites' contain a phase with nanoscale morphology; such materials include nanoparticles, nanotubes, or lamellar nanostructures. Materials used in the dispersion phase and matrix of nanocomposites can be used to categorise these materials [1]. Nowadays, it is feasible to create a variety of intriguing new materials with novel features, thanks to cutting-edge synthesis techniques made possible by this newly developing industry. Nanocomposite materials have recently come to light as viable options for alleviating the shortcomings of many engineering materials [2]. They are reportedly the 21st century's materials.

One of the biggest issues on the planet is water pollution. Due to the influx of significant amounts of contaminants into the water bodies, growing industrialization is increasingly deteriorating the water quality [3]. Water pollution has emerged as a hazard to the entire biosphere, making its elimination crucial. Appropriate materials with high separation capacity, low cost, porosity, and reusability are crucial for water treatment. Researchers have produced a load of work in the last decade, employing membrane technology. It was difficult for researchers to employ membrane technology in various sectors in the past, but they are currently using it in a variety of research fields. In this sense, integration with nanotechnology offers the chance to create new materials with improved hydrophilicity, hydrophobicity, porosity, mechanical strength, and dispersibility qualities for efficient water filtration [4]. Due to their enhanced features, such as high surface area, surface mobility, and improved optical and magnetic properties, nanoparticles (NPs) placed in a polymer matrix or in a thin composite layer to produce nanocomposite membranes have emerged as potential membrane materials [5].

This book chapter examines the most current advancements in membrane technology, emphasizing nanocomposite membrane technology. The focus of this chapter will be on the applications of nanocomposite membrane for water treatment and purification. This book chapter also discusses the crucial aspects of nanocomposites membrane technology for potential future research directions.

5.2 MEMBRANE TECHNOLOGY

Membrane technology is becoming increasingly important in various industrial sectors, including those that deal with the treatment of industrial effluent and wastewater, as well as food, medicine, pharmacy, biotechnology and chemicals [6, 7]. There are several processes that are encompassed with membrane technology, which include membrane distillation, dialysis, electrodialysis, reverse osmosis, pervaporation, gas separation and many more. A chosen thin layer of a semipermeable substance is referred to as a membrane. A membrane isolates unwanted components from a feed solution, depending on their size or affinity, by exerting a suitable gradient, such as pressure, temperature, electric, or concentration differences, as a primary driver. High extraction efficiency, flexibility of operation, and affordability are the strengths of this technique over traditional separation techniques. Membranes are often categorized according to their composition, shape, and average pore size. Both organic and inorganic materials make up membranes. Figure 5.1 shows the types of membrane materials that are commonly used.

The potential of both organic and inorganic membrane types for water treatment applications has been widely researched. Despite having better mechanical, chemical and thermal qualities than polymeric membranes, inorganic membranes composed of metals or ceramics are more costly to produce and are therefore not favoured [8]. Polymeric/organic membranes, on the other hand, are extremely adaptable—their pore diameters are controllable within a certain range. Moreover, by altering the casting circumstances, monomer molecules and their concentrations, additives, and coagulation bath conditions, the membrane characteristics can be changed [9]. Polymeric membranes do, unfortunately, also have a number of significant disadvantages. They are susceptible to fouling because of their innate hydrophobicity, which is a major drawback especially in water treatment [10]. They also have poor chlorine resilience. Depending on the level of fouling, a thorough physical and chemical wash or membrane replacement may be necessary. Enhancements in membrane technology are now possible, thanks to advancements in nanotechnology, an enabling technology at the atomic level. At present, industrial applications of nanocomposite membrane technology are popular. A higher level of control over a polymer base and the incorporation of nanoparticles in the support layer are required for polymer membrane advances employing next generation materials that widen the commercial uses of membrane technology. The concept of incorporating nanoparticles into polymeric membranes was made with the idea that they would help these membranes overcome some of their shortcomings, such as their propensity for fouling, by utilizing the superior qualities of these nanoparticles.

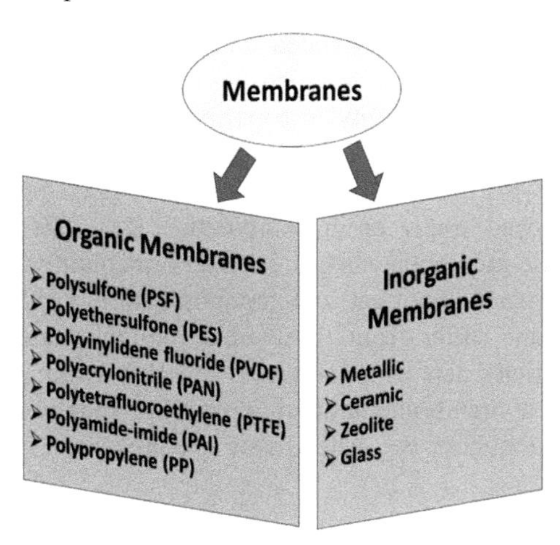

Figure 5.1 Common types of membrane used in various industrial applications.

5.3 NANOCOMPOSITES

Composite materials known as nanocomposites contain a phase with nanoscale morphology, such as nanoparticles, nanotubes, or lamellar nanostructures. The heterogeneous/hybrid substances known as nanocomposites are generated when polymers are combined with inorganic substances (such as oxides or clays) at the nanoscale level. It has been discovered that their structures are more intricate than those of microcomposites. The structure, composition, interfacial interactions, and elements of each specific property have a significant impact [11]. The most common method for creating nanocomposites involves in situ biopolymer and inorganic matrix development and polymerization. These attractive, potentially cutting-edge materials are in high demand and are therefore extremely valuable in a wide range of industries—from small-scale production to very big manufacturing facilities. In comparison with other traditional water filtration techniques, the membrane separation method using nanocomposites has demonstrated excellent outcomes in the removal or recovery of numerous contaminants. In this section, the focus of the chapter will be on nanocomposite membranes for water treatment.

5.3.1 Nanocomposite Membranes

Typically, the creation of nanocomposite membranes involves adding nanoparticle substances (the filler) into a macroscopic substance (the matrix). Before membrane casting, nanoparticles can be distributed in the polymer solution or coated onto the membrane surface [12]. This structure produces polymer-nanocomposite membranes, also known as mixed matrix membranes (MMMs) or nano-enhanced membranes, by incorporating fillers, or additional components, into the primary polymeric matrix [13]. Nanocomposite membranes are a possible substitute to manage the limitations of conventional membranes mentioned above. The inclusion of fillers usually alters the surface characteristics of membranes that affect segregation performance, such as high permeability, stable flux, superior rejection against foulants, and enhanced antifouling behaviour. For example, it is hypothesized that adding a hydrophilic functional group to the surface may enhance the membrane's ability to separate materials while minimizing or controlling the undesirable adhesion and/or adsorption reactions between foulants and the active layer [14]. Numerous techniques have been proposed to do this, which may be employed singly or in combination [15]. The modifications that have been done on the membrane surface include dip coating, grafting, blending, surface chemical, plasma treatment and ion implantation [12]. The features of a composite membrane differ from those of a regular membrane because the inclusion of nanoparticles acts as an enhancer on the membrane surface. Several nanoparticles make a membrane more hydrophilic, reducing the membrane's propensity to foul throughout the water treatment process.

5.3.2 Fabrication and Types of Composite Membranes

Several forms of nanocomposite membranes can be produced depending on the membrane's structure and where the nanoparticles are positioned. Figure 5.2 below depicts the technique of fabricating composite membranes. Popular techniques for the development of composite membranes include (a) covering the membrane with nanoparticles on the surface, (b) entrapping of nanoparticles inside the polymeric framework during fabrication, (c) thin-film composites (TFC) with a nanocomposite substrate, and (d) thin-film composites (TFC) with nanocomposites.

In addition to the techniques mentioned above, there are other blending methods to fabricate nanocomposites; these include combining a variety of organic polymers, organic and organometallic compounds, biological molecules, enzymes, and sol-gel generated polymers with inorganic nanoclusters, fullerenes, clays, metals, oxides, or semiconductors [16]. The distinct features of nanocomposite materials generated by combining two or more different building blocks into one substance are most likely due to their small size, huge surface area, and the interface contact between the phases [11].

5.3.2.1 Non-polymer Nanocomposites

Nanocomposites can be categorised into polymer-based and non-polymer nanocomposites, depending on whether or not the composite contains a polymeric

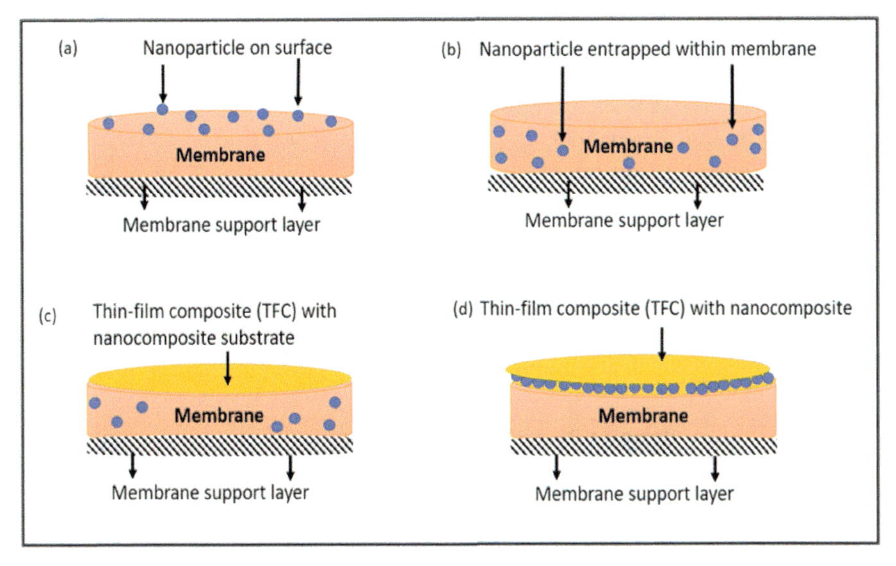

Figure 5.2 Popular techniques for fabricating composite membranes: (a) surface coating of nanoparticles; (b) entrapping nanoparticles in the membrane matrix; (c) thin-film composites (TFC) with a nanocomposite substrate; and (d) thin-film composites (TFC) with nanocomposites

material. Table 5.1 shows the different types of nanocomposites and the advantages of non-polymer nanocomposites. Metal-based nanocomposites can be characterised by the following: low melting temperatures, greater durability and hardness, enhanced magnetic characteristics, high electrical resistivity, and enhanced plasticity. Ceramic composites containing more than one solid phase, with at least one having dimensions in the nanoscale range (50–100 nm), are referred to as ceramic-based nanocomposites. Both layers in these composites integrate the properties of magnetism, chemistry, optics and mechanics. The usage of ceramic/ceramic nanocomposites in the field of artificial joint grafts for fracture failures can immediately lower surgical costs and increase patient mobility [17].

Table 5.1 Different types of non-polymer nanocomposites and their strengths.

Type of Nanocomposites	Composition of Nanocomposites	Advantages	Ref.
Non-polymer	Metal nanocomposites (Metal/Metal nanocomposites)	• Enhanced catalytic properties and advancement in optical properties • Improved toughness and endurance • Enhanced magnetic characteristics	[18]
	Ceramic nanocomposites	• Integrated qualities of magnetics, chemistry, optics and mechanics • Enhanced flexibility, greater strength, and improved durability	[19]
	Ceramic-Ceramic nanocomposites	• High-temperature mechanical behaviour	[20]

5.3.2.2 Polymer-based Nanocomposites

Figure 5.3 displays the different types of polymer-based nanocomposites. Poly nanocomposites are defined as polymers or copolymers with nanoparticles or nanofillers scattered throughout the polymer matrix. Here, one dimension (1D) should lie between 1 and 50 nm in size and have a variety of morphologies, including platelets, fibres, spheroids, etc. The most complicated characteristic of polymer nanocomposites (PNCs) is the complicated interfacial regions between the polymer matrices due to the small-scale huge specific area formed, which emphasizes the significance of polymer-nanoparticle interactions. Examination of the intercalation process between nanoparticles and polymer bases is necessary to obtain desirable mechanical, thermal, optical and electric properties [11].

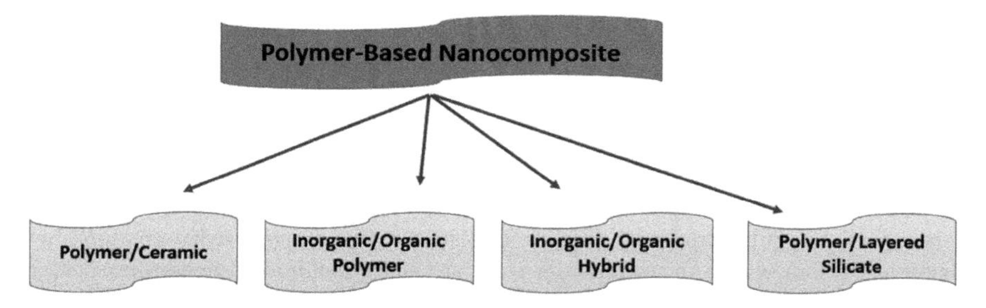

Figure 5.3 Different types of polymer-based nanocomposites.

Mechanical behaviour, clarity, and thermal stability continue to be the limiting factors for diverse applications of polymer nanocomposite membranes. In order to promote realistic results, scientists must create robust, transparent and heat-resistant nanocomposites membranes. The primary inventive goals for developing nanocomposites membranes are to simultaneously achieve high permeability and good selectivity at low costs, combine reactions within pore structures to prevent membrane fouling while preventing additional downstream unit operations, and increase the physical strength of the membranes [21,22].

Over the past three decades, a lot of work has gone into creating advanced membrane technologies to enhance the performance of membranes in terms of flux, solute rejection, and antifouling qualities. Developments will continue to advance in future, and, right now, scientists are concentrating on nanocomposite membranes to enhance the properties of membranes. Researchers have focused on developing advanced thin-film nanocomposite membranes designed to provide high flux, good rejection, and antifouling properties [23,24]. A new technique for integrating functional nanoparticles (10–15 nm) in polymer films was recently devised, and it has helped with the fabrication of thin-film nanocomposite (TFN) membranes [25]. The development of super-hydrophilic nanoparticles is the initial use of TFN-based membranes. The initial study attests that TFN membranes separate water vapour while conserving a large amount of energy and possessing a super-hydrophilic nature. Without significantly altering the TFC membrane manufacturing method, it is feasible to commercialise or produce large quantities of TFN membranes using the produced nanoparticles. The costs involved are also not significantly high. Due to various cutting-edge features, including great flexibility, a wide range of pore sizes and shapes, an easy development process, low costs, and ease of scaling up, polymer membrane technology has drawn significant attention for water treatment applications. The section below will explore the various nanocomposite applications in water treatment and purification.

5.4 APPLICATION OF NANOCOMPOSITES FOR WATER TREATMENT

Due to its effectiveness in addressing pollution and reducing water toxicity, nanotechnology, in the form of nanomaterials or nanoparticles combined with conventional methods, began to emerge as a promising alternative to traditional treatment approaches in the water sector. In addition, throughout the last twenty years, removal strategies utilising nanotechnology have significantly advanced, showing potential to advance water and wastewater treatment procedures, surpassing existing techniques and their shortcomings [26]. The sections below delve into different nanocomposite materials for removing different types of contaminants from water.

5.4.1 Removal of Heavy Metal Ions

A number of researchers are interested in the heavy metal contamination of freshwater bodies because of the significant toxicity and the cancerous effects it has on living things. Nanomaterials are being created and used in water treatment due to their excellent qualities such as strong reactivity, large surface area, and outstanding mechanical capabilities. One of the easiest, most affordable and most effective ways to treat water is adsorption. Numerous nanoadsorbents have been used in many studies to examine the removal of contaminants from wastewater [27]. Graphene, graphene oxide and some other substances related to graphene are some of the most efficient adsorbents for removing organic and inorganic pollutants from wastewater. Carbon-based nanomaterials such as carbon fibres, carbon nanotubes (CNT), activated carbons, graphene, and carbon nanotubes (GNO) are also among these materials. These composites feature high porosity, improved mechanical and thermal stability, a huge specific surface area, and great chemical stability under acid/alkaline environments [28].

Given their adaptability and simplicity in terms of administration and maintenance, membrane technologies, such as nanofiltration, are quickly emerging as a standard method for essential services and industries to purify their water [29]. Nanofiber membranes are notable for their large surface specific area, intrinsic porosity, and selectivity, particularly when functionalized. They have indeed been suggested as a pretreatment step before using ultrafiltration or reverse osmosis. Metals are removed from water using nanofiber membranes. Fouling, unfortunately, still poses a significant problem for membrane processes [30]. Organic, inorganic, or biological fouling are all possible. Fouling results in higher transmembrane pressure, lesser permeate flux, higher operational expenses, more regular membrane clean-up or replacement, and decreased water quality [31,32]. In addition, nanotechnology is also being employed in many contexts to address the fouling issues that normally used membranes in general, and nanofiltration membranes in particular, encounter [15,33]. In a study, composite nanofiber membranes were produced from chitosan and hydroxyapatite (CHAp) for the elimination of Pb^{2+} and Cu^{2+} from water. Chitosan was selected because of its limited solubility, high viscosity, and low chain flexibility as well as the existence of hydroxyl and amine groups that enhance adsorption [34].

For the elimination of heavy metals using nanofiltration (NF), further investigations discovered the cross-linking between graphene oxide framework and ethylenediamine (EDA) [35]. Studies have found that nanochannels of GO-nanosheets are widened and their physical structure is improved by EDA cross-linking. This results in increased water permeability and removal efficiencies [36]. According to a study by Shukla et al. [37], utilising GO combined with a polyphenylsulfone (PPSU) nanofiltration membrane, over 98% of metals are removed from water. Compared to more expensive procedures, the properties of GO improve membrane performance, enable efficient metal ion removal, and offer commercially favourable conditions.

The challenges of membrane fouling and scaling need to be tackled regardless of the positive preliminary findings of such investigations. Another difficulty is the significantly greater energy demand and associated expenses. The elimination of heavy metals from water and wastewater using nanofiltration can be a feasible solution when more research is done to tackle these basic issues that membrane technology faces. In addition, more lab research and pilot-scale research are required to incorporate newly designed nanostructured membranes into current water treatment processes.

5.4.2 Removal of Dyes

Large-scale dye releases into waterways by the textile industry have negative environmental effects. Because of the persistent and stubborn characteristics of these contaminants, removing these colours from contaminated streams is a serious challenge. Because of its high efficiency, compact design and simple scaling up, membrane filtering has become more common. Traditional membrane-based methods do have certain inherent disadvantages, such as significant energy requirements, fouling, and slow water flow and removal rates. The creation of innovative nanocomposite membranes and their usage in the treatment of dyes from contaminated streams are thus the main topics of focus in the current section.

For the removal of dyes, a number of approaches have been devised, including adsorption, coagulation, filtration, photocatalytic, and biochemical degradation. Several investigators have employed polymer nanocomposites in this aspect to remove dyes effectively. For removing dyes, PNC exhibits a number of characteristics, including a great adsorption ability, and catalytic and magnetic capabilities. Multiple hybrid designs have been employed for the removal of dyes as the incorporation of metal produces functional sites. A brand-new class of filtration materials called nanocomposite membranes is made up of nanofillers in a polymeric matrix. The selection of materials for particular separation procedures is a complicated process. Stability, productivity, segregation effectiveness, and mechanical integrity under operating parameters are the main criteria to be considered. The aforementioned polymer is actually imparted with functional characteristics by the nanofiller.

In a latest attempt, mixed matrix membranes (flat sheets) were generated by combining polyimide (P84) with various MOF-2(Cd) compositions (metal organic frameworks based on cadmium). These films were subsequently used to repel methylene blue (99.9%), sunset yellow (68.4%), and eosin Y (81.2%) simultaneously under various processing settings. Interestingly, adding 0.2% of nanofiller (MOF-2(Cd)) increased the membrane separation performance for the dyes eosin yellow and sunset yellow by 213.3% and 110.2% respectively [38].

For effective dye removal, Zhao et al. created an in-situ thin film nanocomposite membrane based on polyvinylpyrrolidone/Zr-MOF [39]. The resultant membrane's remarkable filtration ability, greater water dispersibility, consistent structure, large surface area, and negative charge made it a promising candidate for salt/dye extraction. As much as 0.3 weight percent of PVP-UiO-66-NH_2 nanofillers were incorporated into an optimised membrane, which resulted in impressive rejection (about 99.89%) of four distinct dyes, including Reactive black 5, methyl blue, Direct red 23, and Congo red, as well as effective water permeability [39]. In a recent study, UiO-66-NH_2 membranes with TFN integration were created using the interfacial polymerization of piperazine (PIP) and trimesoyl chloride (TMC). Malachite green dye and phosphate extraction from water are also being studied using a uniquely created nanocomposite membrane. Through testing, researchers were able to successfully separate dyes from a feed solution of a low concentration of 100 ppm. Borpatra and his group investigated the removal of phosphates using a low concentration (10 ppm) solution, which is usually a complicated process, and found that the proposed TFN membrane had a rejection rate of 78% and a permeability of 22.22 Lm^2h^1 [40].

Reverse osmosis and nanofiltration have been compared for their efficacy in the biological remediation of textile wastewater, and the permeate purity generated from each method has also been evaluated [41]. The permeate was analysed for salinity content, permeate flow, and BOD and COD elimination. For cross flow filtering studies of textile wastewater over a wide array of concentration ratios with various hydrodynamic parameters, flat sheet membranes for reverse osmosis (BW30) and nanofiltration (NF90) have been utilised. In both the cases, it has been discovered that the process streams handled meet the criteria for reclamation and produce reusable water of high quality within each membrane [41]. Nanocomposite membranes have been found to be appropriate alternatives for treating textile effluents for water recycling and colour removal, as opposed to traditional methods like coagulation, flocculation, and adsorption. Owing to its characteristics complementing those of the dyes to be removed, nanofiltration is considered to be the most successful method of membrane-based separation [42].

5.4.3 Removal of Other Pollutants

Microorganisms, insecticides, pathogens, and other organic compounds are among the other significant water contaminants. Nanocomposite membranes have been used for the removal of various other pollutants, with the aim of treating water. For the purpose of separating waste oil from sewage, Maphutha et al. created a membrane from a carbon nanotube-polymer composite employing a polyvinyl alcohol barrier stratum [43]. Chemical vapour deposition was used to create the carbon nanotubes, and a phase inversion approach was applied to combine the carbon nanotubes with the composite materials solution that was utilized to create the membrane. The permeate revealed that oil contents were lower than the typical threshold of 10 mgL^{-1}, and higher than 95% oil rejection was attained consequently [43]. In another study, for preventing biofouling on the membrane surface and to achieve zero bacterial cell (*E. coli*) count in the permeate water, silver-nanoparticles were used to modify of the surface of polysulfone membranes. Silver nanoparticle-polyethersulfone membranes showed a steady permeate flow rate (3.45 Lh^{-1}) as an outcome of completely eliminating *Escherichia coli* cells after administering $AgNO_3$, which caused H^+ ions to be replaced by Ag^+ ions. Thus, the enhancement of the antibacterial effect was made possible by the incorporation of silver nanoparticles on the membrane surface [44].

The development of sophisticated disinfection techniques has been motivated by the risks connected with them, including the development of disinfection byproducts and multi-drug-resistant bacterial species. Upon close interaction, the nanostructured composite restores the cell membrane integrity while killing infections by releasing toxic compounds. It can also form reactive oxygen species (ROS) in certain circumstances. Metals have been known to have bactericidal properties since antiquity, but advances in nanotechnology have increased their effectiveness and made it possible to employ them as effective disinfectants [45]. Ag nanoparticles incorporated in cellulose acetate fibres have been discovered to be effective against bacteria [46]. Research findings show that over time, Ag-incorporated polymer filler's ability to disinfect reduces. Pathogenic bacteria can also be effectively removed by a variety of composites with carbon nanostructures, such as carbon nanotubes (CNTs) and activated carbon fibres (ACFs) [47,48]. In practical field applications, the utilisation of nanofibers and composite nanostructure membranes can aid in the degradation of a variety of organic and inorganic pollutants as discussed above.

5.5 CONCLUSION

Global population increase and environmental issues are both major contributors to the long-term evolution of the world's water crisis. The use of nanoparticles in water treatment has significantly increased in recent years. Due to the recent and rapid breakthroughs in nanotechnology, the study of nanocomposites has gained significance in the development of innovative components for cutting-edge uses. Nanocomposites, which are both a diverse class of materials and have a substantial amount of comprehensive interaction, are the greatest choice to address the growing demand for multifunctional materials. In this chapter, exploration of nanocomposites in membrane technology for various applications in water treatment was thoroughly discussed. Nanocomposites seem to be our best option since they have a high degree of integrated connection and are a diverse class of materials. This is a multidisciplinary field that requires expertise in both technological and scientific backgrounds to produce macroscopic designed materials obtained through nanoscale structures. Nanocomposites have been successfully utilised for removing heavy metal ions, dyes and organic compounds, and even eliminating bacteria from water. Therefore, it is anticipated that they will have a significant impact on keeping the environment greener, healthier, and safer in the years to come.

REFERENCES

[1] Wang, J. and Kaskel, S. 2012. KOH activation of carbon-based materials for energy storage, J. Mater. Chem. 22: 23710. doi:10.1039/c2jm34066f.

[2] Markandan, K. and Chai, W.S. 2022. Perspectives on nanomaterials and nanotechnology for sustainable bioenergy generation. Materials (Basel). 15: 1–20. doi:10.3390/ma15217769.

[3] Rashid, R., Shafiq, I., Akhter, P., Iqbal, M.J. and Hussain, M. 2021. A state-of-the-art review on wastewater treatment techniques: the effectiveness of adsorption method, Environ. Sci. Pollut. Res. 28: 9050–9066. doi:10.1007/s11356-021-12395-x.

[4] Daer, S., Kharraz, J., Giwa, A., and Hasan, S.W. 2015. Recent applications of nanomaterials in water desalination: A critical review and future opportunities. Desalination. 367: 37–48. doi:10.1016/j.desal.2015.03.030.

[5] Ingole, P.G. 2009. Application of sustainable nanocomposites in membrane technology, Sustain. Polym. Compos. Nanocomposites. 935–960. doi:10.1007/978-3-030-05399-4_32.

[6] Radjenović, J., Petrović, M. and Barceló, D. 2008. Fate and distribution of pharmaceuticals in wastewater and sewage sludge of the conventional activated sludge (CAS) and advanced membrane bioreactor (MBR) treatment, Water Res. 43: 831–841. doi:10.1016/j.watres.2008.11.043.

[7] Jhaveri, J.H. and Murthy, Z.V.P. 2016. A comprehensive review on anti-fouling nanocomposite membranes for pressure driven membrane separation processes, Desalination. 379: 137–154. doi:10.1016/j.desal.2015.11.009.

[8] Lee, M., Wu, Z. and Li, K. 2015. Advances in ceramic membranes for water treatment. pp. 43-82. *In*: Angelo Basile, Alfredo Cassano and Navin K. Rastogi (eds). Advances in Membrane Technologies for Water Treatment. Materials, Processes and Applications. Woodhead Publishing is an imprint of Elsevier. doi:10.1016/B978-1-78242-121-4.00002-2.

[9] Goh, P.S. and Ismail, A.F. 2018. A review on inorganic membranes for desalination and wastewater treatment. Desalination. 434: 60–80. doi:10.1016/j.desal.2017.07.023.

[10] Lee, A., Elam, J.W. and Darling, S.B. 2016. Membrane materials for water purification: design, development, and application. Environ. Sci. Water Res. Technol. 2: 17–42. doi:10.1039/C5EW00159E.

[11] Sen, M. 2020. Nanocomposite Materials. *In*: Nanotechnol. Environ., IntechOpen, 1–12. doi:10.5772/intechopen.93047.

[12] Al Aani, S., Wright, C.J., Atieh, M.A. and Hilal, N. 2017. Engineering nanocomposite membranes: Addressing current challenges and future opportunities. Desalination. 401: 1–15. doi:10.1016/j.desal.2016.08.001.

[13] Mueller, N.C., van der Bruggen, B., Keuter, V., Luis, P., Melin, T., Pronk, W., et al. 2012. Nanofiltration and nanostructured membranes—Should they be considered nanotechnology or not?, J. Hazard. Mater. 211–212: 275–280. doi:10.1016/j.jhazmat.2011.10.096.

[14] Kochkodan, V., Johnson, D.J. and Hilal, N. 2014. Polymeric membranes: Surface modification for minimizing (bio)colloidal fouling, Adv. Colloid Interface Sci. 206: 116–140. doi:10.1016/j.cis.2013.05.005.

[15] Mohammad, A.W., Teow, Y.H., Ang, W.L., Chung, Y.T., Oatley-Radcliffe, D.L. and Hilal, N. 2015. Nanofiltration membranes review: Recent advances and future prospects. Desalination. 356: 226–254. doi:10.1016/j.desal.2014.10.043.

[16] Markandan, K., Sankaran, R., Tiong, Y.W., Siddiqui, H. and Khalid, M. 2023. A review on the progress in chemo-enzymatic processes for CO_2 conversion and upcycling. Catalysts. 13(3): 1–26.

[17] Thostenson, E.T., Li, C. and Chou, T.W. 2005. Nanocomposites in context. Compos. Sci. Technol. 65: 491–516. doi:10.1016/j.compscitech.2004.11.003.

[18] Fam, D.W.H., Palaniappan, A., Tok, A.I.Y., Liedberg, B. and Moochhala, S.M. 2011. A review on technological aspects influencing commercialization of carbon nanotube sensors. Sensors Actuators B Chem. 157: 1–7. doi:10.1016/j.snb.2011.03.040.

[19] Pandey, J.K., Kumar, A.P., Misra, M., Mohanty, A.K., Drzal, L.T. and Palsingh, R. 2005. Recent advances in biodegradable nanocomposites. J. Nanosci. Nanotechnol. 5: 497–526. doi:10.1166/jnn.2005.111.

[20] Palmero, P. 2015. Structural ceramic nanocomposites: A review of properties and powders' synthesis methods. Nanomaterials. 5: 656–696. doi:10.3390/nano5020656.

[21] Wang, F., Wu, Y., Huang, Y. and Liu, L. 2018. Strong, transparent and flexible aramid nanofiber/POSS hybrid organic/inorganic nanocomposite membranes. Compos. Sci. Technol. 156: 269–275. doi:10.1016/j.compscitech.2018.01.016.

[22] Wang, M., Liu, G., Cui, X., Feng, Y., Zhang, H., Wang, G., et al. 2018. Self-crosslinked organic-inorganic nanocomposite membranes with good methanol barrier for direct methanol fuel cell applications, Solid State Ionics. 315: 71–76. doi:10.1016/j.ssi.2017.12.001.

[23] He, Y., Tang, Y.P., Ma, D. and Chung, T.-S. 2017. UiO-66 incorporated thin-film nanocomposite membranes for efficient selenium and arsenic removal. J. Memb. Sci. 541: 262–270. doi:10.1016/j.memsci.2017.06.061.

[24] He, Y., Liu, J., Han, G. and Chung, T.-S. 2018. Novel thin-film composite nanofiltration membranes consisting of a zwitterionic co-polymer for selenium and arsenic removal, J. Memb. Sci. 555: 299–306. doi:10.1016/j.memsci.2018.03.055.

[25] Ingole, P.G., Choi, W.K., Lee, G.B. and Lee, H.K. 2017. Thin-film-composite hollow-fiber membranes for water vapor separation. Desalination. 403: 12–23. doi:10.1016/j.desal.2016.06.003.

[26] Bhati, M. and Rai, R. 2017. Nanotechnology and water purification: Indian know-how and challenges. Environ. Sci. Pollut. Res. 24: 23423–23435. doi:10.1007/s11356-017-0066-3.

[27] Damiri, F., Andra, S., Kommineni, N., Balu, S.K., Bulusu, R., Boseila, A.A., et al. 2022. Recent advances in adsorptive nanocomposite membranes for heavy metals ion removal from contaminated water: A comprehensive review. Materials (Basel). 15. doi:10.3390/ma15155392.

[28] Kalaitzidou, K., Zouboulis, A. and Mitrakas, M. 2020. Cost evaluation for Se(IV) removal, by applying common drinking water treatment processes: Coagulation/precipitation or adsorption. J. Environ. Chem. Eng. 8: 104209. doi:10.1016/j.jece.2020.104209.

[29] Tambe Patil, B.B. 2015. Wastewater treatment using nanoparticles. J. Adv. Chem. Eng. 5: 1000131. doi:10.4172/2090-4568.1000131.

[30] Zheng, Y., Yao, G., Cheng, Q., Yu, S., Liu, M. and Gao, C. 2013. Positively charged thin-film composite hollow fiber nanofiltration membrane for the removal of cationic dyes through submerged filtration. Desalination. 328: 42–50. doi:10.1016/j.desal.2013.08.009.

[31] Mahendran, B., Lin, H., Liao, B. and Liss, S.N. 2011. Surface properties of biofouled membranes from a submerged anaerobic membrane bioreactor after cleaning. J. Environ. Eng. 137: 504–513. doi:10.1061/(ASCE)EE.1943-7870.0000341.

[32] Malaeb, L., Le-Clech, P., Vrouwenvelder, J.S., Ayoub, G.M. and Saikaly, P.E. 2013. Do biological-based strategies hold promise to biofouling control in MBRs?. Water Res. 47: 5447–5463. doi:10.1016/j.watres.2013.06.033.

[33] Kim, Y.C., Han, S. and Hong, S. 2011. A feasibility study of magnetic separation of magnetic nanoparticle for forward osmosis. Water Sci. Technol. 64: 469–476. doi:10.2166/wst.2011.566.

[34] Aliabadi, M., Irani, M., Ismaeili, J., Piri, H. and Parnian, M.J. 2013. Electrospun nanofiber membrane of PEO/Chitosan for the adsorption of nickel, cadmium, lead and copper ions from aqueous solution. Chem. Eng. J. 220: 237–243. doi:10.1016/j.cej.2013.01.021.

[35] Yunus, H.I.S., Kurniawan, A., Adityawarman, D. and Indarto, A. 2012. Nanotechnologies in water and air pollution treatment. Environ. Technol. Rev. 1: 136–148. doi:10.1080/21622515.2012.733966.

[36] Zhang, Y., Zhang, S. and Chung, T.-S. 2015. Nanometric graphene oxide framework membranes with enhanced heavy metal removal via nanofiltration. Environ. Sci. Technol. 49: 10235–10242. doi:10.1021/acs.est.5b02086.

[37] Shukla, A.K., Alam, J., Alhoshan, M., Arockiasamy Dass, L., Ali, F.A.A., Mishra, M.M.R.U., et al. 2018. Removal of heavy metal ions using a carboxylated graphene oxide-incorporated polyphenylsulfone nanofiltration membrane. Environ. Sci. Water Res. Technol. 4: 438–448. doi:10.1039/C7EW00506G.

[38] Baneshi, M.M., Ghaedi, A.M., Vafaei, A., Emadzadeh, D., Lau, W.J., Marioryad, H., et al. 2020. A high-flux P84 polyimide mixed matrix membranes incorporated with cadmium-based metal organic frameworks for enhanced simultaneous dyes removal: Response surface methodology. Environ. Res. 183: 109278. doi:10.1016/j.envres.2020.109278.

[39] Zhao, P., Li, R., Wu, W., Wang, J., Liu, J. and Zhang, Y. 2019. *In-situ* growth of polyvinylpyrrolidone modified Zr-MOFs thin-film nanocomposite (TFN) for efficient dyes removal. Compos. Part B Eng. 176: 107208. doi:10.1016/j.compositesb.2019.107208.

[40] Borpatra Gohain, M., Karki, S., Yadav, D., Yadav, A., Thakare, N.R., Hazarika, S., Lee, H.K., et al. 2022. Development of antifouling thin-film composite/nanocomposite

membranes for removal of phosphate and malachite green dye. Membranes (Basel). 12. doi:10.3390/membranes12080768.

[41] Liu, M., Lü, Z., Chen, Z., Yu, S. and Gao, C. 2011. Comparison of reverse osmosis and nanofiltration membranes in the treatment of biologically treated textile effluent for water reuse, Desalination. 281: 372–378. doi:10.1016/j.desal.2011.08.023.

[42] Zahid, M., Ahmad, H., Drioli, E., Rehan, Z.A., Rashid, A., Akram, S., et al. 2021. Role of polymeric nanocomposite membranes for the removal of textile dyes from wastewater. pp. 91–103. *In*: Kamel A. Abd-Elsalam, Muhammad Zahid (eds). Aquananotechnology: Applications of Nanomaterials for Water Purification, A volume in Micro and Nano Technologies. Elsevier. https://doi.org/10.1016/B978-0-12-821141-0.00006-9.

[43] Maphutha, S., Moothi, K., Meyyappan, M. and Iyuke, S.E. 2013. A carbon nanotube-infused polysulfone membrane with polyvinyl alcohol layer for treating oil-containing waste water. Sci. Rep. 3: 1509. doi:10.1038/srep01509.

[44] Biswas, P. and Bandyopadhyaya, R. 2017. Biofouling prevention using silver nanoparticle impregnated polyethersulfone (PES) membrane: E. coli cell-killing in a continuous cross-flow membrane module. J. Colloid Interface Sci. 491: 13–26. doi:10.1016/j.jcis.2016.11.060.

[45] Pandey, N., Shukla, S.K. and Singh, N.B. 2017. Water purification by polymer nanocomposites: an overview. Nanocomposites. 3: 47–66. doi:10.1080/20550324.2017.1329983.

[46] Kamanina, N.V. 2014. Advanced optical materials modified with carbon nanoobjects. pp. 275–315. *In*: A. Tiwari and S.K. Shukla (eds). Advanced Carbon Materials and Technology. John Wiley&Sons, Inc. Hoboken, New Jersey, and Scrivener Publishing LLC. doi.org/10.1002/9781118895399.ch7

[47] Lilly, M., Dong, X., McCoy, E. and Yang, L. 2012. Inactivation of bacillus anthracis spores by single-walled carbon nanotubes coupled with oxidizing antimicrobial chemicals. Environ. Sci. Technol. 46: 13417–13424. doi:10.1021/es303955k.

[48] Srinivasan, N.R., Shankar, P.A. and Bandyopadhyaya, R. 2013. Plasma treated activated carbon impregnated with silver nanoparticles for improved antibacterial effect in water disinfection. Carbon N.Y. 57: 1–10. doi:10.1016/j.carbon.2013.01.008.

Chapter **6**

Sustainable Thermoplastic Nanocomposites: Properties and Applications

Akash Vincent and N. Harshavardhana*
Department of Mechanical Engineering,
SRM Institute of Science and Technology,
Kattankulathur, Chengalpattu, Tamil Nadu, India

6.1 INTRODUCTION: SUSTAINABLE THERMOPLASTIC NANOCOMPOSITES

Thermoplastic nanocomposites are a class of advanced materials due to their potential to provide a high level of performance. They are made by incorporating nanoscale filler particles into a thermoplastic matrix, resulting in a material that has enhanced strength, stiffness and durability. The combination of thermoplastics and nanoparticles is used for improvement in mechanical, electrical and optical properties of the polymer matrix. The mechanical properties of thermoplastic nanocomposites can be significantly improved by the addition of nanoparticles in the thermoplastic matrix. The small size of nanoparticles allows for better dispersion throughout the polymer matrix, leading to a more uniform and stronger material. Nanoparticles can act as a reinforcement for the polymer matrix, providing enhanced strength, electrical conductivity and stiffness. This improved performance has significant applications in various industries, such as packaging, construction, and automotive parts [1, 2].

*For Correspondence: harshavardhananatarajan@gmail.com

The term 'sustainable' has gained significant attention in recent years and refers to the environmentally responsible production and disposal of these thermoplastic polymers. Sustainable thermoplastic nanocomposites are materials that are designed to be environmentally friendly, while also having improved physical and mechanical properties compared to traditional thermoplastics. The use of renewable and biodegradable materials, such as bio-based polymers or recycled plastics, can reduce the reliance on hydrocarbon-based thermoplastics resources and decrease the amount of waste in the environment [3]. For example, the use of cellulose nanofibrils and clay nanoparticles can improve the properties (such as flame- retardant ability) of the polymer matrix without the need for harmful additives. The small size of the nanoparticles allows for better dispersion throughout the polymer matrix, leading to a more uniform and stronger material. The use of sustainable thermoplastic nanocomposites in various applications and can also help reduce the carbon footprint of these industries. For example, the lighter weight and improved strength of these materials can lead to reduced fuel consumption in transportation.

In summary, sustainable thermoplastic nanocomposites are materials that offer a high level of performance and sustainability. The use of renewable and biodegradable materials, as well as the incorporation of nanoparticles such as bio-based polymers and nanoparticles can lead to materials that are stronger and more durable, while also being more environmentally friendly. The improved mechanical and physical properties of these materials can have significant applications in various industries, including packaging, construction, and automotive parts, which can help reduce the carbon footprint of these industries.

The following topics have been covered in this chapter:

- Material design for sustainable thermoplastic nanocomposites materials
- Advantages and limitations of conventional thermoplastic nanocomposites
- Processing of sustainable thermoplastic nanocomposites materials
- Healthcare applications of sustainable thermoplastic nanocomposites materials
- Ethical and regulatory issues in sustainable thermoplastic nanocomposites materials
- Future of sustainable thermoplastic nanocomposites materials

6.2 MATERIAL DESIGN FOR SUSTAINABLE THERMOPLASTIC NANOCOMPOSITES

Thermoplastics can be melted and made into any shape without losing their physical properties. Thermoplastics are classified as non-degradable and degradable thermoplastics. Both non-degradable and degradable thermoplastics can be derived from bio-based or petroleum-based materials. The classification of thermoplastics is shown in Figure 6.1. Some of the common examples of widely used degradable thermoplastics are listed in Table 6.1.

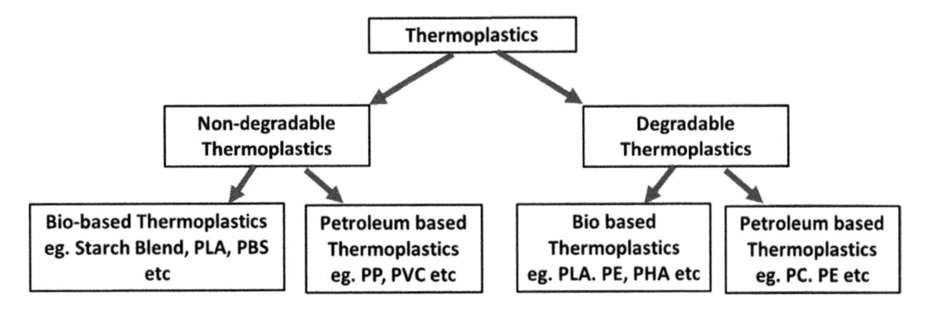

Figure 6.1 Types of plastics available, based on its degradability [4].

Table 6.1 Various non-degradable thermoplastics and their properties [5].

Thermoplastic	Description
Polyethylene (PE)	A versatile, lightweight plastic that is commonly used in packaging, consumer goods, and toys
Polypropylene (PP)	A strong, lightweight plastic that is commonly used in packaging, consumer goods, and automotive parts
Polyvinyl chloride (PVC)	A rigid plastic that is commonly used in pipes, vinyl flooring, and electrical cable insulation
Polystyrene (PS)	A lightweight plastic that is commonly used in packaging materials, insulation, and disposable tableware
Acrylonitrile butadiene styrene (ABS)	A thermoplastic that is commonly used in toys, automotive parts, and electronic housings
Polycarbonate (PC)	A strong, lightweight plastic that is commonly used in eyewear, electronics, and medical devices
Polyethylene terephthalate (PET)	A thermoplastic that is commonly used in beverage containers, food packaging, and textiles
Nylon	A strong, flexible thermoplastic that is commonly used in ropes, gears, and electronic components

On the other hand, degradable or sustainable thermoplastics are made from renewable resources or with a reduced environmental impact compared to traditional petroleum-based thermoplastics. Table 6.2 shows the list of degradable thermoplastics and their properties. One of the most commonly used sustainable thermoplastics is polylactic acid (PLA) made from corn starch or sugarcane. PLA is a biodegradable and compostable material that is used as a replacement for traditional plastic in packaging and consumer goods. This material has a lower carbon footprint than traditional plastics, as it is made from renewable resources and does not produce the same amount of greenhouse gas emissions as traditional plastic manufacturing processes. Another type of sustainable thermoplastic is polyhydroxyalkanoate (PHA), which is made from the bacterial fermentation of plant sugars. PHA is a biodegradable and compostable material that is used in packaging and consumer goods. These materials are biodegradable and compostable and are often used in applications such as packaging and consumer goods. The use of these materials in fabrication helps reduce the environmental impact of plastic production and contribute to a more sustainable future. Some sustainable/degradable thermoplastics are presented in Table 6.2.

Table 6.2 Various degradable thermoplastics and their properties [4].

Sustainable Thermoplastic	Description
Biodegradable polyethylene (PE)	Polyethylene designed to break down in the environment over time
Polylactic acid (PLA)	Biodegradable thermoplastic made from corn starch or sugar cane
Biodegradable polypropylene (PP)	Polypropylene designed to break down in the environment over time
Renewably-sourced polyethylene (PE)	Polyethylene made from plant-based materials rather than petroleum
Bio-based polyvinyl chloride (PVC)	PVC made from renewable resources
Starch-based thermoplastics	Thermoplastics made from renewable resources such as corn starch or potato starch
Bio-based polystyrene (PS)	Polystyrene made from renewable resources
Biodegradable acrylonitrile butadiene styrene (ABS)	ABS designed to break down in the environment over time
Renewably sourced polycarbonate (PC)	Polycarbonate made from renewable resources
Biodegradable polyethylene terephthalate (PET)	PET designed to break down in the environment over time
Bio-based nylon	Nylon made from renewable resources
Eco-friendly polyethylene (PE)	Polyethylene made using a more sustainable production process, such as reducing greenhouse gas emissions or using renewable energy
Recycled thermoplastics	Thermoplastics made from recycled materials, reducing waste and conserving resources

Sustainable thermoplastic nanocomposite materials aim to develop advanced materials that have high strength, toughness and stiffness, while also being environmentally friendly and biodegradable or recyclable. This involves the use of (a) recyclable nanoscale filler particles such as nanoclays, nanotubes and nanofibers (b) bio-based nano-fillers, which are added to sustainable thermoplastic matrices to enhance the properties of materials. These sustainable nanocomposites have several advantages over conventional materials, including mechanical strength, enhanced thermal stability, improved barrier properties, and reduced weight. Some of the widely used thermoplastic nanocomposites materials are [6]:

(i) Thermoplastic nanocomposites based on layered silicates.
(ii) Thermoplastic nanocomposites based on carbon nanotubes.
(iii) Thermoplastic nanocomposites based on inorganic nanoparticles.

(i) Thermoplastic Nanocomposites Based on Layered Silicates: Thermoplastic nanocomposites based on layered silicates are produced by incorporating small amounts of layered silicates into a thermoplastic matrix, resulting in enhanced properties compared to pure polymers. There are various types of thermoplastics used for the preparation of nanocomposites with layered silicates, which include vinyl polymers, polyolefin biodegradable polymers, etc. The manufacturing process includes melt compounding, solution blending, in-situ polymerization, and intercalation/exfoliation. Figure 6.2 shows the schematic of various types of composites generated on mixing layered silicates with polymer matrices. Initially, the layered silicate and thermoplastics do not mix and form micro composites. In intercalated nanocomposites, the layered silicate mixes with the polymer matrix at a crystalline level. In exfoliated nanocomposites, the individual clay layers get separated to form a continuous polymer matrix. The high strength of these materials is due to the interaction between the polymer and the layered silicates, as well as the arrangement and dispersion of the silicates within the polymer matrix. Nucleation and crystallization behavior can also be affected by the presence of layered silicates [7].

Characterization techniques such as X-ray, TEM and SEM are commonly used to analyze the morphology of the resulting nanocomposites. Mechanical properties, such as tensile strength and modulus, can be improved by the addition of layered silicates. Furthermore, thermal stability and flammability can be improved, while photo-oxidative stability can be reduced. Ionic conductivity, optical transparency, and electro-rheology are some additional properties that can be influenced by the presence of layered silicates.

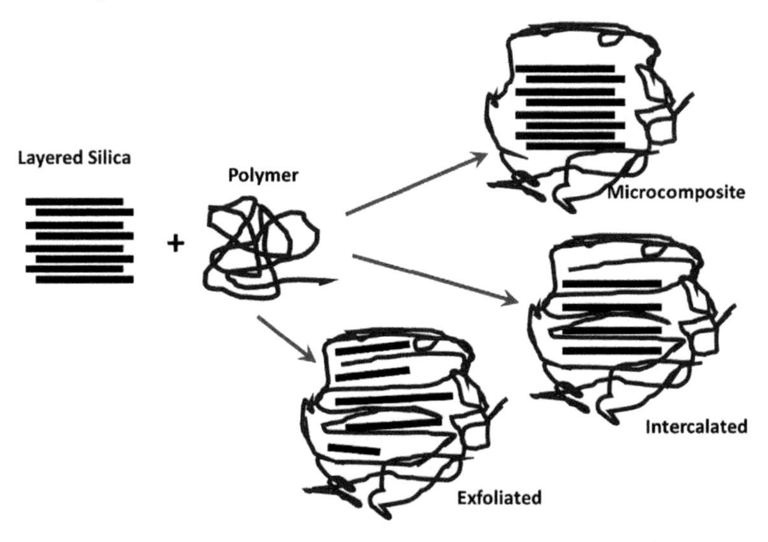

Figure 6.2 Schematic representation of the different types of composites generated by mixing layered silicates with polymer matrices: (a) microcomposites, (b) intercalated nanocomposite, and (c) exfoliated nanocomposites [7].

(ii) Thermoplastic Nanocomposites Based on Carbon Nanotubes: Thermoplastic nanocomposites based on carbon nanotubes (CNTs) and nanofibers (CNFs) are synthesized by incorporating small amounts of CNTs or CNFs into a polymer matrix. These materials have unique properties that make them attractive for a range of applications, including electronics, sensors, energy storage, and structural materials. The addition of CNTs or CNFs to a polymer matrix can influence the crystallization behavior and morphology of the resulting nanocomposite. CNTs and CNFs act as nucleation sites for the polymer matrix, which can cause changes in the crystallization behavior and the morphology of the material. The degree of influence depends on factors such as the concentration and type of the CNTs or CNFs used, as well as the processing conditions.

There are several methods for preparing polymer nanocomposites based on CNTs and CNFs, including solution blending, melt blending, and in-situ polymerization. In solution blending, CNTs or CNFs are dispersed in a solvent along with the polymer matrix, and the resulting solution is then cast into films or fibers. In melting blending, CNTs or CNFs are directly mixed with the polymer matrix in a molten state, and the resulting mixture is then molded or extruded into the desired shape. In-situ polymerization involves the polymerization of monomers in the presence of CNTs or CNFs, resulting in a thermoplastic matrix with uniformly dispersed CNTs or CNFs. Figure 6.3 shows the mechanism of dispersion of CNT or CNFs in thermoplastics. The dispersion of CNTs in thermoplastics is a sequential process starting with (a) wetting of the CNT agglomerates, (b) infiltration of the thermoplastic melt into the CNT agglomerates, (c) disintegration of agglomerate fractions weakened by the infiltration process, and (d) distribution of CNTs in the polymer host.

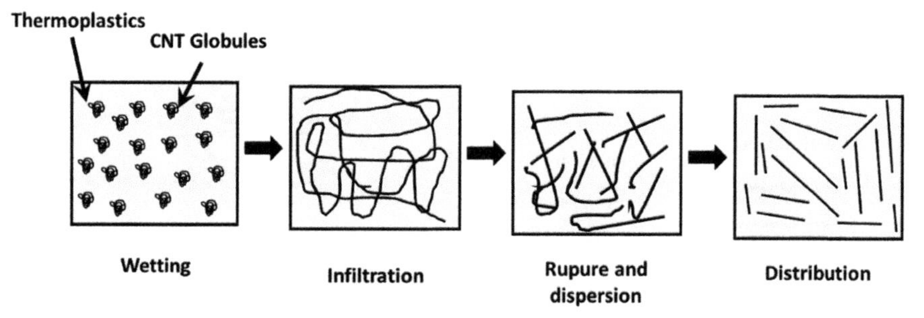

Figure 6.3 Route for the dispersion of CNTs or CNFs in thermoplastics [6].

Morphological Characterization: Transmission electron microscopy (TEM) is commonly used to visualize the morphology of the nanocomposites, as it can provide information about the dispersion, alignment, and interaction of CNTs or CNFs with the polymer matrix. Scanning electron microscopy (SEM) is also used to observe the surface morphology of nanocomposites. X-ray diffraction (XRD) and Raman spectroscopy are other techniques that can provide information on the crystal structure and orientation of CNTs or CNFs in the polymer matrix.

Mechanical Properties: CNTs and CNFs have high strength and stiffness, and the addition of these materials to a polymer matrix can result in significant improvements in the mechanical properties of the resulting nanocomposites. Tensile strength, Young's modulus, and toughness can all be improved by incorporating CNTs or CNFs into the polymer matrix. The degree of improvement depends on factors such as the type and concentration of the CNTs or CNFs used, the interaction between CNTs or CNFs and the polymer matrix, and the alignment of CNTs or CNFs in the polymer matrix.

Electrical Properties: CNTs and CNFs are highly conductive materials, and the addition of these materials to a polymer matrix can result in significant improvements in the electrical conductivity of the resulting nanocomposites. The conductivity of nanocomposites can be tuned by varying the concentration of CNTs or CNFs, and their alignment in the polymer matrix to form percolation. Other electrical properties, such as dielectric constant and dielectric loss, can also be influenced by the presence of CNTs or CNFs in the polymer matrix.

Thermal Properties: CNTs and CNFs have high thermal conductivity, and the addition of these materials to a polymer matrix can result in significant improvements in the thermal conductivity of the resulting nanocomposites. The thermal stability of nanocomposites can also be improved by incorporating CNTs or CNFs into the polymer matrix, as these materials are highly resistant to thermal degradation. Other thermal properties, such as the glass transition temperature and the coefficient of thermal expansion, can also be influenced by the presence of CNTs or CNFs in the polymer matrix.

Optical Properties: The addition of CNTs or CNFs to a polymer matrix can result in changes in the optical properties of the resulting nanocomposite. For example, the addition of CNTs or CNFs to a polymer matrix can alter the refractive index, absorbance, and transmittance of the material. These changes can be used to develop materials with unique optical properties that are useful in a range of applications.

Field Emission Properties: CNTs and CNFs have high aspect ratios, which makes them attractive for use as field emission materials. The addition of these materials to a polymer matrix can result in significant improvements in the field emission properties of the resulting nanocomposite. Field emission properties such as turn-on field, emission current density, and field enhancement factor can all be improved by incorporating CNTs or CNFs into the polymer matrix.

Fiber Surface Properties: The surface properties of CNTs and CNFs can be modified to improve their compatibility with the polymer matrix. This can be done by functionalizing the surface of the CNTs or CNFs with functional groups such as carboxyl, hydroxyl or amine groups. This modification improves the interaction between the CNTs or CNFs and the polymer matrix, resulting in better dispersion and improved mechanical and electrical properties.

Structure-Property Relationship: The properties of polymer nanocomposites based on CNTs and CNFs are highly dependent on the structure of the resulting material. Factors such as the dispersion, alignment and interaction of CNTs or CNFs with the polymer matrix can all influence the resultant properties of the nanocomposites. Understanding the relationship between the structure and the properties of these materials is important for the development of new materials with improved performance.

Thermal Stability and Flammability: Addition of CNTs or CNFs to a polymer matrix can improve the thermal stability and flammability of the resulting nanocomposite. CNTs and CNFs have high thermal stability and are highly resistant to thermal degradation. They also have low flammability, which makes them useful in developing fire-resistant materials.

In conclusion, thermoplastic nanocomposites based on CNTs and CNFs have unique properties that make them attractive for a range of applications. The properties of these materials can be tuned by varying factors such as the type and concentration of CNTs or CNFs used, the interaction between CNTs or CNFs and the polymer matrix, and the alignment of CNTs or CNFs in the polymer matrix. Understanding the relationship between the structure and the properties of these materials is important for the development of new materials with improved performance.

(iii) Thermoplastic Nanocomposites Based on Inorganic Nanoparticles: The preparation of thermoplastic nanocomposites based on inorganic nanoparticles typically involves the dispersion of these nanoparticles in a polymer matrix, using various methods such as melt blending, solution blending, in-situ polymerization, and electrospinning. Some examples of discrete inorganic nanocomposites include nanoparticles, nanorods, nanofibers, etc. The choice of method depends on the type of nanoparticles and the polymer matrix used, as well as the desired properties of the resulting nanocomposites. Figure 6.4 shows the dispersion of inorganic nanoparticles in the thermoplastic matrix. The agglomerated particles pass through two rollers, which are rotating in opposite directions and positioned close to each other. Due to the rolling action, the agglomerated particles break down and disperse within the thermoplastic matrix At last, the characterization of thermoplastic nanocomposites based on inorganic nanoparticles is essential to understand the structure-properties correlation of these materials [8][9]. X-ray diffraction (XRD), transmission electron microscopy (TEM), scanning electron microscopy (SEM), dynamic mechanical analysis (DMA), thermal gravimetric analysis (TGA), and differential scanning calorimetry (DSC) are commonly used to analyze the structure and properties of these materials.

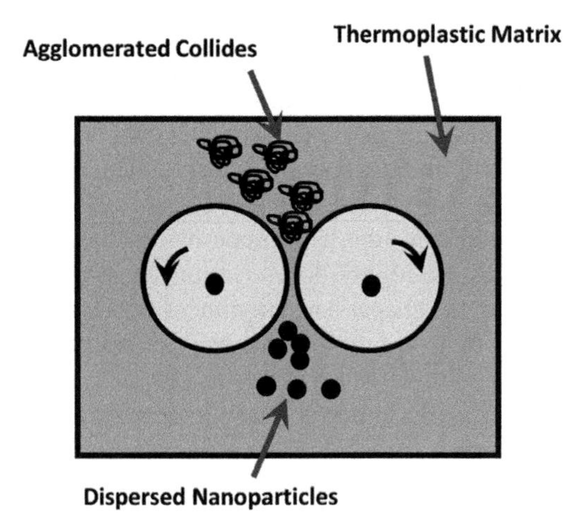

Figure 6.4 Dispersion of inorganic nanoparticles in the thermoplastic matrix through rolling [8, 9].

6.3 APPLICATION OF CONVENTIONAL THERMOPLASTIC NANOCOMPOSITES

Automobiles: Thermoplastic nanocomposites are widely used in the automobile industry due to their light weight, high strength, and improved mechanical properties. These materials are used to produce various automobile parts such as body panels, bumpers, interior components and engine parts. The use of polymer nanocomposites in automobiles can result in improved fuel efficiency, reduced emissions, and improved safety.

Aerospace Applications: Thermoplastic nanocomposites are also used in the aerospace industry due to their high strength, light weight, and improved thermal properties. These materials are used to produce aircraft parts such as wing structures, landing gear, and interior components. The use of polymer nanocomposites in the aerospace industry can result in enhanced fuel efficiency, reduced emissions, and improved safety.

Injection Molded Products: Thermoplastic nanocomposites are used in injection molded products due to their improved mechanical and thermal properties. These materials are used to produce various products such as electrical and electronic components, household appliances, and automotive parts. The use of polymer nanocomposites in injection molded products can result in improved product performance, reduced costs, and improved environmental sustainability.

Coatings: Thermoplastic nanocomposites are used in coatings due to their superior mechanical and thermal properties. These materials are used to produce coatings for various applications such as corrosion protection, wear resistance, and antifouling properties. The use of polymer nanocomposites in coatings can result in improved product performance, reduced costs, and improved environmental sustainability.

Adhesives: Thermoplastic nanocomposites are used in adhesives due to their improved mechanical and thermal properties. These materials are used to produce adhesives for various applications such as structural bonding, electronic packaging, and medical devices. The use of polymer nanocomposites in adhesives can result in improved product performance, reduction in cost, and improved environmental sustainability.

Fire-retardants: Thermoplastic nanocomposites are used as fire-retardants due to their ability to improve the thermal stability and flammability of materials. These materials are used to produce various products such as building materials, textiles, and automotive parts. The use of polymer nanocomposites as fire retardants can result in improved product safety, reduced costs, and improved environmental sustainability.

Packaging Materials: Thermoplastic nanocomposites are used in packaging materials due to their enhanced mechanical and thermal properties. These materials are used to produce various products such as food packaging, medical packaging, and industrial packaging. The use of polymer nanocomposites in packaging materials can result in improved product performance, reduced costs, and improved environmental sustainability.

Microelectronic Packaging: Thermoplastic nanocomposites are used in microelectronic packaging due to their improved mechanical, thermal, and electrical properties. These materials are used to produce various products such as printed circuit boards, microprocessors, and sensors. The use of polymer nanocomposites in microelectronic packaging can result in better product performance, reduced costs, and improved environmental sustainability.

Optical Integrated Circuits: Thermoplastic nanocomposites are used in optical integrated circuits due to their superior optical properties. These materials are used to produce various products such as optical waveguides, photonic crystals, and sensors. The use of thermoplastic nanocomposites in optical integrated circuits can result in improved product performance, reduced costs, and improved environmental sustainability.

Drug Delivery: Thermoplastic nanocomposites are used in drug delivery due to their ability to improve drug solubility, stability, and bioavailability. These materials are used to produce drug delivery systems such as nanoparticles, micelles, and hydrogels. The use of polymer nanocomposites in drug delivery can result in improved therapeutic efficacy, reduced side effects, and improved patient compliance.

Sensors: Thermoplastic nanocomposites are used in sensors due to their enhanced electrical, mechanical, and optical properties. These materials are used to produce sensors for various applications such as environmental monitoring, medical diagnosis, and security systems. The use of polymer nanocomposites in sensors can result in improved sensitivity, selectivity, and response time.

Membranes: Thermoplastic nanocomposites are used in membranes due to their ability to improve membrane performance such as permeability, selectivity, and stability. These materials are used to produce various types of membranes such as gas separation membranes, water purification membranes, and fuel cell membranes. The use of polymer nanocomposites in membranes can result in better product performance, reduced costs, and improved environmental sustainability.

Medical Devices: Thermoplastic nanocomposites are used in medical devices due to their enhanced mechanical, thermal, and biocompatible properties. These materials are used to produce medical devices such as implants, catheters, and drug delivery systems. The use of polymer nanocomposites in medical devices can result in better product performance, reduced costs, and improved patient outcomes.

Consumer Goods: Thermoplastic nanocomposites are used in consumer goods due to their superior mechanical and thermal properties. These materials are used to produce various products such as sporting goods, electronics, and home appliances. The use of polymer nanocomposites in consumer goods can result in improved product performance, reduced costs, and improved environmental sustainability.

In conclusion, thermoplastic nanocomposites have a wide range of applications in various fields such as aerospace, automobiles, biomedical, medical devices, drug delivery, packaging, optical, sensors and consumer goods [8][10]. The unique properties of these materials make them attractive for many applications, and their use can result in better product performance, reduced costs, and improved environmental sustainability.

6.4 ADVANTAGES AND LIMITATIONS OF SUSTAINABLE THERMOPLASTIC NANOCOMPOSITES

Advantages

Thermoplastic nanocomposites have gained popularity in recent years and have exceptional mechanical, thermal and electrical properties. They are made by incorporating nanoscale fillers, such as nanoparticles, nanotubes, and nanofibers, into a thermoplastic matrix [11, 12]. In this section, the advantages of CTN will be discussed in detail.

(i) **Improved Electrical Properties:** Conventional thermoplastic nanocomposites exhibit improved electrical properties compared to pure thermoplastic materials. The addition of conductive fillers, such as carbon nanotubes or graphene, can significantly enhance the electrical conductivity of thermoplastic nanocomposites. This makes them suitable for applications in electrical and electronics industries, where high conductivity is required.

(ii) **Improved Thermal Stability:** Thermoplastic nanocomposites exhibit better thermal stability compared to pure thermoplastic materials. The nanoscale fillers act as nucleation sites for the formation of a crystalline structure, which enhances the thermal stability of the material. This results in a reduced rate of thermal degradation and an increased resistance to thermal aging.

(iii) **Enhanced Barrier Properties:** Thermoplastic nanocomposites are known for their exceptional barrier properties which prevent the migration of gases and liquids through the material. The barrier properties are a result of the combination of the thermoplastic matrix and the nanoscale fillers, making them ideal for involving protection within materials. Thermoplastic nanocomposites have been shown to be effective barriers to water vapor, oxygen and other gases.

(iv) **Processing Advantages:** Thermoplastic nanocomposites can be processed using the conventional thermoplastic route and have several processing advantages over pure thermoplastic materials.

(v) **Enhanced Mechanical Properties**: The addition of nanoscale fillers to the thermoplastic matrix results in a significant improvement in the mechanical properties of thermoplastic nanocomposites. The fillers serve to reinforce the matrix and increase its strength, fatigue resistance and stiffness, leading to an overall improvement in the mechanical performance of the material. This has been shown to exhibit higher tensile strength, flexural strength and impact resistance compared to the pure thermoplastic counterparts. As a result, thermoplastic nanocomposites are ideal for applications that require high strength and fatigue resistance, such as aerospace and automotive components.

(vi) **Improved Damping Properties:** Thermoplastic nanocomposites exhibit superior damping properties compared to pure thermoplastic materials. The nanoscale fillers in the material serve to dissipate energy, which leads to an overall reduction in the magnitude of vibrations and increased damping. This makes them ideal for applications where damping is a critical factor, such as shock absorbers, vibration dampers, and noise reduction.

(vii) **Increased Flame Retardation:** Thermoplastic nanocomposites have been shown to exhibit increased flame retardation compared to pure thermoplastic materials. The addition of flame-retardant fillers, such as aluminium hydroxide or magnesium hydroxide, can significantly enhance the flame retardation property of thermoplastic nanocomposites. This makes them suitable for applications in construction, transportation, and electrical industries, where flame retardation is a critical factor.

(viii) **Recyclability:** Thermoplastic nanocomposites are recyclable and can be melted and molded multiple times, which makes them environmentally friendly. The recyclability of thermoplastic nanocomposites is a result of the thermoplastic matrix, which can be melted and molded without degradation. This makes thermoplastic nanocomposites an attractive alternative to other materials that are not recyclable and are harmful to the environment.

Limitations

Despite their numerous advantages over traditional composites, there are also several limitations associated with conventional thermoplastic nanocomposites that need to be addressed [13]

(i) **Compatibility Issues and Poor Mechanical Properties:** One of the major limitations of conventional thermoplastic nanocomposites is the compatibility between the polymer matrix and the nanofillers. This can lead to poor dispersion of nanofillers, resulting in degraded mechanical properties and decreased thermal stability. This can result in reduced tensile strength, modulus, and toughness, making it difficult to use these materials in high-stress applications.

(ii) **Processing Difficulty:** Another major limitation of conventional thermoplastic nanocomposites is the difficulty in processing these materials. This is due to the high viscosity and the low melt flow index of the polymer matrix, which can lead to processing difficulties, particularly during injection molding.

(iii) **Poor Thermal Stability:** Conventional thermoplastic nanocomposites are often limited in terms of thermal stability due to the low thermal stability of the polymer matrix. This can result in the degradation of the mechanical properties at elevated temperatures, making these materials unsuitable for high-temperature applications.

(iv) **Lack of Scalability:** Conventional thermoplastic nanocomposites are often limited in terms of scalability due to the difficulty in processing these materials on a large scale. This can result in high production costs and limited commercialization of these materials.

(v) **Cost:** The high cost of raw materials, particularly the nanofillers, is another major limitation of conventional thermoplastic nanocomposites. This can make it difficult to produce these materials in large quantities, which can limit their commercialization.

(vi) **Health and Environmental Concerns:** There are also several health and environmental concerns associated with the use of conventional thermoplastic nanocomposites, particularly with regard to the release of nanoparticles from the materials. This can result in potential exposure to toxic and harmful substances, which can pose a risk to human health and the environment.

(vii) **Lack of Standardization:** There is currently a lack of standardization in the production of conventional thermoplastic nanocomposites, which can result in variability in the quality of the materials. This can make it difficult to ensure consistent performance of these materials in different applications.

(viii) **Limited Commercialization:** Despite the several advantages of conventional thermoplastic nanocomposites, they are yet to be widely commercialized, due to the limitations discussed above. This can make it difficult to fully realize the potential of these materials in a wide range of industries.

Despite the countless advantages of conventional thermoplastic nanocomposites, there are also several limitations associated with these materials; these limitations need to be addressed in order to fully realize their potential of thermoplastic nanocomposites. Some of the key limitations include compatibility issues and sustainability. To overcome these limitations, it is important to continue research and development to find the optimum process parameters for improving the performance and scalability of these materials.

6.5 PROCESSING OF SUSTAINABLE THERMOPLASTIC NANOCOMPOSITE MATERIALS

Development of sustainable thermoplastic nanocomposite materials is a complex and multi-step process that involves several stages including material selection, design, processing, and characterization. One of the key challenges in the processing of thermoplastic nanocomposites is to ensure that the nanofillers are well dispersed within the matrix, and that the interactions between the matrix and the filler particles are optimized. This can be achieved through several processing techniques, including melt mixing, solution blending, and melt compounding. Melt mixing is a common processing method that is used to produce thermoplastic nanocomposites. In this method, the nanofillers are added to the thermoplastic matrix while it is in a molten state, and the two components are subsequently mixed together using an extruder. The resulting mixture is then cooled and solidified, producing a nanocomposite material with well-dispersed nanofillers. Solution blending is another common processing method that is used to produce thermoplastic nanocomposites. In this method, the nanofillers are dispersed in a solvent, and then blended with the thermoplastic matrix. After this, the resulting mixture is dried, producing a nanocomposite material with well-dispersed nanofillers. Once the thermoplastic nanocomposites have been produced, they can be fabricated into various products using various fabrication techniques.

(i) **Material Selection:** The first step in the processing and fabrication of sustainable thermoplastic nanocomposites is to select appropriate materials. The selection of materials for nanocomposites is based on several factors such as mechanical properties, thermal stability, and processing conditions. The matrix material must have good thermal stability and should be compatible with the nanofiller. In addition, the mechanical properties of the matrix material should be improved by the addition of the nanofiller.

(ii) **Material Design:** The next stage involves the development of a suitable design for the nanocomposite. Various factors such as properties of the materials, the processing conditions, and the desired end use properties of the nanocomposite have to be considered for material design. For example, if the nanocomposite is intended to be used in applications where high strength is required, the design should be such that the nanofiller is well dispersed and well interconnected within the matrix.

(iii) **Processing:** The processing stage involves the preparation of the nanocomposite material. There are several methods for processing sustainable thermoplastic nanocomposites, including extrusion, injection molding, compression molding, and thermoforming. Every method has its own advantages and disadvantages and depends on the desired end use properties and the processing conditions. It is important to carefully control the processing conditions to ensure that the final product is of high quality and has the desired properties.

(a) **Extrusion Molding:** In the extrusion process, the thermoplastic matrix material and the nanofiller are mixed together and then extruded to form a continuous profile. The extrusion process is well suited for the production of long continuous lengths of nanocomposites. Figure 6.5 shows the schematic diagram of the extrusion molding process [13]. The process starts with material selection, where the manufacturer selects the appropriate thermoplastic material based on the product requirements such as strength, stiffness, and heat resistance. The material is then loaded into an extruder, where it is melted and formed into a tube known as a parison. The parison is then transferred to a blow molding machine, where it is placed in a mold. The mold is then closed and air is injected into the extruder, causing it to expand and take the shape of the mold. Once the material has expanded to fill the mold, it is allowed to cool and solidify. Post solidification, the mold is opened and the product is removed. Extrusion molding is a highly automated process that minimizes the need for manual labor, reducing the risk of human error. The process is ideal for producing products with consistent quality and dimensional accuracy, making it suitable for mass production runs. The process is also well-suited for producing products with complex shapes and intricate details.

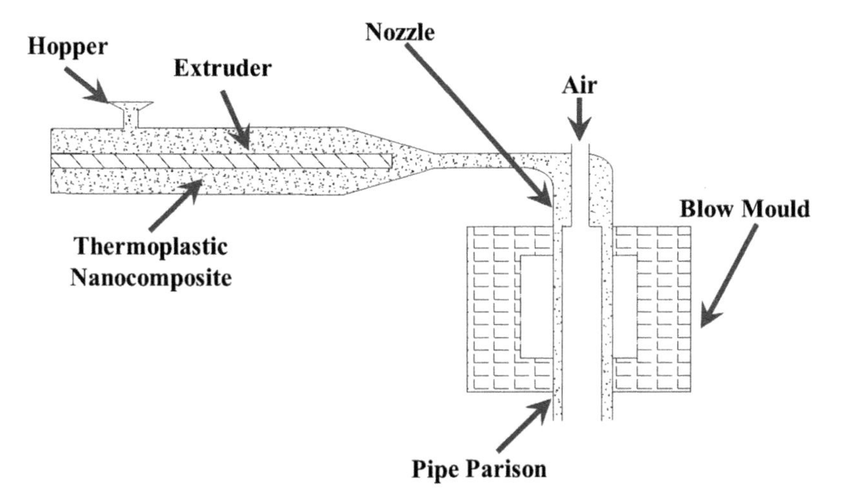

Figure 6.5 Schematic diagram of extrusion molding [13].

(b) **Injection Molding:** Injection molding is a process where the thermoplastic material is melted and injected into a mold to get solidified [14]. Figure 6.6 shows the schematic diagram of the injection molding process. This process is well suited for the production of complex geometries and high-volume production runs. The process starts with melting the plastic material, which is then fed into an injection molding machine through a hopper. The molten plastic is then injected into the mold under high pressure, through a rotary screw, and the mold is closed to keep the plastic material inside. The plastic material cools and solidifies in the mold, taking the shape of the mold cavity. Once the plastic material is solidified, the mold is opened and the part is removed. The part is then typically trimmed and finished to remove any flash or excess material and is inspected to ensure that it meets quality and design requirements. One of the key benefits of injection molding is its ability to produce large numbers of identical parts with complex shapes, precise dimensional tolerances, and high detail. Injection molding also has a low environmental impact, as the process produces very little waste material. The scrap generated during the process can be recycled and used to produce new parts, reducing the overall environmental impact of the process.

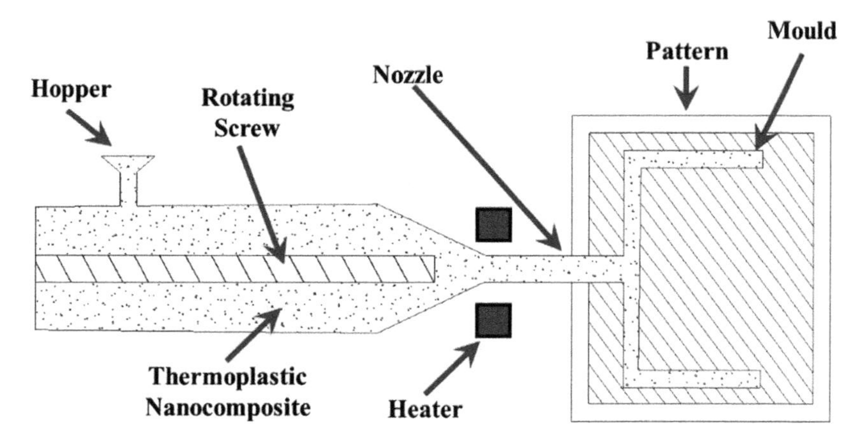

Figure 6.6 Injection molding process for thermoplastic nanocomposites [14].

(c) **Compression Molding:** Compression molding is a process in which a preformed thermoplastic material is placed into a mold and subjected to pressure and heat [15]. This process is well suited for the production of complex geometries and high-volume production runs. It is a method in which thermoplastic material is placed into a heated mold, and subsequently subjected to pressure to form the desired shape. This process is widely used for the production of a wide range of parts, including large automotive components, household goods, and industrial components. Figure 6.7 shows the schematic diagram of the compression molding process. Compression molding starts with the heating of a thermoplastic material to its melting point. The melted material is then placed between the top and bottom dies, which is then closed and subjected to high pressure. The pressure forces the material to fill the entire mold cavity and get the shape of the mold. The mold is then cooled, and the solidified plastic part is removed.

The process is versatile and can be used to produce parts with a high degree of accuracy and consistency. The pressure applied during the molding process helps in eliminating the air pockets or voids in the material, if any, resulting in a more uniform and consistent part. This makes it ideal for the production of parts with tight tolerance and precise dimensions.

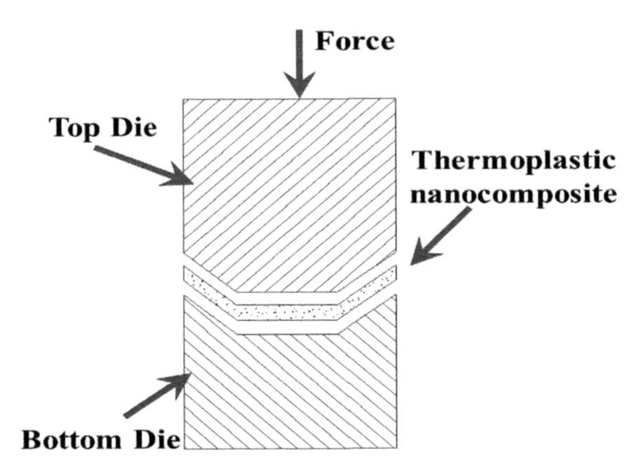

Figure 6.7 Compression molding process for thermoplastic nanocomposites [15].

(d) **Thermoforming:** Thermoforming is a process in which a sheet of thermoplastic material is heated and formed into a desired shape using a mold. This process is well suited for the production of complex geometries and low-volume production runs. The molded part is cooled to solidify the material into its final form. This process is commonly used for the large-scale manufacture of products such as packaging containers, consumer goods, and automotive parts.

(iv) **Characterization:** The final stage in the processing and fabrication of sustainable thermoplastic nanocomposites is characterization. The characterization stage involves the evaluation of the properties of the nanocomposite material. The properties that are usually evaluated include mechanical properties, thermal stability, and processing conditions. It is important to carefully control the processing conditions to ensure that the final product is of high quality and has the desired properties. For example, the temperature and pressure during the injection molding process must be carefully controlled to ensure that the product is molded with the correct shape and has the desired mechanical properties.

Plastic materials produced using environmentally friendly methods are biodegradable or recyclable. The processing of sustainable thermoplastic materials involves several stages, including problem identification, design, material selection, and object fabrication. Each of these stages is crucial in ensuring that the final product is of high quality, and is environmentally friendly and economically feasible.

(i) **Problem Identification:** This stage involves identifying a problem or a need for a specific product and determining how to address it using sustainable thermoplastic materials. For example, a company might need to create a product that is biodegradable, durable and cost-effective. The problem identification stage is critical because it sets the direction for the rest of the design and production process.

(ii) **Design:** This stage involves developing a design for the product, including its shape, size, and functionality. It is essential to consider the end use of the product, the production process, and the materials that will be used during the fabrication process. For example, a designer may choose to use a particular type of thermoplastic material that is biodegradable and can be easily molded into a specific shape.

(iii) **Material Selection:** In this stage, the designer selects the appropriate sustainable thermoplastic materials that will be used in the fabrication process. It is critical to consider the mechanical properties, cost, and availability of the materials. The designer must also consider the processing conditions that will be used to fabricate the product, such as temperature, pressure, and cooling rate. For example, a designer may choose a sustainable thermoplastic material that is strong and can withstand high temperatures, such as polylactic acid (PLA).

(iv) **Object Fabrication:** This stage involves the actual production of the product using the sustainable thermoplastic materials that were selected in the previous stage. The fabrication process can use a variety of techniques, including injection molding, blow molding, thermoforming, and extrusion. It is important to carefully control the processing conditions to ensure that the final product is of high quality and has the desired properties. For example, the temperature and pressure during the injection molding process must be carefully controlled to ensure that the product is molded into the correct shape and has the desired mechanical properties.

Various steps in the fabrication of sustainable thermoplastic nanocomposite materials include:

1. **Raw material selection:** The selection of sustainable and environmentally friendly raw materials is the first step in the fabrication process. This includes selecting the appropriate thermoplastic polymer and the desired nanofillers, such as nanoparticles or nanofibers, which will be used to reinforce the material.
2. **Material preparation:** Once the raw materials have been selected, they must be prepared for processing. This typically involves blending the thermoplastic polymer and nanofillers together, either through physical mixing or through chemical modification of the surface of the nanofillers to improve their compatibility with the polymer matrix.
3. **Compounding:** The compounded mixture of a thermoplastic polymer and nanofillers is then processed into a homogeneous material. This can be done through a variety of methods, such as melt blending, solution blending, or in situ polymerization.

4. **Molding:** The compounded material is then molded into the desired shape, using a variety of processing techniques such as injection molding, blow molding, thermoforming, or compression molding.
5. **Characterization:** After the material has been molded, it is important to characterize its properties to assess the effectiveness of the processing and fabrication processes. This typically involves testing the mechanical, thermal, electrical, and other properties of the material to determine its performance.
6. **Optimization:** Based on the results of characterization, the processing and fabrication conditions can be optimized to improve the performance of the sustainable thermoplastic nanocomposite material.

By following these steps, sustainable thermoplastic nanocomposite materials can be fabricated with improved properties and performance. These materials have the potential to be used in a wide range of applications, including packaging, construction, automotives, and electronics.

It is important to note that the processing and fabrication of sustainable thermoplastic nanocomposite materials is a complex and multi-disciplinary field, and it is essential to work with experts in materials science, engineering, and processing to achieve the best results.

In addition, it is also important to consider the sustainability of the entire life cycle of the material, including the sourcing of raw materials, the processing and fabrication of the material, and its disposal at the end of its life. This requires a comprehensive and holistic approach for the development of sustainable thermoplastic nanocomposite materials.

6.6 APPLICATIONS OF SUSTAINABLE THERMOPLASTIC MATERIALS IN HEALTHCARE

Healthcare applications of sustainable thermoplastic nanocomposite materials are on a rise, for these materials have the potential to revolutionize the medical industry. These materials are becoming increasingly popular due to their unique properties, such as high strength, toughness, biocompatibility, and resistance to wear and tear, which make them ideal for a wide range of medical applications [16].

One of the main applications of sustainable thermoplastic nanocomposite materials in healthcare is the development of implantable devices such as pacemakers, stents and artificial joints. These devices are designed to be implanted into the human body and remain there for an extended period of time; therefore, the materials used in the manufacture of such devices must be biocompatible, durable and non-toxic. Sustainable thermoplastic nanocomposite materials have been shown to possess these properties, making them ideal for use in implantable devices. Other applications of sustainable thermoplastic nanocomposite materials involve the development of surgical instruments and tools. Surgical instruments are required to be strong and durable, yet flexible enough to maneuver around delicate tissues. Sustainable thermoplastic nanocomposite materials offer the perfect combination of strength and flexibility, making them ideal for use in surgical instruments and tools. In addition to these applications, sustainable thermoplastic nanocomposite materials are also finding use in the development of medical devices and equipment, such as diagnostic instruments, drug delivery systems, and dialysis machines. These devices often require materials that are lightweight, durable, and capable of resisting corrosion, wear and tear; this makes sustainable thermoplastic nanocomposite materials the ideal choice for such applications. Sustainable thermoplastic nanocomposite materials have high strength-to-weight ratios. This means that these materials can be used to produce lightweight and strong medical devices, which can reduce the risk of injury and increase patient comfort. Some of the key applications of sustainable thermoplastic nanocomposites in healthcare are discussed below.

Medical Devices and Equipment: Thermoplastic materials are extensively used in the production of medical devices and equipment, including blood collection tubes, syringes, catheters, dialysis equipment, and surgical instruments. These materials offer excellent mechanical strength, chemical resistance, and biocompatibility, making them suitable for use in medical applications. For example, Polycarbonate (PC) is widely used in the production of medical devices such as centrifuges and pipettes. This material offers excellent clarity and is also impact resistant, which is important in laboratory settings where devices are frequently handled.

Packaging: Thermoplastic materials are also used in the production of medical packaging, such as blister packs for medications and medical devices, as well as containers for medical devices and equipment. These materials offer excellent barrier properties, making them ideal for protecting sensitive medical products from moisture, light, and other environmental factors. For example, Polyethylene Terephthalate (PET) is commonly used in the production of medical packaging due to its high clarity, transparency and strength.

Implants: Thermoplastic materials are also used in the production of medical implants such as artificial joints, spinal implants, and other orthopedic devices. These materials offer excellent biocompatibility, making them suitable for use in the human body. For example, Polyetheretherketone (PEEK) is widely used in the production of spinal implants due to its high mechanical strength, toughness and biocompatibility.

Disposable Products: Thermoplastic materials are also used in the production of disposable medical products such as gloves, gowns, face masks, and drapes. These materials offer excellent barrier properties, making them ideal for use in medical applications where infection control is critical. For example, Polyethylene (PE) is commonly used in the production of disposable medical products due to its low cost and excellent barrier properties.

In conclusion, thermoplastic materials play a critical role in promoting sustainability in the healthcare industry. These materials offer excellent mechanical strength, chemical resistance, and biocompatibility, making them suitable for the aforementioned applications. Sustainable thermoplastic nanocomposite materials offer a wide range of benefits to the healthcare industry, including biocompatibility, durability, strength and flexibility. These materials are being increasingly used in the development of implantable devices, surgical instruments, medical devices and equipment, etc. With the ongoing advancements in nanotechnology and materials science, it is expected that sustainable thermoplastic nanocomposite materials will play an even more important role in the healthcare industry in the years to come.

6.7 ETHICAL AND REGULATORY ISSUES WITH SUSTAINABLE THERMOPLASTIC COMPOSITES

The use of thermoplastic materials has become widespread in many industries due to their versatility, ease of processing and low cost. However, the production and disposal of these materials have raised significant environmental and ethical concerns. The development of sustainable thermoplastic materials is an attempt to address these issues and provide a more responsible alternative to traditional thermoplastics. In this section, we explore the ethical and regulatory issues involved in the use of sustainable thermoplastic materials, and their future [17].

Ethical Issues

(a) **Environmental Impact:** Although sustainable thermoplastic composites are designed for less harmful environmental impact compared to traditional materials, worries regarding the overall ecological footprint of these materials may persist. Ethical considerations includes, assessing the materials' life cycle and environmental impact of sustainable thermoplastic composites, ensuring minimal resource depletion, and mitigating pollution throughout the production process.

(b) **End-of-Life Management:** Ethical considerations extend up to the end-of-life management of sustainable thermoplastic composites. Companies should design products with recyclability or biodegradability in mind and invest in infrastructure for recycling or composting. The company must promote appropriate disposal techniques which helps to reduce pollution to the environment and damage to ecosystems.

(c) **Greenwashing:** There is a risk of greenwashing, where companies overstate or falsify the environmental advantages the environmental benefits of their products. Transparency and accuracy in reporting the sustainability credentials of sustainable thermoplastic composites, supported by reliable certifications and independent verification, are essential components of ethical responsibility.

(d) **Supply Chain Transparency:** Ethical sourcing of raw materials is crucial to ensuring that sustainable thermoplastic composites do not contribute to environmental degradation or human rights abuses. Companies must track the origins of their commodities and confirm that they originate from sustainably managed and sustainable sources.

Regulatory Issues

- Material Safety and Standards: Standards and laws pertaining to the performance and safety of sustainable thermoplastic composites may be established by regulatory organizations. By adhering to these standards, materials are guaranteed to satisfy basic specifications for durability, strength, and chemical make-up as well as health and safety criteria for uses.
- Environmental Regulations: Throughout the life cycle of sustainable thermoplastic composites, governments may establish legislation to address the impact these materials have on the environment. Regulations pertaining to waste disposal, end-of-life management, and emissions during production may fall under this category. In order to maintain sustainable operations and reduce environmental impact, companies are required to adhere to certain standards.
- Product Labeling and Certification: To assure transparency and educate customers about the environmental qualities of sustainable thermoplastic composites, regulatory bodies may mandate the labeling or certification of these materials. Claims about sustainability, recycled content, or biodegradability can be verified with the use of certification programs like eco-labels or third-party certifications.

6.8 THE FUTURE OF SUSTAINABLE THERMOPLASTIC NANOCOMPOSITES

- **Renewable Resources:** The future of sustainable thermoplastics is likely to be based on renewable resources such as corn starch, sugarcane and cellulose. These materials can be produced more efficiently and at a lower cost than traditional thermoplastics, and are biodegradable and compostable, which reduces their impact on the environment [18].
- **Biodegradable Plastics:** Biodegradable plastics, which break down into harmless substances in the environment, are an important part of the future of sustainable thermoplastics. These materials can be used in applications where traditional thermoplastics would persist in the environment for hundreds of years, such as food packaging, disposable cutlery, and shopping bags.
- **Recycling:** The recycling of thermoplastics will become increasingly important as we look to reduce waste and conserve resources. Companies are developing technologies to make thermoplastics easier to recycle, such as materials that can be sorted by color or chemical composition.
- **Public Policy:** Governments will play an important role in promoting the development and use of sustainable thermoplastics. This can include regulations to reduce waste, promote recycling, and incentivize the use of renewable resources. Additionally, governments can provide funding for

research and development in this area, and promote education and public awareness on the importance of sustainability in the use of thermoplastic materials.

6.9 CONCLUSIONS

Sustainable thermoplastic nanocomposites are composite materials that are being currently used in a wide range of industrial and medical applications due to their superior properties and biodegradable nature. The production of traditional thermoplastic nanocomposites is a significant contributor to greenhouse gas emissions; on the other hand, sustainable thermoplastic materials have lower greenhouse gas emissions and carbon footprint. Biopolymers derived from renewable resources like corn, sugarcane and potato starch can be used to produce sustainable thermoplastic materials. These materials have a lower carbon footprint, as their production requires less energy and results in lower greenhouse gas emissions. Furthermore, sustainable thermoplastic nanocomposites can be recycled or biodegraded at the end of their life. This significantly reduces the amount of waste that ends up in landfills, in turn reducing the environmental impact.

In summary, the design and development of sustainable thermoplastic nanocomposites requires a comprehensive understanding of the desired properties for the intended application, the environmental impact of the materials, and the processing techniques employed. This chapter provides an overview of the possibilities offered by these materials towards improving the performance of a wide range of products, while minimizing their environmental impact.

REFERENCES

[1] Haider, S., Khan, Y., Almasry, W.A. and Haider, A. 2012. Thermoplastic nanocomposites and their processing techniques, thermoplastic—composite materials. *In*: Intech. 113–130.

[2] Markandan, K. and Quan, C. 2023. Fabrication, properties and applications of polymer composites additively manufactured with filler alignment control : A review. Compos. Part B. 256: 110661. https://doi.org/10.1016/j.compositesb.2023.110661.

[3] Cetiner, B., Dundar, G.S., Yusufoglu, Y. and Okan, B.S. 2023. Sustainable engineered design and scalable manufacturing of upcycled graphene reinforced polylactic acid/ polyurethane blend composites having shape memory behavior. Polymers 15(5): 1085. https://doi.org/10.3390/polym15051085.

[4] Tokiwa, Y., Calabia, B.P., Ugwu, C.U. and Aiba, S. 2009. Biodegradability of plastics. Int. J. Mol. Sci. 10: 3722–3742. https://doi.org/10.3390/ijms10093722.

[5] Azeez, T.O. 2019. We are IntechOpen, the world's leading publisher of Open Access books Built by scientists, for scientists TOP 1%. *In*: Intech. 1–19.

[6] Cholleti, E.R. and Gibson, I. 2018. ABS nano composite materials in additive manufacturing. IOP Conf. Ser. Mater. Sci. Eng. 455: 012038. DOI 10.1088/1757-899X/455/1/012038.

[7] Müller, K., Bugnicourt, E., Latorre, M., Jorda, M., Sanz, Y.E. and Lagaron, J.M. 2017. Review on the processing and properties of polymer nanocomposites and nanocoatings and their applications in the packaging, automotive and solar energy fields. Nanomaterials. 7(4): 74. https://doi.org/10.3390/nano7040074.

[8] Chi, H., Gu, Y., Xu, T. and Cao, F. 2017. Multifunctional organic—inorganic hybrid nanoparticles and nanosheets based on chitosan derivative and layered double hydroxide: Cellular uptake mechanism and application for topical ocular drug delivery. Int. J. Nanomedicine. 12: 1607–1620.

[9] Tanahashi, M. 2010. Development of fabrication methods of filler/polymer nanocomposites: With focus on simple melt-compounding-based approach without surface modification of nanofillers. Materials (Basel). 3: 1593–1619.

[10] Puglia, D., Luzi, F. and Torre, L. 2022. Preparation and applications of green thermoplastic and thermosetting nanocomposites based on nanolignin. Polymers (Basel). 14(24): 5470. https://doi.org/10.3390/polym14245470.

[11] Modi, V.K., Shrives, Y., Sharma, C. and Sen, P.K. 2015. Review on green polymer nanocomposite and their applications. Int. J. Innov. Res. Sci. Eng. Technol. 3: 17651–17656. https://doi.org/10.15680/IJIRSET.2014.0311079.

[12] Manral, A., Kumar Bajpai, P., Ahmad, F. and Joshi, R. 2021. Processing of sustainable thermoplastic based biocomposites: A comprehensive review on performance enhancement. J. Clean. Prod. 316: 128068. https://doi.org/10.1016/j.jclepro.2021.128068.

[13] Khalifa, M., Anandhan, S., Wuzella, G., Lammer, H. and Mahendran, A.R. 2020. Thermoplastic polyurethane composites reinforced with renewable and sustainable fillers–A review. Polym. Technol. Mater. 59: 1751–1769. https://doi.org/10.1080/25740881.2020.1768544.

[14] Mondragón, M., Hernández, E.M., Rivera-Armenta, J.L. and Rodríguez-González, F.J. 2009. Injection molded thermoplastic starch/natural rubber/clay nanocomposites: Morphology and mechanical properties, Carbohydr. Polym. 77: 80–86.

[15] Quadrini, F., Bellisario, D., Santo, L., Stan, F. and Catalin, F. 2017. Compression moulding of thermoplastic nanocomposites filled with MWCNT, Polym. Polym. Compos. 25: 611–620. https://doi.org/10.1177/096739111702500806.

[16] Hosseini, E.S., Dervin, S., Ganguly, P. and Dahiya, R. 2021. Biodegradable materials for sustainable health monitoring devices. ACS Appl. Bio Mater. 4: 163–194. https://doi.org/10.1021/acsabm.0c01139.

[17] Norizan, M.N., Shazleen, S.S., Alias, A.H., Sabaruddin, F.A., Asyraf, M.R.M., Zainudin, E.S. et al. 2022. Nanocellulose-based nanocomposites for sustainable applications: A review. Nanomaterials. 12: 1–51.

[18] Thakur, M., Chandel, M., Rani, A. and Sharma, A. 2022. Introduction to biorenewable nanocomposite materials: Methods of preparation, current developments, and future perspectives. ACS Symp. Ser. 1411: 1–24. https://doi.org/10.1021/bk-2022-1411.ch001.

Chapter 7

Natural 2D Material Polymer Composites: Processing, Properties, and Applications

Naimul Arefin[1], Ragavanantham Shanmugam[2] and Minxiang Zeng*[1]
[1]Department of Chemical Engineering, Texas Tech University, Lubbock, TX 79409, USA
[2]Department of Engineering Technology, Fairmont State University, WV 26554, USA

7.1 INTRODUCTION

A nanocomposite is a multiphase system containing materials with one, two, or three dimensions <100 nanometers (nm) [1]. Compositions and nanostructures are essential to composite properties, such as preferential reinforcement of one phase or targeted reinforcement of the interphase, filler-induced changes in phase morphology, or phase alignment [2,3]. In this chapter, we will focus on clays and other natural two-dimensional materials and discuss the processing, properties, and applications of polymer matrix-nanoclay composites. For clay-based composites, early investigations involve studies such as polymer/layered silicate clay mineral composites [1,4,5]. Nanocomposites are gaining popularity as they have several advantages over traditional micro-composites or monolithic materials, including a high surface area-to-volume ratio, improved stability, and enhanced mechanical properties [6].

*For Correspondence: minzeng@ttu.edu

Typically, appropriately adding nanoclays to a polymer matrix can improve its performance by leveraging the intrinsic properties of the nanoscale filler [7]. As natural clays are composed of 2D silicate layers that could be intercalated by small molecules, they are often used as additives/fillers for various composite systems [8, 9]. Two unique specific properties of layered silicates are utilized in preparing nanocomposites. The first is that the silicate particles can be exfoliated into individual thin layers, and the second is that their surface chemistry can be fine-tuned via exchange reactions with ions or functionalized using ligands [10, 11]. Silicates typically form two types of nanocomposites. In the first type, the distance between the layered silicates is increased. On the other hand, in some of the structures, the silicate thin layers are dispersed throughout the polymer [11, 12]. The former is called an intercalated structure, and the latter is an exfoliated structure. Several nanocomposite preparation methods have been used to date, resulting in both intercalated and exfoliated nanostructures, as discussed in this chapter.

The popularity of nanocomposites has increased tremendously in the past decades [13–15]. The manufacturing process of nanocomposites influences the morphology, structures, and properties of clay/polymer nanocomposites [16]. Clay-based nanocomposites have enhanced barrier properties as the gas permeability is decreased [17]. Clay-based nanocomposites are now widely used in various environmental and biomedical applications. One emergying research topic is electrospinning to create bioresorbable nanofiber scaffolds for tissue engineering applications [18]. Clay-based nanocomposites have the potential to be used for drug delivery/release applications as well [18]. These nanocomposites are also suitable adsorbents for removing pollutants from water. Furthermore, nanocomposites have promising applications in other fields as well. Polymer nanocomposites have emerged as promising candidates for developing materials capable of effectively attenuating photon or particle radiation [19]. Nanocomposites are also considered favorable materials for additive manufacturing [20–22]. The applications of nanocomposites can be further expanded to develop gas sensors/biosensors, which are vital in the clinical sectors, bioengineering, and environmental applications [23].

7.2 PROCESSING OF 2D MATERIAL-BASED COMPOSITES

Natural clay minerals containing layered silicate structures have been utilized over the past decade in various applications, from industrial materials and environmental pollution control to applications in biomedical devices [24]. The performance of these composites is primarily attributed to the natural properties of clay, including its high specific surface area, cation exchange capacity (CEC), selectivity, surface hydrophilicity, and surface electronegativity. Recent investigations have focused on integrating these inherent characteristics into materials through the incorporation of various polymers (vinyl polymers, condensation polymers, polyolefins) [25, 26] with clay to fabricate enhanced composites [27]. These advanced composites demonstrate superior mechanical and thermal properties, enhanced biodegradability, controlled gas permeability, and reduced flammability [25, 28, 29]. In pursuit of the synthesis and processing of clay-based composites, tremendous efforts have been made to improve the composite performance (Figure 7.1), and the main approaches have been discussed in this section.

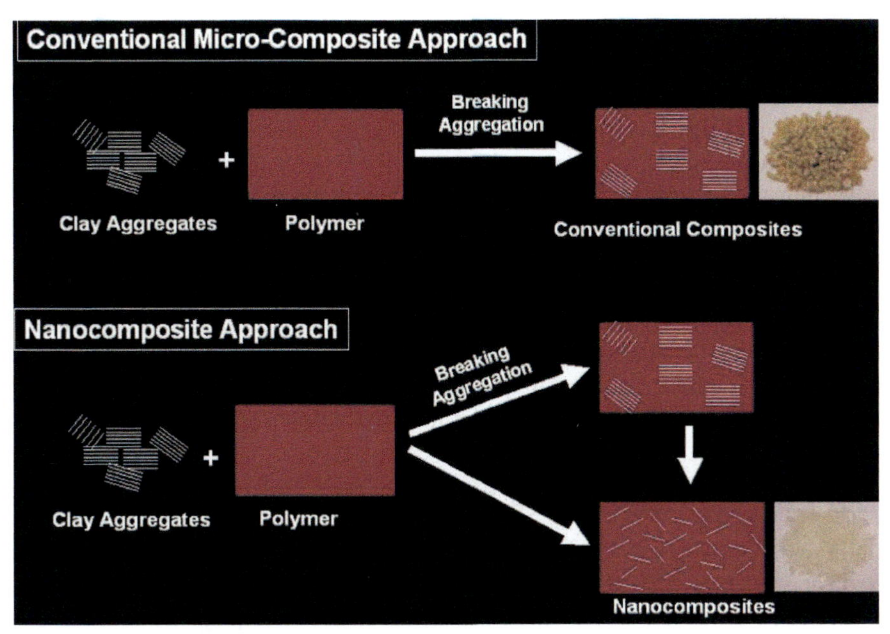

Figure 7.1 The general approach in the fabrication of conventional micro-composites and nanocomposites. Adapted from ref. [30] with permission from Science Direct.

7.2.1 *In Situ* Template Synthesis

Typically, the fabrication of stereospecific polymers has been accomplished through the *in-situ* polymerization process, which involves the polymerization of monomers within the confines of molecular-sized gaps, utilizing either a gel or solution to integrate the mineral clay into the polymer matrix. Here, the polymer serves as a template for the formation of layers by aiding the inorganic host crystal's growth and nucleation and fixes itself when the layers rise [31]. This technique has been successfully applied with polymers such as polyvinylpyrrolidone (PVP), hydroxypropylmethylcellulose (HPMC), polyacrylonitrile (PAN), poly(diallyldimethylammonium chloride) (PDDA), and polyaniline (PANI) [12]. One example for the *in situ* hydrothermal crystallization of nanocomposites involves a two-day reflux process of the clay with 2 wt% gel composed of the selected polymer, lithium fluoride (LiF), magnesium hydroxide ($Mg(OH)_2$), and silica sol (SiO_2) in water [32]. The main challenge of this method to scale up is the slow growth rate of the crystals, often taking hours to days to complete [31]. Additionally, the expansion of silicate layers during the process may result in the aggregation of molecules, consequently restricting the elongation of the layers, which is a critical factor to consider in synthesizing nanocomposites [32–34].

7.2.2 Intercalation of Polymer from Solution

This process employs a dual-stage approach involving the dispersion of the polymer with suitably intercalated clays [35]. The initial stage introduces silicate layers into a solvent aimed at their expansion. Following this, the second stage facilitates the amalgamation of polymers with the intercalated clays, where the polymer is either diffused, dispersed, or dissolved into the expandable layers of silicate, wherein solvent selection is critical for achieving both a high solubility of the matrix and an optimal distribution of the layered silicates. Subsequently, the solvent is effectively removed from the composite material via methods such as vaporization or vacuum extraction [36]. The swelling behavior of montmorillonite (MMT) with alkyl chains differs noticeably [15, 37]. Employing this methodology, Chang et al. have successfully fabricated poly(lactic acid) (PLA) nanocomposites, incorporating various organically modified clays such as cloisite 15A, 25A, and 30B in their study [38]. A crucial discovery from their research highlighted the enthalpic interactions between the polymer and clay as one of the pivotal factors in achieving an efficient dispersion of clays within polymer composites, enhancing the composite's overall performance and material properties [38, 39].

7.2.3 *In Situ* Intercalative Polymerization

Traditional methods of *in situ* intercalation-based polymerization involve the expansion of layered silicates in a liquid monomer or monomer solution, which facilitates polymer formation within the intercalated sheets [40]. As a result, the silicate layers undergo exfoliation, leading to a disordered structure, attributed to the equalized rate of polymerization occurring both within and outside the interlayers [40]. The polymer chain length can be engineered according to the desired applications by various polymerization techniques [29, 32, 41–43]. Greesh et al. examined the impact of sodium montmorillonite (Na–MMT) on the synthesis of poly(styrene-*co*-butyl acrylate) copolymers via free-radical polymerization in an emulsion, resulting in an intercalated structural arrangement [44]. Additionally, the study noted that the reactivity of the polymer groups imposed limitations on the penetration of polymers into the clay galleries [44]. Conversely, Oral et al. were able to synthesize poly(cyclohexene oxide) (PCHO)/clay (cloisite 30B) nanocomposites through the *in situ* photoinitiated activated monomer cationic polymerization, selecting an organically modified clay, cloisite 30B, as the layered silicate [45]. Initial X-ray diffraction (XRD) results indicated a basal spacing expansion from 1.84 to 4.86 nm, signifying an exfoliated structure [45]. Despite these advancements, this process is limited by the time-intensive nature, the less stable exfoliation phase, and the tendency to produce agglomerates with layered silicates. These challenges present avenues for further research, indicating the need for continued investigation to overcome these limitations and enhance the efficiency of this approach in the synthesis of nanocomposites [46–48].

7.2.4 Melt Intercalation

Melt intercalation has emerged as one of the favorable techniques in the fabrication of thermoplastic polymer nanocomposites, involving heating the polymer matrix with the fillers to elevated temperatures followed by mixing to ensure their homogeneous distribution throughout the composite [49]. The formation of intercalated or exfoliated nanocomposites relies on the polymer's compatibility with the silicate layers' surface. This method stands out because of its environmentally friendly nature, attributed to being solvent-free. Additionally, its compatibility with modern industrial and manufacturing techniques enhances its applicability [32, 46, 49]. Zheng et al. demonstrated the effectiveness of melt intercalating polystyrene with an oligomerically modified clay (e.g., hectorite) containing maleic anhydride and vinyl-benzyl trimethylammonium chloride to produce nanocomposites [50]. Further, Li et al. [51] explored the impact of three different organically modified clays (cloisite 6A, cloisite 10A, cloisite 30B) on the properties of nanocomposites. The melt intercalation process has benefits like lower cost, easy utilization of traditional mechanical techniques, no competition between the polymer and the solvent, etc., underscoring its utility and efficiency in nanocomposite production [49, 52, 53].

7.2.5 Direct Mixing of Polymers and Particles

This method is characterized by its simplicity, eliminating the necessity for external solvents, high temperatures, or catalysts. Within this process, the polymer and layered silicate undergo mechanical stirring at a specific speed [54]. Employing this technique of direct mixing, Velmurugan and Mohan have adeptly synthesized an MMT-based epoxy nanocomposite [55]. This approach facilitated the dispersion of clay layers within a polymer matrix, leading to a composite material where the interplay between the clay and the epoxy polymer is finely tuned. Comparative analysis revealed that composites infused with organoclay exhibit superior tensile strength to those filled with unmodified clay. Similar enhancements in mechanical properties are observed in glass fiber-reinforced epoxy composites that incorporate organoclay, demonstrating the potential of this method in improving the structural integrity and performance of composite materials [55].

7.3 PROPERTIES OF 2D MATERIAL-BASED COMPOSITES

7.3.1 Mechanical Properties

Integrating layered silicate materials or clays into a polymer matrix often improves the material's mechanical characteristics, including elastic modulus, creep resistance, fracture toughness, and overall strength. The extent of enhancement in these mechanical properties is intricately linked to the microstructure resulting from the dispersion of clay layers within the polymer matrix [56]. Notably, homogeneous dispersion of clay particles within the matrix is essential for enhancing the tensile modulus, storage modulus, and tensile strength [56]. Chan et al. [57] investigated interactions between MMT nanoclay and the surrounding matrix. It was found that adding 5% MMT to the epoxy matrix increased Young's modulus and tensile strength by 34% and 25%, respectively, compared to an unmodified sample. Cauvin et al. [58] further studied the effect of clay concentrations on mechanical properties. The increase of elastic modulus was seen to be directly proportional to the clay amount in a certain range, and it reached the maximum at 7 wt% MMT content. Zhang et al. [59] observed significant enhancements in impact strength (88%) and tensile strength (21%) in epoxy clay nanocomposites enriched with 3% organoclay (Na-MMT). Furthermore, investigations into exfoliated epoxy nanocomposite structures have demonstrated that a 5% clay (O-MMT) loading can elevate the strength and modulus by 10% and 130%, respectively, highlighting the impact of clay incorporation on the mechanical performance of polymer composites [60].

7.3.2 Thermal Stability

The inclusion of clay layers within a polymer matrix acts as an insulating barrier that impedes the mass transport of volatile substances produced during thermal degradation and also aids in the formation of char. These mechanisms collectively contribute to enhanced thermal stability and flame retardancy of the composite materials [61–63]. Studies on polymer-clay nanocomposites have demonstrated significant improvements in thermal stability across a diverse array of organoclays and polymer matrices. Zhang et al. [59] highlighted that epoxy clay nanocomposites containing 5% clay (Na-MMT) exhibited notable increases in heat distortion temperature and thermal decomposition compared to an unmodified epoxy matrix. Furthermore, the integration of flame-retardant polymers or fillers into composites has been shown to improve flame-retardance properties [61, 64]. The improved flame-retardance characteristics of nanocomposite materials primarily arise from char formation, which serves as a protective barrier for the matrix, enhancing its resistance to combustion [65–68].

7.3.3 Barrier Properties

The incorporation of even minimal amounts of layered silicate into nanocomposites elevates their barrier capabilities. Herrera-Alonso et al. [69] studied the barrier properties of polyacrylate-layered silicate nanocomposites. Their findings revealed an inverse relationship between gas permeability and the amount of modified clay (Na–MMT, cloisite) within the composite, with a notable reduction in gas permeability (28%) achieved at a clay concentration of 5 wt% [69]. Similarly, Jawahar et al. [70] reported improvements in the barrier properties of polyester polymers upon the addition of bentonite, which consequently enhanced the composite's wear resistance by forming a protective silicate layer over the matrix. Additional studies have also demonstrated that an increase in clay content improves the barrier properties of nanocomposites, underlining the critical role of layered silicates in the functional performance of these materials [71–74].

7.3.4 Rheological Properties

Rheological properties serve as a fundamental metric for characterizing the flow and deformation behavior of fluids [42, 63, 75]. Hwang et al. [76] observed that the integration of organoclay (MMT) into nanocomposites not only led to a reduction in cell size but also resulted in an increase in viscosity with higher organoclay concentrations. Similarly, Franchini et al. [77] showed that high clay/modifier contents induced shear thinning behavior. Kaushik et al. [78] highlighted the impact of adding nanoclays to cementitious mixtures, noting an increase in yield stress, which is particularly advantageous for the material's structural integrity. As shown in Figure 7.2, Jin et al. [79] proposed that laponite nanoclay-based nanocomposite hydrogels can be used to prepare extrusion-based 3D printable hydrogel inks. They studied the rheological implications of nanoclay addition, illustrating how a suspension of laponite nanoclay transitions into a shear-thinning fluid upon extrusion through a nozzle, a behavior that is essential for extrusion-based 3D printing processes [79].

7.3.5 Dynamic Mechanical Properties

Dynamic mechanical analysis (DMA) provides insight into the viscoelastic properties of materials by applying a small sinusoidal deformation, which can be in the form of tension, bending, or twisting under a controlled temperature setting. This technique measures three critical parameters: the storage modulus, loss modulus, and tan delta, which are pivotal for characterizing the viscoelasticity of a material. Fukushima et al. [80] have demonstrated that integrating MMT into a polymer matrix improves its dynamic mechanical properties. This enhancement is mainly due to the immobilization of polymer chains above the glass transition temperature (T_g) due to the presence of layered silicates within the matrix, contributing to the observed improvement in the material's viscoelastic performance [80].

7.3.6 Toxicity Properties

As the use of nanoclays is expanding worldwide, evaluation of their toxicity properties is important, particularly in scenarios involving human exposure. This exposure can occur at any stage during development, manufacture, use, or disposal [81, 82]. Schmidt et al. [83] investigated the total and specific migration of all the individual components in Mg-Al-layered double hydroxide composites (PLA-LDH-C12). The results complied with the migration limits for total migration and specific lauric acid migration established by European legislation for food contact materials. Simon et al. [84] investigated the thermodynamics governing the migration of nanoparticles from packaging into food, deducing that such migration would continue until the nanoparticles reached an equilibrium distribution between the packaging material and the food content [84]. These studies illustrate the importance of evaluating the safety and regulatory compliance of nanoclay applications, particularly in contexts where they may come into direct contact with consumable products.

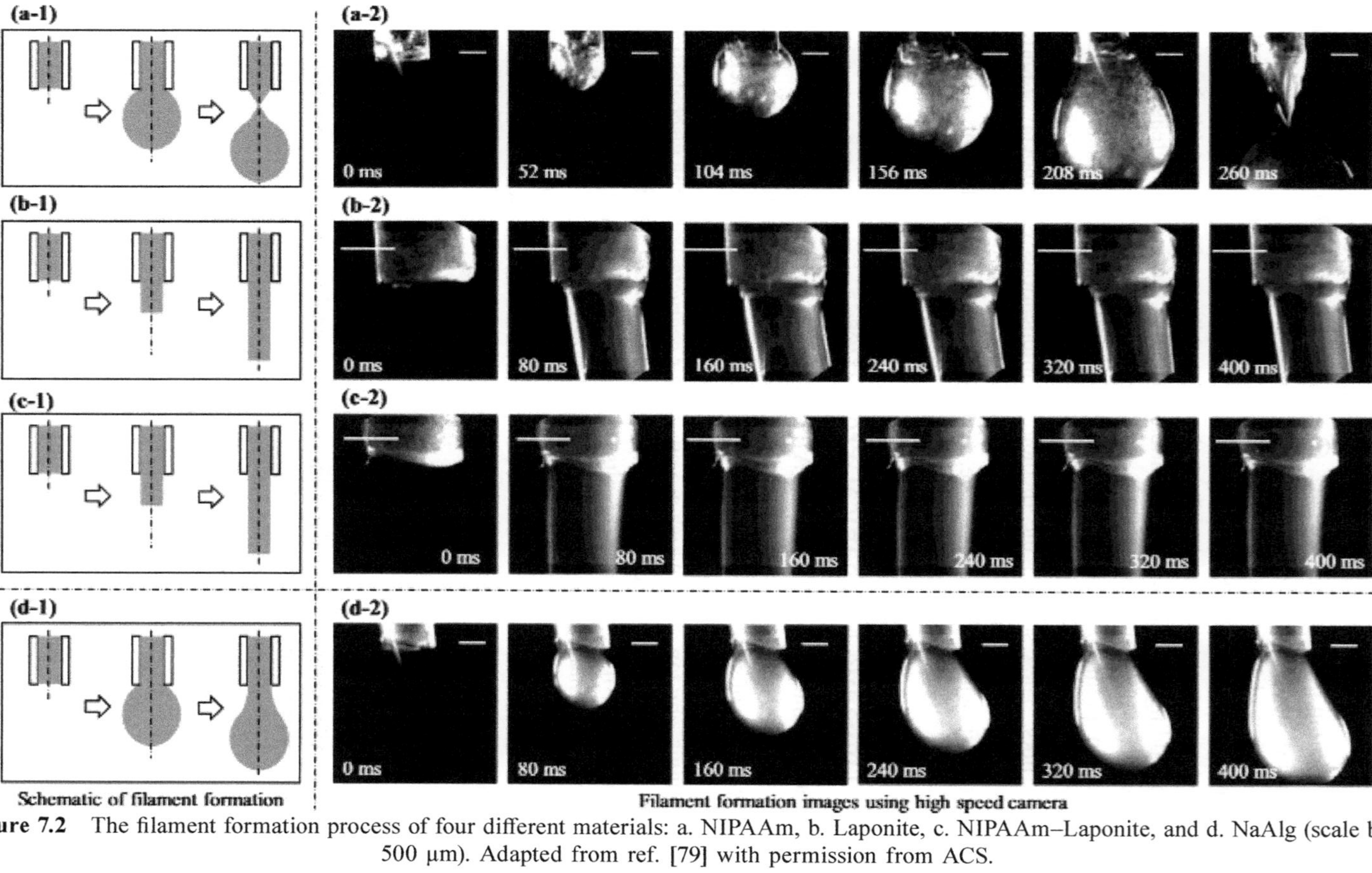

Figure 7.2 The filament formation process of four different materials: a. NIPAAm, b. Laponite, c. NIPAAm–Laponite, and d. NaAlg (scale bars: 500 μm). Adapted from ref. [79] with permission from ACS.

7.4 APPLICATIONS

7.4.1 Biomedical Applications

The interaction between nanoclays and cells enables the optimization of cellular functions for a range of tissue engineering applications, including those focused on bone, cartilage, osteochondral regeneration, cell adhesion, and immunomodulation. The generally non-toxic nature of nanoclays toward cells has been the reason behind extensive research, where the role of nanoclay in promoting osteogenic differentiation in stem cells has been one of the prime focuses [85, 86]. Scaffolding, cells, and bioactive signals are the three main components of regenerative engineering approaches [87]. Utilizing nanoclays, a diverse array of scaffolds such as hydrogels, nanofibrous electrospun scaffolds, and porous structures have been developed, showcasing the versatility of nanoclays in facilitating tissue regeneration and engineering.

The direct application of nanoclay in cellular environments has been recognized for its osteogenic potential, further extending to its incorporation within scaffolds for bone tissue engineering [88]. One particular study (Figure 7.3) demonstrated the capability of nanocomposites to facilitate cellular and tissue functionalities by applying a diffusion gelation method, where the laponite dispersions exhibited instant gelation after being in contact with buffered saline or blood serum. This method was used to generate injectable bioactive microenvironments for osteogenesis [88]. The resulting nanoclay diffusion gels effectively supported the synthesis of osteogenic matrices, enhancing the cell-driven mineralization of the matrix [88]. Wang et al. [89] reported that sintering laponite (LAP) powder yielded bioceramics for bone tissue engineering applications. The LAP bioceramics demonstrated an innate ability to stimulate osteoblast differentiation from rat mesenchymal stem cells (rMSCs) within a growth medium, eliminating the need for external inducing factors [89].

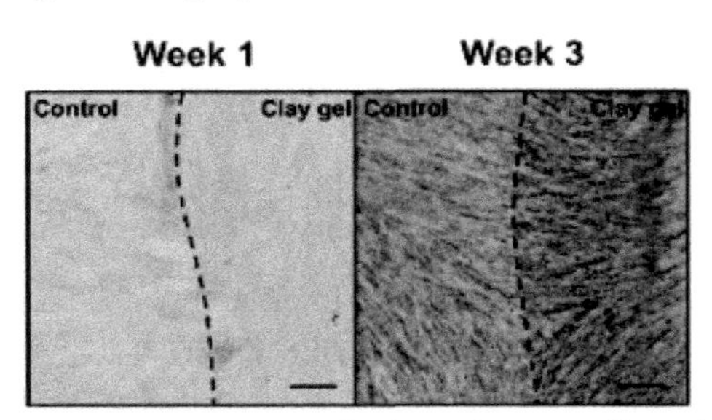

Figure 7.3 Nanoclay gel supports the adhesion of stem cells and induces osteogenic differentiation as determined by alizarin red staining. Adapted from ref. [88] with permission from Advanced Healthcare Materials.

Recent research has unveiled that varying the clay content within the laponite-polyethylene oxide (PEO) system impacts cellular behaviors, such as adhesion, spreading, and proliferation [90]. The properties of the material, including its hydration, dissolution, and mechanical properties, play a crucial role in enhancing cell adhesion and facilitating functional qualities like bioactivity while concurrently optimizing the material's overall performance [90]. Furthermore, nanocomposites have been employed in the development of wound-healing applications. One particular study (Figure 7.4) demonstrated the synthesis of a zwitterionic poly(sulfobetaine acrylamide) (pSBAA)/laponite nanocomposite hydrogel, which exhibited potential as a dressing for chronic wounds [91]. Nanoclay has a unique ability to form a gel phase that can be combined with natural and synthetic polymers to formulate shear-thinning biomaterial inks. This property is particularly advantageous for various additive manufacturing techniques [92]. Chimene et al. [93] developed bioactive inks with good printability and mechanical properties using a novel reinforcement technique called nanoengineered ionic covalent entanglement (NICE). Additionally, the applications of nanoclays can be extended into the medical field as vaccine adjuvants, where they have been shown to provoke potent immune responses, underscoring the multifaceted applications of nanoclays from tissue engineering to immunology [94–96].

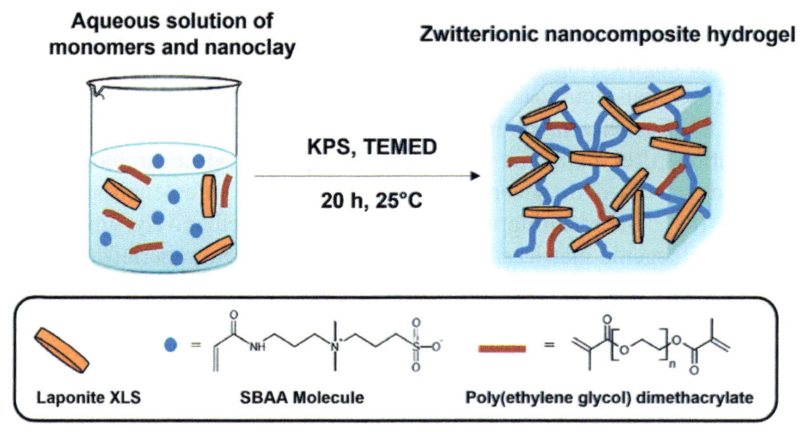

Figure 7.4 The chemical structure of SBAA and the network of zwitterionic nanocomposite hydrogel after free-radical polymerization. Adapted from ref. [91] with permission from Journal of Materials Chemistry.

7.4.2 Environmental Applications

Clays, owing to their minimal toxicity, affordability, and superior adsorption capacity, are widely employed as adsorbents for purifying water by removing pollutants [97]. One important application of clays in environmental remediation includes the removal of heavy metals, dyes, pesticides, and pharmaceutical compounds [98, 99]. For example, Rafati et al. [100] used a nanoclay (cloisite 15A) composite functionalized by mono-tosyl-β-cyclodextrin (β-CD) to remove ibuprofen from an aqueous solution. The research highlighted that factors such as pH and contact time influence the adsorption performance. Rafati et al. [101] conducted another study to find out the optimal conditions for the removal of naproxen from water using cloisite 15A, showcasing the potential of clays to address environmental pollution through water purification.

Similarly, the contamination of natural water bodies with dye-containing wastewater presents a complex array of environmental challenges. A variety of remediation strategies have been explored for the removal of dyes from wastewater, encompassing techniques such as coagulation, chemical oxidation, biological degradation, and membrane separation. El Haouti et al. [102] investigated the kinetics and isotherms of two cationic dyes, toluidine blue (TB) and crystal violet (CV), on a Na–MMT clay. The investigation delineated that factors including adsorbent quantity, duration of contact, and the solution's pH level exhibited a proportional relationship to the performance of CV and TB dye removal [102]. Similarly, Shirsath et al. [103] assessed the effectiveness of synthesized nanocomposites in dye adsorption processes. Their findings indicated that the synergistic application of hydrogel coupled with ultrasound technology enhanced the removal rate of dyes, surpassing the outcomes achieved with hydrogel alone.

Additionally, nanocomposites can be an effective tool for removing heavy metal contaminants from water. In one particular study, a novel nanocomposite was used to investigate the removal of Pb (II) from water [104]. The authors discovered that the adsorbent could remove 60% of lead with an initial concentration of 4 mg/L in 15 minutes. The removal efficiency increased to 90% after 90 minutes [104]. Additionally, Chen et al. [105] investigated the use of soil clays to remove Cu^{2+} from aqueous solutions.

Moreover, clays have demonstrated considerable potential in the adsorption of gaseous pollutants, offering a sustainable and low-cost solution for improving air quality by capturing harmful emissions. Thermal oxidation, while effective for diminishing gaseous pollutants, presents significant economic and environmental costs due to its high energy demands. In this context, clays emerge as a promising and energy-efficient alternative. These materials have demonstrated the ability to capture gaseous molecules from the air, addressing pollutants like hydrogen sulfide (H_2S), which poses severe respiratory risks to humans at elevated concentrations [106]. Mohamadalizadeh et al. [107] conducted a comprehensive analysis of the adsorptive capabilities of modified bentonites for H_2S removal, exploring the influence of various parameters, such as metal ion presence and ambient temperature, on the efficiency of pollutant extraction.

Carbon dioxide (CO_2) stands as a principal contributor to the greenhouse effect, posing substantial environmental challenges and accelerating climate change. CO_2 capture and storage strategies have been recognized as effective measures to mitigate CO_2 emissions and their impact on the environment [108, 109]. Recent research has explored the dynamics of CO_2 adsorption using modified MMT clay, aiming to predict the conditions for maximum adsorption [110]. This investigation revealed that the critical factors affecting the capacity for CO_2 capture were the weight percentage of NaOH, pressure, and temperature, as demonstrated by an analysis of variance (ANOVA). Roth et al. [111] introduced MMT clay as an economical option for CO_2 adsorption. The enhancement of this clay adsorbent involved a two-step modification process incorporating N-dimethylformamide, 3-aminopropyltrimethoxysilane, and polyethyleneimine. Initial observations indicated that while pristine clay exhibited limited CO_2 adsorption capabilities, the chemically treated adsorbents demonstrated enhanced performance [111].

7.5 CONCLUSION

A range of routes can be used to prepare clay-based composites with high homogeneity, depending on the structure and chemistry of the clay used (natural or modified) and the properties to be achieved. The main goal is to identify the optimal conditions for improving the interaction between the polymer matrix and the layered silicates. To enhance the interactions between the two materials, the layered silicates must be pre-treated in most cases. Variations in factors like polymer, clay source, particle dimensions, and preparation methods can lead to the formation of nanocomposites with either exfoliated or intercalated structures. Regardless of which structure is obtained, most nanocomposites demonstrate enhanced mechanical properties like storage modulus, creep properties, and fracture toughness. Moreover, the clay layer acts as an insulator in the final product, which enhances thermal stability and flame retardancy. Adding clay contents to the polymer matrix also improves barrier properties, which is essential in various purification and separation applications. The interaction of clay with biological cells has unveiled its bioactive potential, facilitating its use in biomedical fields such as tissue engineering. Moreover, the high surface-area-to-volume ratio and ionic charges of clays render them suitable for drug delivery and therapeutics. However, there are still open questions regarding the comprehensive toxicity-related properties of 2D clay-based composites, which require further studies. Apart from the biomedical potential, clays' internal and external surfaces can interact with dissolved species due to their unique porous structure and high specific surface area, rendering them suitable adsorbents to treat waste streams and remove heavy metals and industrial pollutants from water. Other unique mechanisms, such as co-precipitation and substitution, may also contribute to environmental applications. Future research is needed to understand the underlying mechanisms of clay composite design and optimization and to utilize their benefits across scientific research and industrial applications.

REFERENCES

[1] Omanović-Mikličanin, E., Badnjević, A., Kazlagić, A. and Hajlovac, 2020. M. Nanocomposites: A brief review. Health Technol. 10(1): 51–59.

[2] Bockstaller, M.R., Mickiewicz, R.A. and Thomas, E.L. 2005. Block copolymer nanocomposites: perspectives for tailored functional materials. Adv. Mater. 17(11): 1331–1349.

[3] Kim, B.J., Bang, J., Hawker, C.J. and Kramer, E.J. 2006. Effect of areal chain density on the location of polymer-modified gold nanoparticles in a block copolymer template. Macromolecules. 39(12): 4108–4114.

[4] LeBaron, P.C., Wang, Z. and Pinnavaia, T.J. 1999. Polymer-layered silicate nanocomposites: an overview. Appl. Clay Sci. 15(1–2): 11–29.

[5] Okpala, C.C. 2013. Nanocomposites–an overview. Int. J. Eng. Res. Deve. 8(11): 17–23.

[6] Moya, J.S., Lopez-Esteban, S. and Pecharroman, C. 2007. The challenge of ceramic/metal microcomposites and nanocomposites. Progr. Mater. Sci. 52(7): 1017–1090.

[7] Manias, E. 2007. Stiffer by design. Nat. Mater. 6(1): 9–11.

[8] Salahuddin, N.A. 2004. Layered silicate/epoxy nanocomposites: synthesis, characterization and properties. Polym. Adv. Technol. 15(5): 251–259.

[9] Vaia, R.A., Jandt, K.D., Kramer, E.J. and Giannelis, E.P. 1996. Microstructural evolution of melt intercalated polymer-organically modified layered silicates nanocomposites. Chem. Mater. 8(11): 2628–2635.

[10] Gonsalves, K. and Chen, X. 1996. In materials research soc symposium proceedings. Mater. Res. Soc.: Warrendale PA. 435: 55.

[11] Ishida, H., Campbell, S. and Blackwell, J. 2000. General approach to nanocomposite preparation. Chem. Mater. 12(5): 1260–1267.

[12] Carrado, K.A. and Xu, L. 1998. In situ synthesis of polymer—Clay nanocomposites from silicate gels. Chem. Mater. 10(5): 1440–1445.

[13] Okada, A. and Usuki, A. 1995. The chemistry of polymer-clay hybrids. Mater. Sci. Eng.: C. 3(2): 109–115.

[14] Usuki, A., Koiwai, A., Kojima, Y., Kawasumi, M., Okada, A., Kurauchi, T., et al. 1995. Interaction of nylon 6-clay surface and mechanical properties of nylon 6-clay hybrid. J. Appl. Polym. Sci. 55(1): 119–123.

[15] Usuki, A., Kojima, Y., Kawasumi, M., Okada, A., Fukushima, Y., Kurauchi, T., et al. 1993. Synthesis of nylon 6-clay hybrid. J. Mater. Res. 8(5): 1179–1184.

[16] Albdiry, M., Yousif, B., Ku, H. and Lau, K. 2013. A critical review on the manufacturing processes in relation to the properties of nanoclay/polymer composites. J. Compos. Mater. 47(9): 1093–1115.

[17] Choudalakis, G. and Gotsis, A. 2009. Permeability of polymer/clay nanocomposites: A review. Eur. Polym. J. 45(4): 967–984.

[18] Paul, D.R. and Robeson, L.M. 2008. Polymer nanotechnology: Nanocomposites. Polymer. 49(15): 3187–3204.

[19] Nambiar, S. and Yeow, J.T. 2012. Polymer-composite materials for radiation protection. ACS Appl. Mater. Interfaces. 4(11): 5717–5726.

[20] Chimene, D., Kaunas, R. and Gaharwar, A.K. 2020. Hydrogel bioink reinforcement for additive manufacturing: a focused review of emerging strategies. Adv. Mater. 32(1): 1902026.

[21] Hmeidat, N.S., Pack, R.C., Talley, S.J., Moore, R.B. and Compton, B.G. 2020. Mechanical anisotropy in polymer composites produced by material extrusion additive manufacturing. Addit. Manuf. 34: 101385.

[22] Kasraie, M. and Abadi, P.P.S.S. 2021. Additive manufacturing of conductive and high-strength epoxy-nanoclay-carbon nanotube composites. Addit. Manuf. 46: 102098.

[23] Palanisamy, P., Chavali, M., Kumar, E.M. and Etika, K.C. 2020. Hybrid nanocomposites and their potential applications in the field of nanosensors/gas and biosensors. pp. 253–280. *In*: K. Pal and F. Gomes (eds). Nanofabrication for Smart Nanosensor Applications: A volume in Micro and Nano Technologies. Elsevier.

[24] Uddin, F. 2008. Clays, nanoclays, and montmorillonite minerals. Metall. Mater. Transa. A. 39(12): 2804–2814.

[25] Ray, S.S. and Okamoto, M. 2003. Polymer/layered silicate nanocomposites: a review from preparation to processing. Prog. Polym. Sci. 28(11): 1539–1641.

[26] Wypych, F. and Satyanarayana, K.G. 2005. Functionalization of single layers and nanofibers: A new strategy to produce polymer nanocomposites with optimized properties. J. Colloid Interface Sci. 285(2): 532–543.

[27] Han, H., Rafiq, M.K., Zhou, T., Xu, R., Mašek, O. and Li, X. 2019. A critical review of clay-based composites with enhanced adsorption performance for metal and organic pollutants. J. Hazard. Mater. 369: 780–796.

[28] Cui, H.W. and Du, G.B. 2012. Preparation and characterization of exfoliated nanocomposite of polyvinyl acetate and organic montmorillonite. Adv. Polym. Technol. 31(2): 130–140.

[29] Okamoto, K., Sinha Ray, S. and Okamoto, M. 2003. New poly (butylene succinate)/ layered silicate nanocomposites. II. Effect of organically modified layered silicates on structure, properties, melt rheology, and biodegradability. J. Polym. Sci., Part B: Polym. Phys. 41(24): 3160–3172.

[30] Gao, F. 2004. Clay/polymer composites: the story. Mater. Today. 7(11): 50–55.

[31] Alateyah, A., Dhakal, H. and Zhang, Z. 2013. Processing, properties, and applications of polymer nanocomposites based on layer silicates: a review. Adv. Polym. Technol. 32(4): 21368.

[32] Alexandre, M. and Dubois, P. 2000. Polymer-layered silicate nanocomposites: preparation, properties and uses of a new class of materials. Mater. Sci. Eng.: R: Rep. 28(1–2): 1–63.

[33] Pavlidou, S. and Papaspyrides, C. 2008. A review on polymer–layered silicate nanocomposites. Prog. Polym. Sci. 33(12): 1119–1198.

[34] Zanetti, M., Lomakin, S. and Camino, G. 2000. Polymer layered silicate nanocomposites. Macromol. Mater. Eng. 279(1): 1–9.

[35] Theng, B.K.G. 2012. Formation and Properties of Clay-polymer Complexes. Elsevier.

[36] Romero, R.B., Leite, C.A.P. and do Carmo Gonçalves, M. 2009. The effect of the solvent on the morphology of cellulose acetate/montmorillonite nanocomposites. Polymer. 50(1): 161–170.

[37] Kojima, Y., Usuki, A., Kawasumi, M., Okada, A., Fukushima, Y., Kurauchi, T., et al. 1993. Mechanical properties of nylon 6-clay hybrid. J. Mater. Res. 8(5): 1185–1189.

[38] Chang, J.H., An, Y.U. and Sur, G.S. 2003. Poly (lactic acid) nanocomposites with various organoclays. I. Thermomechanical properties, morphology, and gas permeability. J. Polym. Sci., Part B: Polym. Phys. 41(1): 94–103.

[39] Krikorian, V. and Pochan, D.J. 2003. Poly (L-lactic acid)/layered silicate nanocomposite: fabrication, characterization, and properties. Chem. Mater. 15(22): 4317–4324.

[40] Vo, V.-S., Mahouche-Chergui, S., Nguyen, V.-H., Naili, S., Singha, N.K. and Carbonnier, B. 2017. Chemical and photochemical routes toward tailor-made polymer–Clay nanocomposites: recent progress and future prospects. Clay-Polymer Nanocomposites. 145–197.

[41] Beyer, G. 2002. Nanocomposites: A new class of flame retardants for polymers. Plast. Addit. Compound. 4(10): 22–28.

[42] Solomon, M.J., Almusallam, A.S., Seefeldt, K.F., Somwangthanaroj, A. and Varadan, P. 2001. Rheology of polypropylene/clay hybrid materials. Macromolecules. 34(6): 1864–1872.

[43] Tasdelen, M.A., Kreutzer, J. and Yagci, Y. 2010. In situ synthesis of polymer/clay nanocomposites by living and controlled/living polymerization. Macromol. Chem. Phys. 211(3): 279–285.

[44] Greesh, N., Hartmann, P.C., Cloete, V. and Sanderson, R.D. 2008. Impact of the clay organic modifier on the morphology of polymer–clay nanocomposites prepared by in

situ free-radical polymerization in emulsion. J. Polym. Sci., Part A: Polym. Chem. 46(11): 3619–3628.

[45] Oral, A., Tasdelen, M.A., Demirel, A.L. and Yagci, Y. Poly (cyclohexene oxide)/clay nanocomposites by photoinitiated cationic polymerization via activated monomer mechanism. J. Polym. Sci., Part A: Polym. Chem. 47(20): 5328–5335.

[46] Becker, O., Varley, R. and Simon, G. 2002. Morphology, thermal relaxations and mechanical properties of layered silicate nanocomposites based upon high-functionality epoxy resins. Polymer. 43(16): 4365–4373.

[47] Kornmann, X., Lindberg, H. and Berglund, L.A. 2001. Synthesis of epoxy–clay nanocomposites: influence of the nature of the clay on structure. Polymer. 42(4): 1303–1310.

[48] Le Pluart, L., Duchet, J. and Sautereau, H. 2005. Epoxy/montmorillonite nanocomposites: Influence of organophilic treatment on reactivity, morphology and fracture properties. Polymer. 46(26): 12267–12278.

[49] Vaia, R.A., Jandt, K.D., Kramer, E J. and Giannelis, E.P. 1995. Kinetics of polymer melt intercalation. Macromolecules. 28(24): 8080–8085.

[50] Zheng, X., Jiang, D.D. and Wilkie, C.A. 2006. Polystyrene nanocomposites based on an oligomerically-modified clay containing maleic anhydride. Polym. Degrad. Stab. 91(1): 108–113.

[51] Li, X., Kang, T., Cho, W.J., Lee, J.K. and Ha, C.S. 2001. Preparation and characterization of poly (butyleneterephthalate)/organoclay nanocomposites. Macromol. Rapid Commun. 22(16): 1306–1312.

[52] Huang, J.-C., Zhu, Z.-k., Yin, J., Qian, X.-f. and Sun, Y.-Y. 2001. Poly (etherimide)/montmorillonite nanocomposites prepared by melt intercalation: morphology, solvent resistance properties and thermal properties. Polymer. 42(3): 873–877.

[53] Vaia, R.A. and Giannelis, E.P. 1997. Polymer melt intercalation in organically-modified layered silicates: model predictions and experiment. Macromolecules. 30(25): 8000–8009.

[54] Camargo, P.H.C., Satyanarayana, K.G. and Wypych, F. 2009. Nanocomposites: Synthesis, structure, properties and new application opportunities. Mater. Res. 12: 1–39.

[55] Velmurugan, R. and Mohan, T. 2009. Epoxy-clay nanocomposites and hybrids: synthesis and characterization. J. Reinf. Plast. Compos. 28(1): 17–37.

[56] Azeez, A.A., Rhee, K.Y., Park, S.J. and Hui, D. 2013. Epoxy clay nanocomposites–processing, properties and applications: A review. Composites, Part B. 45(1): 308–320.

[57] Chan, M.-l., Lau, K.-t., Wong, T.-t., Ho, M.-p. and Hui, D. 2011. Mechanism of reinforcement in a nanoclay/polymer composite. Composites. Part B: Engineering. 42(6): 1708–1712.

[58] Cauvin, L., Kondo, D., Brieu, M. and Bhatnagar, N. 2010. Mechanical properties of polypropylene layered silicate nanocomposites: Characterization and micro-macro modelling. Polym. Test. 29(2): 245–250.

[59] Zhang, K., Wang, L., Wang, F., Wang, G. and Li, Z. 2004. Preparation and characterization of modified-clay-reinforced and toughened epoxy-resin nanocomposites. J. Appl. Polym. Sci. 91(4): 2649–2652.

[60] Li, X., Zhan, Z.-J., Peng, G.-R. and Wang, W.-K. 2012. Nano-disassembling method—A new method for preparing completely exfoliated epoxy/clay nanocomposites. Appl. Clay Sci. 55: 168–172.

[61] Becker, O., Varley, R.J. and Simon, G.P. 2004. Thermal stability and water uptake of high performance epoxy layered silicate nanocomposites. Eur. Polym. J. 40(1): 187–195.

[62] Hussain, F., Hojjati, M., Okamoto, M. and Gorga, R.E. 2006. Polymer-matrix nanocomposites, processing, manufacturing, and application: an overview. J. Compos. Mater. 40(17): 1511–1575.

[63] Ray, S.S. and Bousmina, 2005. Biodegradable polymers and their layered silicate nanocomposites: in greening the 21st century materials world. Progr. Mater. Sci. 50(8): 962–1079.

[64] Porter, D., Metcalfe, E. and Thomas, M. 2020. Nanocomposite fire retardants: A review. Fire Mater. 24(1): 45–52.

[65] Kaynak, C., Nakas, G.I. and Isitman, N.A. 2009. Mechanical properties, flammability and char morphology of epoxy resin/montmorillonite nanocomposites. Appl. Clay Sci. 46(3): 319–324.

[66] Szustakiewicz, K., Kiersnowski, A., Gazińska, M., Bujnowicz, K. and Pigłowski, J. 2011. Flammability, structure and mechanical properties of PP/OMMT nanocomposites. Polym. Degrad. Stab. 96(3): 291–294.

[67] Tai, Q., Yuen, R.K., Song, L. and Hu, Y. 2012. A novel polymeric flame retardant and exfoliated clay nanocomposites: preparation and properties. Chem. Eng. J. 183: 542–549.

[68] Wu, G.M., Schartel, B., Bahr, H., Kleemeier, M., Yu, D. and Hartwig, A. 2012. Experimental and quantitative assessment of flame retardancy by the shielding effect in layered silicate epoxy nanocomposites. Combust. Flame. 159(12): 3616–3623.

[69] Herrera-Alonso, J.M., Sedláková, Z. and Marand, E. 2010. Gas transport properties of polyacrylate/clay nanocomposites prepared via emulsion polymerization. J. Membr. Sci. 363(1–2): 48–56.

[70] Jawahar, P., Gnanamoorthy, R. and Balasubramanian, M. 2006. Tribological behaviour of clay–Thermoset polyester nanocomposites. Wear. 261(7–8): 835–840.

[71] Bagherzadeh, M. and Mahdavi, F. 2007. Preparation of epoxy–clay nanocomposite and investigation on its anti-corrosive behavior in epoxy coating. Prog. Org. Coat. 60(2): 117–120.

[72] Ke, Z. and Yongping, B. 2005. Improve the gas barrier property of PET film with montmorillonite by in situ interlayer polymerization. Mater. Lett. 59(27): 3348–3351.

[73] Liu, W., Hoa, S.V. and Pugh, M. 2005. Fracture toughness and water uptake of high-performance epoxy/nanoclay nanocomposites. Compos. Sci. Technol. 65(15–16): 2364–2373.

[74] Ogasawara, T., Ishida, Y., Ishikawa, T., Aoki, T. and Ogura, T. 2006. Helium gas permeability of montmorillonite/epoxy nanocomposites. Compos. Part A: Appl. Sci. Manuf. 37(12): 2236–2240.

[75] Cho, J. and Paul, D. 2001. Nylon 6 nanocomposites by melt compounding. Polymer. 42(3): 1083–1094.

[76] Hwang, S.-s., Liu, S.-p., Hsu, P.P., Yeh, J.-m., Yang, J.-p. and Chen, C.-l. 2012. Morphology, mechanical, and rheological behavior of microcellular injection molded EVA–clay nanocomposites. Int. Commun. Heat Mass Transfer. 39(3): 383–389.

[77] Franchini, E., Galy, J. and Gérard, J.-F. 2009. Sepiolite-based epoxy nanocomposites: Relation between processing, rheology, and morphology. J. Colloid Interface Sci. 329(1): 38–47.

[78] Kaushik, S., Sonebi, M., Amato, G., Perrot, A. and Das, U.K. 2022. Influence of nanoclay on the fresh and rheological behaviour of 3D printing mortar. Mater. Today: Proc. 58: 1063–1068.

[79] Jin, Y., Shen, Y., Yin, J., Qian, J. and Huang, Y. 2018. Nanoclay-based self-supporting responsive nanocomposite hydrogels for printing applications. ACS Appl. Mater. Interfaces. 10(12): 10461–10470.

[80] Fukushima, K., Tabuani, D. and Camino, G. 2012. Poly (lactic acid)/clay nanocomposites: effect of nature and content of clay on morphology, thermal and thermo-mechanical properties. Mater. Sci. Eng.: C. 32(7): 1790–1795.

[81] Lordan, S., Kennedy, J.E. and Higginbotham, C.L. 2011. Cytotoxic effects induced by unmodified and organically modified nanoclays in the human hepatic HepG2 cell line. J. Appl. Toxicol. 31(1): 27–35.

[82] Seaton, A., Tran, L., Aitken, R. and Donaldson, K. 2010. Nanoparticles, human health hazard and regulation. J. R. Soc. Interface. 7(suppl_1): S119–S129.

[83] Schmidt, B., Katiyar, V., Plackett, D., Larsen, E.H., Gerds, N., Koch, C.B., et al. 2011. Migration of nanosized layered double hydroxide platelets from polylactide nanocomposite films. Food Addit. Contam.: Part A. 28(7): 956–966.

[84] ŠIMON, P., Chaudhry, Q. and BAKOŠ, D. 2008. Migration of engineered nanoparticles from polymer packaging to food—a physicochemical view. J Food Nutr. Res. 47(3).

[85] Carrow, J.K., Cross, L.M., Reese, R.W., Jaiswal, M.K., Gregory, C A., Kaunas, R., et al. 2018. Widespread changes in transcriptome profile of human mesenchymal stem cells induced by two-dimensional nanosilicates. Proc. Natl. Acad. Sci. 115(17): E3905–E3913.

[86] Gaharwar, A., Avery, R., Assmann, A., Paul, A. and McKinley, G. 2013. ACS Nano. 2014. 8: 9833; (c) Gaharwar, A.K., Mihaila, S.M., Swami, A., Patel, A., Sant, S., Reis, R.L. AP Marques, ME Gomes, A. Khademhosseini. Adv. Mater. 25: 3329.

[87] Vacanti, J.P. and Langer, R. 1999. Tissue engineering: the design and fabrication of living replacement devices for surgical reconstruction and transplantation. The Lancet. 354: S32–S34.

[88] Shi, P., Kim, Y.H., Mousa, M., Sanchez, R.R., Oreffo, R.O. and Dawson, J.I. 2018. Self-assembling nanoclay diffusion gels for bioactive osteogenic microenvironments. Adv. Healthcare Mater. 7(15): 1800331.

[89] Wang, C., Wang, S., Li, K., Ju, Y., Li, J., Zhang, Y., et al. 2014. Preparation of laponite bioceramics for potential bone tissue engineering applications. PloS one. 9(6): e99585.

[90] Gaharwar, A.K., Schexnailder, P.J., Kline, B.P. and Schmidt, G. 2011. Assessment of using Laponite® cross-linked poly (ethylene oxide) for controlled cell adhesion and mineralization. Acta Biomater. 7(2): 568–577.

[91] Huang, K.-T., Fang, Y.-L., Hsieh, P.-S., Li, C.-C., Dai, N.-T. and Huang, C.-J. 2016. Zwitterionic nanocomposite hydrogels as effective wound dressings. J. Mater. Chem. B. 4(23): 4206–4215.

[92] Gaharwar, A.K., Cross, L.M., Peak, C.W., Gold, K., Carrow, J.K., Brokesh, A., et al. 2019. 2D nanoclay for biomedical applications: regenerative medicine, therapeutic delivery, and additive manufacturing. Adv. Mater. 31(23): 1900332.

[93] Chimene, D., Peak, C.W., Gentry, J.L., Carrow, J.K., Cross, L.M., Mondragon, E., et al. 2018. Nanoengineered ionic–covalent entanglement (NICE) bioinks for 3D bioprinting. ACS Appl. Mater. Interfaces. 10(12): 9957–9968.

[94] Chen, W., Zhang, B., Mahony, T., Gu, W., Rolfe, B. and Xu, Z.P. 2016. Efficient and durable vaccine against intimin β of diarrheagenic *E. coli* induced by clay nanoparticles. Small. 12(12): 1627–1639.

[95] Chen, W., Zuo, H., Mahony, T.J., Zhang, B., Rolfe, B. and Xu, Z.P. 2017. Efficient induction of comprehensive immune responses to control pathogenic *E. coli* by clay nano-adjuvant with the moderate size and surface charge. Sci. Rep. 7(1): 1–12.

[96] Chen, W., Zuo, H., Rolfe, B., Schembri, M.A., Cobbold, R.N., Zhang, B., et al. 2018. Clay nanoparticles co-deliver three antigens to promote potent immune responses against pathogenic *Escherichia coli*. J. Controlled Release. 292: 196–209.

[97] Uddin, M.K. 2017. A review on the adsorption of heavy metals by clay minerals, with special focus on the past decade. Chem. Eng. J. 308: 438–462.

[98] Zhao, S., Huang, W., Wang, X., Fan, Y. and An, C. 2019. Sorption of Phenanthrene onto Diatomite under the Influences of Solution Chemistry: A Study of Linear Sorption based on Maximal Information Coefficient. J. Environ. Inf. 34(1).

[99] Zhao, Y., Huang, G., An, C., Huang, J., Xin, X., Chen, X., et al. 2020. Removal of Escherichia coli from water using functionalized porous ceramic disk filter coated with Fe/TiO_2 nano-composites. J. Water Process Eng. 33: 101013.

[100] Rafati, L., Ehrampoush, M., Rafati, A., Mokhtari, M. and Mahvi, A. 2018. Removal of ibuprofen from aqueous solution by functionalized strong nano-clay composite adsorbent: kinetic and equilibrium isotherm studies. Int. J. Environ. Sci. Technol. 15(3): 513–524.

[101] Rafati, L., Ehrampoush, M.H., Rafati, A.A., Mokhtari, M. and Mahvi, A.H. 2016. Modeling of adsorption kinetic and equilibrium isotherms of naproxen onto functionalized nano-clay composite adsorbent. J. Mol. Liq. 224: 832–841.

[102] El Haouti, R., Ouachtak, H., El Guerdaoui, A., Amedlous, A., Amaterz, E., Haounati, R., et al. 2019. Cationic dyes adsorption by Na-Montmorillonite Nano Clay: Experimental study combined with a theoretical investigation using DFT-based descriptors and molecular dynamics simulations.J. Mol. Liq. 290: 111139.

[103] Shirsath, S., Hage, A., Zhou, M., Sonawane, S. and Ashokkumar, M. 2011. Ultrasound assisted preparation of nanoclay Bentonite-FeCo nanocomposite hybrid hydrogel: a potential responsive sorbent for removal of organic pollutant from water. Desalination. 281: 429–437.

[104] Matei, E., Rapa, M., Covaliu, C.I., Predescu, A.M., Turcanu, A., Predescu, C., et al. 2020. Sodium alginate-cellulose-nano-clay composite adsorbent applied for lead removal from wastewater. Rev. Chim. 71: 416–424.

[105] Chen, Y.M., Tsao, T.M., Wang, M.K., Yu, S., Liu, C.C., Li, H.C., et al. 2015. Kinetic and thermodynamic studies on removal of Cu (II) from aqueous solutions using soil nanoclays. Water Environ. Res. 87(1): 88–95.

[106] Adib, F., Bagreev, A. and Bandosz, T.J. 2000. Analysis of the relationship between H_2S removal capacity and surface properties of unimpregnated activated carbons. Environ. Sci. Technol. 34(4): 686–692.

[107] Mohamadalizadeh, A., Towfighi, J., Rashidi, A., Manteghian, M., Mohajeri, A. and Arasteh, R. 2011. Nanoclays as nano adsorbent for oxidation of H_2S into elemental sulfur. Korean J. Chem. Eng. 28(5): 1221–1226.

[108] Dong, C., Huang, G., Cheng, G., An, C., Yao, Y., Chen, X., et al. 2019. Wastewater treatment in amine-based carbon capture. Chemosphere. 222: 742–756.

[109] Feng, Q., An, C., Chen, Z. and Wang, Z. 2020. Can deep tillage enhance carbon sequestration in soils? A meta-analysis towards GHG mitigation and sustainable agricultural management. Renewable Sustainable Energy Rev. 133: 110293.

[110] Khajeh, M. and Ghaemi, A. 2020. Exploiting response surface methodology for experimental modeling and optimization of CO_2 adsorption onto NaOH-modified nanoclay montmorillonite. J. Environ. Chem. Eng. 8(2): 103663.

[111] Roth, E.A., Agarwal, S. and Gupta, R.K. 2013. Nanoclay-based solid sorbents for CO_2 capture. Energy Fuels. 27(8): 4129–4136.

Chapter **8**

Energy Harvesting Applications of Sustainable Structural Materials

Lillian J.A. Olule
Department of Electrical and Computer Engineering,
University of New Brunswick, Canada
Email: lillian.olule@unb.ca

8.1 STRUCTURAL MATERIALS

Structural materials are materials with the primary purpose of conveying or bearing a force [1]. As such, they are used to construct the load-bearing components of a structure. Their key application, therefore, depends on their mechanical properties rather than their electronic, magnetic, chemical or optical properties [2]. They are (in many instances) required to be durable, strong and stiff in order to support the loads that the structure is designed to bear. They can generally be classified into metals, ceramics, polymers, and wood and natural fibers [3]. Metals used as structural materials, such as aluminum and steel, are typically strong and durable, and have high thermal conductivity [4]. Ceramics include materials such as concrete and bricks. They are hard (but brittle), wear resistant, and demonstrate good insulation properties [5]. Polymers like plastics and composites have good corrosion resistance in addition to being lightweight and flexible [6]. Wood and natural fibers are versatile, strong, lightweight and environmentally friendly [7].

8.1.1 Sustainable Structural Materials

Sustainable structural materials refer to structural materials that are eco-friendly or have a low carbon footprint. These materials are gaining importance as the world focuses on lowering carbon emissions and achieving sustainability goals, especially given the rapid depletion of natural resources. The goal with sustainable structural materials is to optimize their structural characteristics and durability while ensuring low economic and environmental costs [3]. Sustainable structural materials can be broadly classified into natural structural materials such as wood (timber, plywood) [8], stone, natural fibers (coir, sisal, flax, straw, jute and cane), metals, earth, and artificial structural materials such as metallic, ceramic, polymeric or other material composites as illustrated in Figure 8.1.

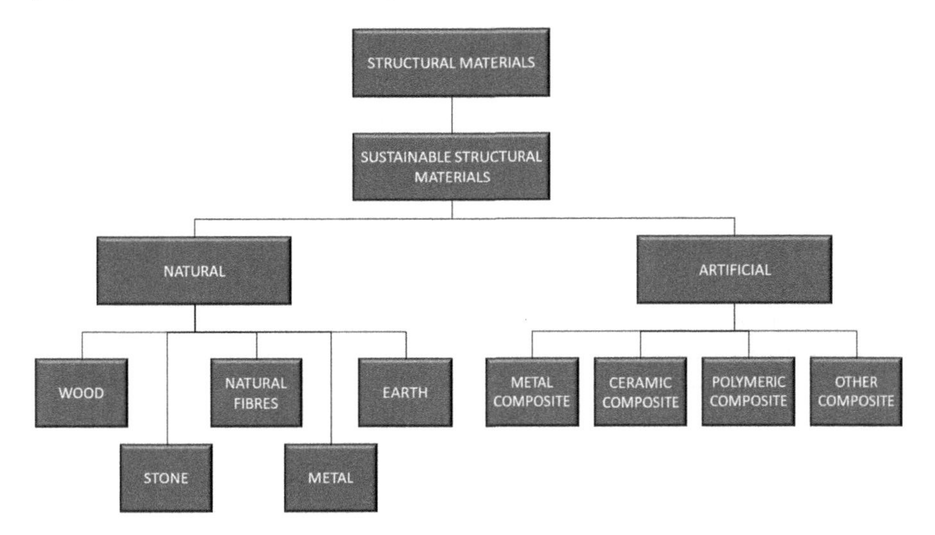

Figure 8.1 General classification of sustainable structural materials.

It is important to mention at this point that there are three major themes to sustainability: social, economic, and environmental [9]. This chapter will focus more on the economic and environmental aspects of sustainable structural materials. We will consider materials which have low environmental impact, are renewable (can be recycled or reused at the end of their life cycle) are non-toxic, and have low embodied energy (require minimal energy for extraction, processing and transport). For instance, bamboo is a highly renewable resource because it grows rapidly and can be harvested within a few years. It can therefore be a sustainable alternative to wood [10].

Whereas natural materials such as bamboo, rammed earth [11] and timber [12, 13] have shown much promise as sustainable resources, research efforts have also been geared towards the development of artificial materials such as metallic, ceramic, polymeric and other material composites which, in addition to maintaining their structural properties, can be engineered to achieve multifunctionality [14].

One of the additional functions that we will explore in this chapter is their ability to be used for energy harvesting applications.

8.2 OVERVIEW OF ENERGY HARVESTING

Energy harvesting (also known as 'energy scavenging' or 'power harvesting') is the process of extracting energy from sources available in the environment and converting it into electrical energy that can be used to power devices [15]. The driving force behind the advancement of energy harvesting technologies is the need to meet the world's increasing energy demands (it is anticipated to reach about 886 quadrillion British thermal units (Btu) by 2050 [16]), in applications such as consumer devices and in the medical, transportation, industrial and military fields. Additionally, energy harvesting can potentially serve as a backup or even replacement to traditional energy storage systems such as batteries, allowing for longer device lifespan, reduction of the environmental impact caused by battery disposal and elimination of the inconvenience of battery replacement especially in cases where it may pose a hazard [17, 18].

There are several methods of energy harvesting, each with its own distinct advantages and disadvantages.

- *Solar energy harvesting:* This method involves using photovoltaic cells to convert sunlight into electricity [19]. This is one of the most widely used methods of energy harvesting and is commonly used in solar panels and solar-powered devices. Advantages of this method include its reliability, scalability, and the fact that it does not produce emissions or pollutants. One of its major disadvantages is that the efficiency of the cells decreases with increasing temperature, which can be a major issue in hot environments. This type of energy harvesting is also not effective at night or in rainy or dark environments [20].
- *Thermal energy harvesting:* This is the process of converting waste heat or heat produced by nature into electric energy [21]. Studies have shown that over 70% of primary energy waste is lost in transportation, industrial and commercial applications, and this waste is mainly dissipated in the form of heat [22]. In addition, natural heat sources such as geothermal heat, volcanic heat and solar heat are the potential energy resources that remain largely untapped [23]. Therefore, thermal energy harvesting can provide a cost-effective and reliable way to make use of this waste. It can be used in a variety of applications, including powering sensors in automotive or industrial systems [23], or in personal electronic devices. The major advantage of this method is that it does not require any external energy source, and the main disadvantages are that it is not as efficient as some other methods [24] and is not effective in environments with unpredictable temperature fluctuations/gradients [20].
- *Kinetic (or mechanical) energy harvesting:* This method utilizes the motion present in the application environment or the human body to

generate electricity [25–27]. It can be used to power wearable devices such as fitness trackers, or to monitor industrial equipment. The main advantage of this method is that it is readily available in the environment due to human activities, structural and machinery vibrations, wind flows, water, and electromagnetic waves [20], however, this method is highly dependent on the type of motion and the specific implementation; therefore, it suffers from low efficiency, low output power, low reliability and poor environmental adaptability [27].

- *Electrochemical energy harvesting:* This method uses the chemical reactions between different materials/biomaterials to generate electricity [28] [29]. It can be used in a variety of applications, such as powering sensors in biomedical devices or powering small electronic devices [28]. The major advantages of this method are that it is biocompatible, can be used in a wide range of environments, and is inexpensive, scalable and relatively easy to implement [28]. However, the efficiency of this method is highly dependent on the specific materials used and the specific implementation. Hence, it is characterized by low output energy [29].

The summary of the energy harvesting methods and their respective advantages and disadvantages is shown in Figure 8.2.

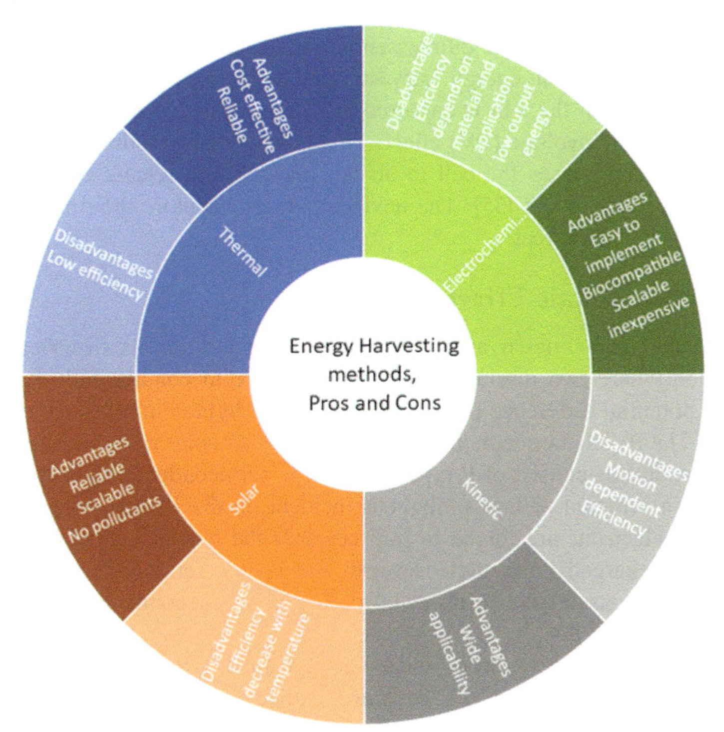

Figure 8.2 Energy harvesting methods and their respective advantages and disadvantages.

8.3 SUSTAINABLE STRUCTURAL MATERIALS USED IN ENERGY HARVESTING

This section will highlight the energy harvesting techniques where sustainable structural materials have been employed. For each technique, a brief theoretical background of the energy harvesting principle will be reviewed first. Following this, a discussion will ensue regarding various key sustainable structural materials that have been utilized, along with an exploration of their applications. It should be noted that the emphasis of this chapter will be on singular energy harvesting systems, that is, systems where a single energy harvesting technique is employed. However, more than one energy harvesting technique can be employed in what is called a 'hybrid energy harvesting system'. Furthermore, since many of the works reported on electrochemical energy harvesting involve hybrid energy systems [30–33], sustainable structural materials used in electrochemical energy harvesting will not be reviewed in this chapter. The reader is encouraged to explore the abundantly available literature on hybrid energy harvesting systems.

8.3.1 Applications in Solar Energy Harvesting

Solar energy is considered the quintessential renewable energy source because of its abundance [34]. Solar energy harvesting is the process through which solar energy radiated from the sun is captured and converted into electrical energy. It is based on the photovoltaic (PV) effect. This is where solar radiation incident on a photovoltaic cell causes the cell to absorb photons and release electrons, thus generating electric current [35]. The next section will explore the finer details of the theory of the PV effect.

8.3.1.1 Photovoltaic Effect Theory

The PV effect (also known as the Becquerel Effect, after its discoverer) is a phenomenon where two dissimilar materials that are in contact with each other produce a potential difference between them when a light (sunlight) is incident on them [36]. The PV effect occurs in photovoltaic cells or solar cells. In PV cells, the dissimilar materials consist of P-type and N-type semiconductors joined together to form a p-n junction. When sunlight is incident on the PV cell, the negatively charged free electrons are forced to flow towards the N-type semiconductor and the positively charged holes move towards the P-type semiconductor. The flow of moving electrons results in an electric current when connected to a load as illustrated in Figure 8.3.

An individual PV cell can produce 1 to 2 Watt of power [37]. Therefore, to increase output power, several PV cells are connected to form modules. In turn, several modules can be connected to form arrays which can supply an electricity grid. For environmental protection, PV cells are sandwiched between protective layers of glass or plastics [37].

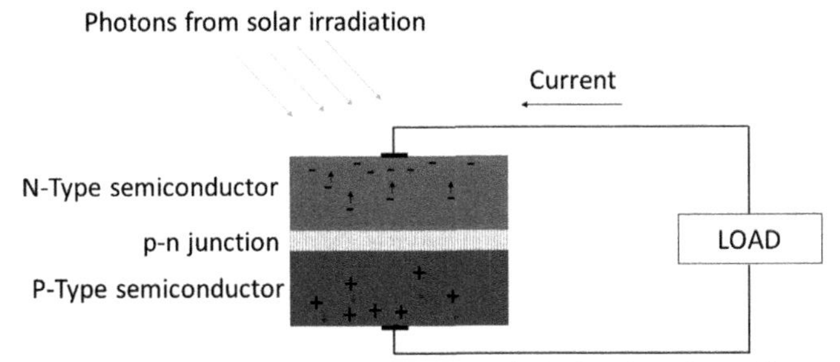

Figure 8.3 Principle of Photovoltaic Effect in a PV cell.

8.3.1.2 Sustainable Structural Materials used in Solar Energy Harvesting

Before delving into a discussion on sustainable structural materials that have been used in solar energy harvesting, it is worth noting and defining two terms that have emerged recently. The two terms are Building Integrated Photovoltaics (BIPV) and Building Applied Photovoltaics (BAPV). BIPV systems are systems where solar cells are integrated into different building components, that is to say they become functional architectural elements, substituting traditional building materials [38]. On the other hand, BAPV systems refer to the ones where the PV system is added to a structurally finished building. Building structures such as roofs, facades and pavements can provide critical avenues for setting up local energy harvesting systems, close to the site of consumption, thereby removing the cost of building new infrastructure and minimizing transmission losses that would occur to transmit energy from remote PV plants [38]. Sustainable structural materials that are used to support PV systems in BAPV and those used to implement PV energy harvesting in BIPV will be discussed in this section.

8.3.1.2.1 Glass

Glass is a well-known structural material [39] and is considered to be a sustainable material because it can be infinitely recycled and it can also be simply reused [40]. When considering its application in solar energy harvesting, it can serve several purposes:

- It is usually the protective cover for the solar cell [34] [37];
- It can be integrated with PV cells or provide the surface onto which the PV cell is built [41] [42];
- It can be modified to accommodate energy harvesting technologies [43].

Glass is one of the materials commonly used as a protective layer for solar cells because it is sustainable, mechanically stable and highly transparent. Typically, 3-mm thick float glass (soda-lime glass composed primarily of silica, sodium oxide and calcium oxide, and may contain iron, magnesium, titanium potassium and aluminium oxides [41]) is employed [34]. It provides environmental protection to solar cells against water, dirt and vapor, and also provides structural support [41]. It should be noted that the cover glass accounts for anywhere from 10% (for crystalline silicon PV modules) [44] to 25% (in silicon PV modules) [41] of the cost of the modules. Therefore, any effort towards reducing the material cost of glass greatly improves the sustainability of these modules.

Glass can be integrated with PV technology or provide the substrate or superstrate onto which the PV cells are built. Multicolored dye sensitized cells have been integrated into glazed surfaces to form translucent and colored photovoltaic panels covering 200 m^2, also known as 'Grätzel cells' for the SwissTech Convention Center [42]. Each of the 65 panels has a micro-converter and is connected to a low-voltage electrical backbone, allowing for modular maintenance [45]. Compared to their silicon-based photovoltaic counterparts, their operating temperatures show less increase, and they have better performance at angles far from optimum. In thin-film PV modules, the modules are fabricated by depositing the silicon layers onto a glass sheet in a superstrate configuration. The glass is selected such that it is transparent enough to allow sufficient light to transmit through it to the solar cell [41]. Float glass available currently can transmit approximately 90% of the light incident on it [34]. A glass back plate provides the protective covering for the back face.

Glass in architectural structures has been modified to accommodate energy harvesting technologies. A water-filled prismatic glass lover was proposed by researchers [43] for solar energy harvesting. The louver is comprised of three uniform silica glasses that form a hollow equilateral triangle as shown in Figure 8.4. Water is made to flow through the hollow section. The ultraviolet and infrared lights incident on the louver are absorbed and used for energy harvesting. Through simulation and experiment, the device has been evaluated for several cases of solar spectra of different air mass (AM) and showed feasibility of harvesting 51–54% of the total incident solar energy (about 87% absorption of ultraviolet radiation and about 81% absorption of infrared radiation). The aim of the concerned study was to develop a 7-band spectral model that can predict the performance of the louver under different spectra conditions. Therefore, the full study with integrated solar energy harvesting was not conducted.

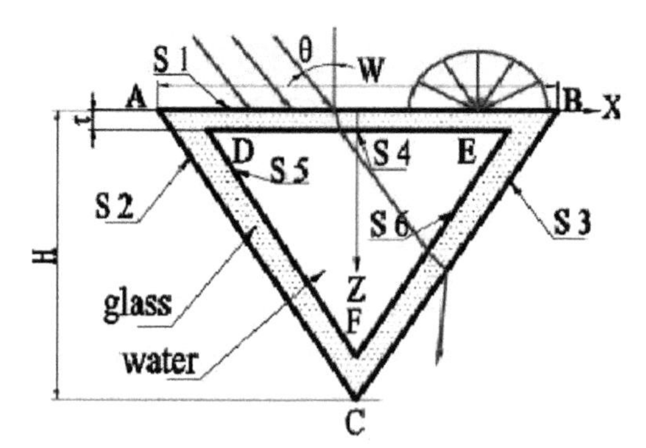

Figure 8.4 Cross section of the water-filled glass louver through which UV and IR lights are harvested. Reprinted from [43] with permission from Elsevier.

8.3.1.2.2 Glass Fiber Reinforced Polymer (GFRP)

Glass Fiber Reinforced Polymers (GFRP) are sustainable structural materials comprised of a combination of polymer matrix with reinforcing glass fibers [46]. GFRP have been employed in PV systems as support structures [47] and also integrated with PV cells, as substrates, due to their low cost, high resistance to corrosion, electrical insulation properties and high strength-to-weight ratio [48].

Han [47] investigated, numerically as well as experimentally, the mechanical properties of GFRP with different densities and orientations for applications in aquatic environments as a structural material for the construction of floating PV power stations. GFRP were selected due to their superior corrosion resistance compared to other FRP composites such as carbon fiber reinforced polymer (CFRP). GFRP was built as a primary beam (Figure 8.5) with a polyurethane (PU) foam core to enable sufficient buoyancy and mechanical support for the photovoltaic modules housed in the superstructure. The material properties, and bending and transverse compression performance of the GFRP were investigated. A 9.5 × 10.2 m floating PV power station was built that supported 36 double-glazed PV modules. The system provided an annual power output of approximately 10,692 kWh over a period of four years, without any maintenance requirements.

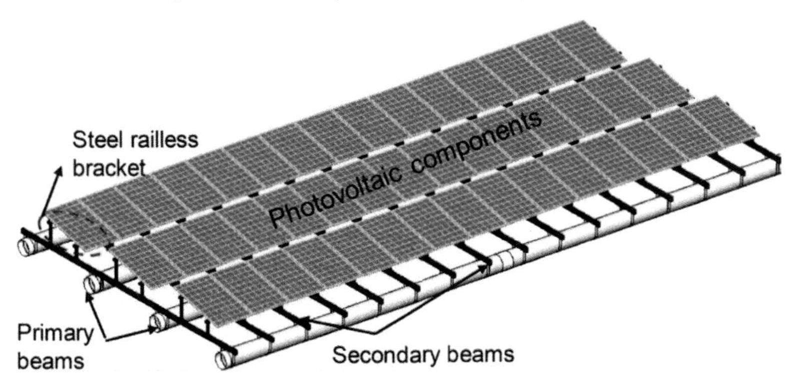

Figure 8.5 Illustration of the location of primary beams in floating photovoltaic power stations. Reprinted from [47] with permission from Elsevier.

8.3.1.2.3 Amorphous Silicon

Amorphous silicon (a-silicon) is a non-crystalline allotropic form of semiconductor silicon, used commonly used for thin-film solar cells, due to its low cost and low material requirements for the fabrication of PV cells [49, 50]. It is employed in PV cells for its optical rather than mechanical properties. Therefore, it is not considered as a sustainable structural material in this context. However, it has been employed in the support structures that hold the PV systems.

BIPV have been demonstrated with PV roof tiles made from amorphous silicon, mounted on a rigid insulation which was in turn resting on a waterproof roofing membrane [51]. The roof tiles were built into a 3.0 kW prototype. Although the gross installation cost of the BIPV was higher than that for a conventional

PV system, due to its dual functional requirement as a roof and as a PV energy harvester, it eliminated the need for a PV support structure, was quick and easy to install, could be reusable if the building was dissembled, had a longer life than a conventional PV, and was less than half the weight of aggregate ballasted roofs. The system was shown to have about 8% reduction in cost compared to the conventional means of structurally mounting a PV system on a roof.

8.3.1.2.4 Cadmium Telluride (CdTe)

Cadmium telluride (CdTe) is the second most commonly used material in PV cells after crystalline silicon [52]. It is used in thin-film PV cells because it has high absorption owing to its large band gap that can be tuned from 1.4 to 1.5 (eV), which allows for highly efficient sunlight absorption [53]. Additionally, they are cheap to manufacture [52] and are the least carbon intensive in production [54]. Another key advantage of CdTe is the feasibility for manufacturing large area solar cells [55]. Te is considered sustainable since they are the by-product of mining copper, steel and gold, and would otherwise be wasted [56]. The CdTe thin-film layer converts the absorbed sunlight into electricity. While CdTe provides good electronic properties for PV cell performance, it also serves as a mechanical support for the cell.

In a study, CdTe-based Semi-Transparent Photovoltaic (STPV) solar cell integrated window was reported [57]. Four of these glazed windows with different levels of transparency were investigated. Considering that radiation can be transmitted, reflected or absorbed, these properties were measured for the four windows. It was observed that higher the glazing, lower was the amount of heat transferred to the interior of the building, resulting in lower AC power consumption. Also, the higher the glazing, the more the absorption of radiation. The window with the least amount of glazing produced 30 kJ of energy, while the window with the heaviest amount of glazing produced about 51 kJ of energy. The higher glazing, however, reduced the amount of light transmitted to the interior and increased the need for artificial lighting during the day. The net energy performance was therefore calculated to determine the energy savings from employing STPV windows. It was found that for a 40% transparent window, net energy savings of 5% were achieved, while almost 20% savings were achieved from a 10% transparent window.

8.3.1.2.5 Graphene and Related Materials

Graphene consists of a single layer of carbon atoms organized in a hexagonal structure [58]. Since it has superior optical and electronic properties such as transparency of up to 97.7% in the near-infrared and visible spectrum [59] and carrier mobility of up to 10^4 cm^2 v^{-1} s^{-1} at room temperature [60], it is suitable as a transparent electrode in solar cells [61]; it also has superior mechanical strength due to the covalent bonds between all atoms [62] and its single layer structure gives it good mechanical flexibility. In terms of sustainability, graphene can enhance the durability of PV cells.

Yoon et al. [63] used a single layer of graphene to build a flexible transparent electrode for a perovskite solar cell. The transparent flexible graphene electrode exhibited superior mechanical stability against bending deformation by maintaining >90% of its efficiency after 1000 bending cycles with a bend radius of 4 mm, and 85% efficiency after 1000 bending cycles with a bend radius of 2 mm, as compared to a traditional flexible indium tin oxide electrode.

8.4 APPLICATIONS IN THERMAL ENERGY HARVESTING

Thermal energy harvesting can be performed based on several different effects: thermoelectric, pyroelectric, thermomagnetic and thermoelastic [23] [35]. Since the focus of this chapter is conversion to electric power, the two effects of concern are thermoelectric and pyroelectric effects. Thermoelectric effect is the most common of the two and is based on the Seebeck effect. Pyroelectric energy harvesting converts a time-varying temperature change into a current or a voltage [18]. This technique is not very well developed in practice due to its dependence on temperature fluctuations [23].

8.4.1 Thermoelectric Energy Harvesting Theory

Thermoelectric energy harvesting generates electrical energy from temperature gradients. It makes use of thermoelectric (TE) materials which are a special class of semiconductors which can generate electric energy when thermal energy migrates through them from a hotter side to a colder side. Thermoelectric materials are compact, fairly lightweight, require minimal maintenance, do not emit any carbon dioxide, and are robust since they do not have moving parts [14].

TE materials can be described by the Seebeck, Peltier and Thompson effects [14]. The Seebeck coefficient (S) defines the solid-state conversion of thermal energy to electrical energy. It describes the generation of electric current when electrons and holes move between two different semiconductor materials due to the temperature difference between them. Conversely, when two different semiconductor materials are in contact, the reverse effect can occur where a temperature difference is produced between them instead. This is also known as the Peltier effect [64] and is used in heating/cooling applications [65]. The Seebeck effect is an intrinsic property of TE materials and is independent of the material's geometry. It is given by:

$$S = \frac{\Delta V}{\Delta T} \tag{1}$$

Where ΔV is the potential difference generated due to the temperature difference ΔT. S is positive for p-type semiconductors and negative for n-type semiconductors [66].

The efficiency of TEs can be compared using the thermoelectric power factor (PF) defined as:

$$PF = \sigma S^2 \quad (2)$$

where σ is the electrical conductivity of the material. A second method of comparing the thermoelectric efficiency of TEs is using the dimensionless figure of merit (ZT), which is defined as:

$$ZT = (\sigma S/\kappa)T \quad (3)$$

where κ is the thermal conductivity of the material and T is the absolute temperature [67]. An efficient thermoelectric material has a high value of *ZT*. Historically, the maximum value obtained for this parameter is 1. However, recently reported materials have shown *ZT* values in the range of 2 to 2.8 [68]–[70]. A newly developed material which comprises a thin layer of iron, vanadium, tungsten and aluminum applied to a silicon crystal has been reported with a *ZT* value between 5 and 6 [71]. From (3), it can be seen that a high *ZT* requires high PF (high σ and S) and low κ. Electrical conductivity depends on the properties of the material. In general, semiconductors possess a higher Seebeck coefficient than metals [14] (although metals have the highest electrical conductivity, their Seebeck coefficient is quite low), and are therefore preferred for fabricating thermoelectric generators. This is due to the doping-induced local distortion in the electronic density of states or band gap in semiconductors, which limits the diffusion of electrons from the cold side to the hot side of the material. In fact, studies have shown the impact of semiconductor doping on the Seebeck coefficient [72]. The working principle of the Seebeck effect is shown in Figure 8.6.

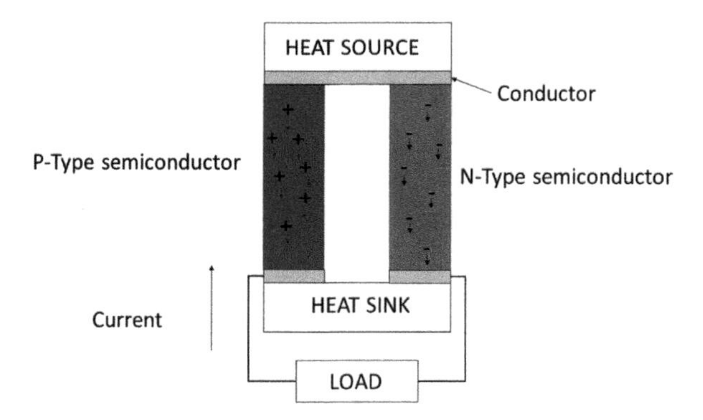

Figure 8.6 Working principle of the Seebeck effect used in thermoelectric energy harvesting.

Thermoelectric materials can be built into thermoelectric generators (TEGs) to increase their energy output. TEGs comprise alternating p-type and n-type thermocouples that are electrically in series and thermally in parallel to allow for all elements to be subject to the same temperature gradient while directly adding their individual thermoelectric voltage contributions (Figure 8.7).

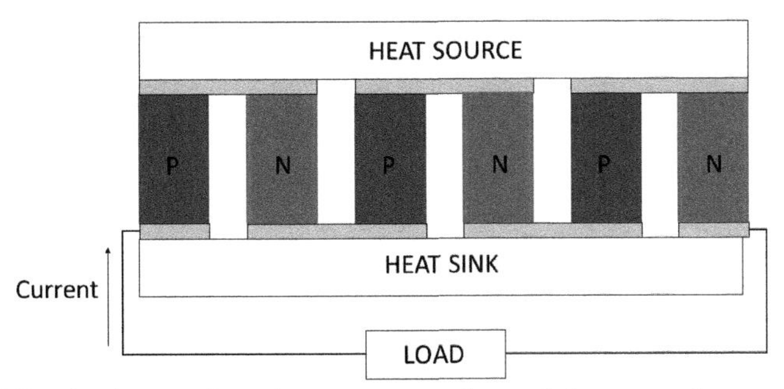

Figure 8.7 A thermoelectric generator formed by a series connection of thermoelectric materials.

8.4.2 Sustainable Structural Materials used in Thermoelectric Energy Harvesting

8.4.2.1 Asphalt

Asphalt or bituminous is one of the essential components used to produce hot mix asphalt [73] and is commonly used as a structural material in road construction. It is considered sustainable because it can be recycled and reused [74]. It should be noted that to improve sustainability, additives can be added to traditional asphalt mixtures to yield more efficient and greener versions [73] [74]. Since asphalt covers a large percentage of urban surface areas and is exposed to large amounts of thermal energy and solar radiation, it is an effective solution for thermal energy harvesting. The use of asphalt for thermoelectric energy harvesting can reduce the temperature of the road surface and minimize degradation caused by high temperatures [75]. The temperature gradient between that of the road surface and deeper layers can be exploited for energy harvesting. It should be noted that the surface temperatures of asphalt pavements can reach up to 70°C, and is dependent on environmental factors like solar radiation, air temperature, wind speed, humidity, and thermal properties of the asphalt pavement [76].

Liang [77] experimented with applying infrared heat from a lamp on asphalt samples to cause a temperature gradient. For different temperature gradients, it was found that the highest system efficiency and the best durability was exhibited when the TEG was 20 mm to 30 mm below the surface. From the experiments, it was shown that the system was able to generate 2592 J in a day.

8.4.2.2 Cement and Cement Composites

Cement is a well-known structural material in the construction industry. While production of cement is extremely energy intensive and leads to considerable carbon dioxide emissions [78], its composites are highly durable (they can last for 50 to 100 years); therefore, the environmental impact is spread over a long period of time [79]. Cement and its composites can serve dual functions as structural materials and thermoelectric materials.

Tao et al. created cement composites using zinc oxide (ZnO) and iron oxide (Fe_2O_3) nanoparticles. By studying variations in the percentage addition and testing the electrical, thermal and thermoelectric properties of the composites, it was found that the addition of 1% to 5% of nano-powders significantly improved the electrical properties and the Seebeck coefficient [80].

8.4.2.3 Graphene and Related Materials

Graphene and its related materials have been used as additives in other building materials to form composites that can provide improved hardness, mechanical strength, durability and flexibility [81, 82]. Hence, graphene is considered sustainable since its introduction into the host material may require less amount of the host material and therefore lower the carbon footprint for the final application. Moreover, graphene is also considered sustainable since it has green production methods [83, 84].

A graphene-based cement composite has been proposed by researchers [85]. The said composite was formed by introducing graphene nanoparticles (GNP) and showed p-type semiconductor behavior. For 15 wt.% (weight percentage) GNP, an electrical conductivity of 16.2 Scm^{-1} and a Seebeck coefficient of 34 μVK^{-1} were observed, along with a maximum ZT of 0.44×10^{-3} at 70°C.

Lin et al. [86] fabricated a p-n all-graphene fiber (GF) TEG for flexible/wearable thermoelectric energy harvesting applications. The prototype was fabricated such that it did not require the typical conductive adhesive used to ensure a good ohmic contact between the p-n legs, thereby simplifying the assembly process. Since tensile properties are a critical performance indicator for wearables, the prototype was tested with 1000 times bend and release cycles. The measured electrical resistance was maintained within 5% of the original value of 75 kΩ, exhibiting good mechanical reliability. The prototype gave a stable output power of about 1.3 pW, using the temperature difference between the human body and the air (70 K temperature difference) over a period of about 5 minutes.

In another study, graphene oxide was introduced into an aerogel to improve its mechanical properties [87]. The addition of graphene provided structural protection by limiting volume shrinkage during the infiltration process of the phase change material. The performance of the graphene/cysteamine aerogel-supported phase change material was tested using a solar light. Energy harvesting efficiencies of 67.92 and 41.72% were reported for the light-on/light-off process respectively.

8.4.2.4 Carbon Nanotubes and Related Materials

Carbon nanotubes are cylindrical molecules that are an allotropic form of carbon [88]. They exhibit exceptional electrical conductivity, tensile strength and thermal conductivity [89]. Whereas in some texts they are described as rolled up sheets of graphene, here they are discussed in an independent section due to their large variety and distinct applications. The classification of carbon nanotubes as sustainable depends on a few factors: green production, disposal at the end of life, and the life cycle impact assessment.

Maliheh et al. [90] developed a novel thermoelectric material from waste-recycled alkali-activated construction material and 1.0 wt.% single-walled carbon nanotubes. The nanocomposite demonstrated decent electrical and thermoelectric properties, with an electrical conductivity of 1660 Sm^{-1} and a Seebeck coefficient of 15.5 μVK^{-1}; the material also showed good structural load bearing. The composite was built into a TEG which gave a maximum power output of 0.695 μW for a temperature gradient of 60 K.

8.4.3 Pyroelectric Energy Harvesting Theory

With the pyroelectric effect, electricity is generated from temperature fluctuations [91]. Pyroelectric materials are inherently polarized. Their pyroelectric property stems from their temperature dependent polarization state. If the rate of change of temperature increases, the polarization level reduces since the dipole orientation is distorted and the number of unbound charges on the surface reduces. An electric current is produced as a result. If the rate of temperature change is reduced, the dipoles re-orient and a reverse current low is produced [20]. The relationship between the temperature gradient (dT/dt) and the current produced (i_p) is given by [20]:

$$i_p = \frac{dQ}{dt} = \frac{Ap.dT}{dt} \tag{4}$$

Where A is the surface area of the material, and p is the pyroelectric coefficient expressed as:

$$p = \frac{dP_s}{dT} \tag{5}$$

Where P_s is spontaneous polarization [92]. Pyroelectric harvesters are typically modelled as current sources. The net charge generated as a result of a temperature variation ΔT is determined as:

$$Q = pA\Delta T \tag{6}$$

Pyroelectric materials are usually dielectric, and their internal capacitance can be calculated as:

$$C = \frac{A\varepsilon_{33}^{\sigma}}{h} \tag{7}$$

Where e_{33}^{σ} is the material permittivity in the polarization direction. The stored energy due to temperature variation is given by:

$$E = \frac{Q^2}{(2C)} \tag{8}$$

Hence, the stored energy for a pyroelectric material is given as:

$$E = 0.5(p^2\varepsilon_{33}^{\sigma})Ah(\Delta T)^2 \tag{9}$$

Although pyroelectric materials have a high conversion efficiency, they require temperature oscillations—natural temperature oscillations are rare [20].

The working principle of the pyroelectric effect used in energy harvesting is shown in Figure 8.8.

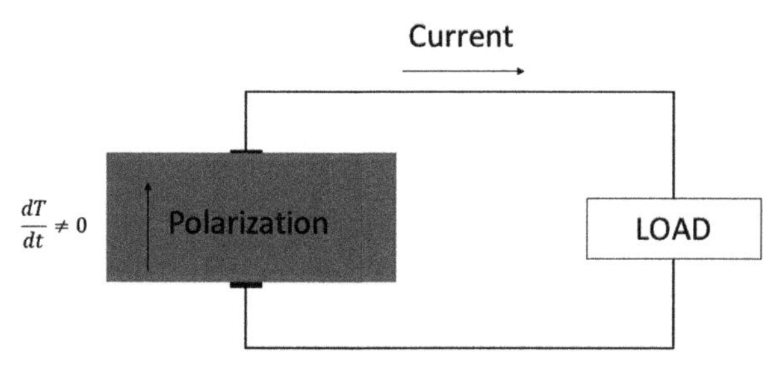

Figure 8.8 Working principle of pyroelectric energy harvesting that utilizes time-varying temperature fluctuations.

8.4 SUSTAINABLE STRUCTURAL MATERIALS USED IN PYROELECTRIC ENERGY HARVESTING

8.4.4.1 Ceramics

Ceramics are inorganic metallic or non-metallic solid compounds with mixed ionic and covalent bonds [93]. They have unique properties such as high-temperature stability, high melting points and good chemical inertness. Therefore, they are useful in a wide variety of applications [93]. Ceramics are considered sustainable as they can be made from waste materials.

In one study [94], a novel hybrid pyroelectric ceramic material was reported. It was formed by combining hexagonal boron nitride (hBN) nanosheets with lead magnesium niobate-lead antimony-manganese-lead zirconate titanate (PMN-PMS-PZT) ceramic. BN was selected as the dopant due to its low-cost, non-toxic and chemically stable nature. Addition of hBN improved the pyroelectric coefficient, the thermal conductivity and the rate of temperature change by 16.2%, 12.5% and 8.2% respectively. An energy harvesting system was designed using two Peltier cells to provide the temperature fluctuation, thermocouples to monitor the temperature, as well as a heat sink and a fan. The results showed that a 65.6% increase in the power harvested was recorded, compared to the power harvested from an unfilled PMN-PMS-PZT ceramic.

8.4.4.2 Carbon Nanotubes

Qingping et al. [95] developed a porous pyroelectric ceramic by introducing 0.3 wt.% of carbon nanotubes (CNT) into PMN-PMS-PZT. The CNT acted as a pore forming agent. The introduced pores decreased the permittivity (since the effective permittivity includes the permittivity of the 'air' in the pores) and the specific heat capacity of the ceramic. The defects altered the electric field distribution and impacted the domain reversal and local piezo response, resulting in a 208% increase in energy harvested. In this context, sustainability is seen by the introduction of pores by the CNT, which effectively reduces the amount of active material required.

8.4.4.3 Piezoelectric Polymers

Piezoelectric polymers are materials that generate electrical charges when subjected to mechanical stress. These materials have been used in pyroelectric energy harvesting due to their pyroelectric properties, ability to withstand high temperatures, high mechanical flexibility, and biocompatibility [91] [96]. Additionally, some piezoelectric polymers such as polyvinylidene fluoride (PVDF) are derived from renewable sources, making them a sustainable option [87].

Xue et al. demonstrated a pyroelectric nanogenerator using a flexible PVDF film installed on an N95 mask. The film was exposed to temperature variations at a rate of 13°C s^{-1} due to subject respiration. During exhalation, the temperature of the air in the mask was close to that of the body, while the temperature of the air reduced during inhalation. The prototype was tested over a period of 10 days and showed a maximum output performance of 8.31 μW with a load resistance of 50 MΩ.

8.5 APPLICATIONS IN KINETIC ENERGY HARVESTING

Kinetic energy harvesting refers to the conversion of kinetic energy (usually present in the form of random displacements, vibrations or forces) into electrical energy [20]. Kinetic energy harvesters are typically classified as electromagnetic, piezoelectric, electrostatic or triboelectric harvesters [20] [97].

8.5.1 Piezoelectric Energy Harvesting Theory

The piezoelectric effect describes the phenomenon where an electric potential is produced across a material when a force is applied. Conversely, reverse piezoelectric effect occurs when the material is exposed to an electric field and results in a mechanical strain [20]. The relationship between forward and reverse piezoelectric effects can be given by the constitutive equations:

$$[\delta D] = [s^E \; d^t \; d \; \varepsilon^T][\sigma \; E] \tag{10}$$

where σ and δ are the stress component and the strain component respectively, E is the electric field, D is the electric flux density, s is the elastic compliance, ε is the dielectric constant and d is the piezoelectric coefficient. Superscript E denotes evaluation performed under constant electric field, superscript T under constant stress, and superscript t stands for transpose. The working principle of the piezoelectric effect is shown in Figure 8.9.

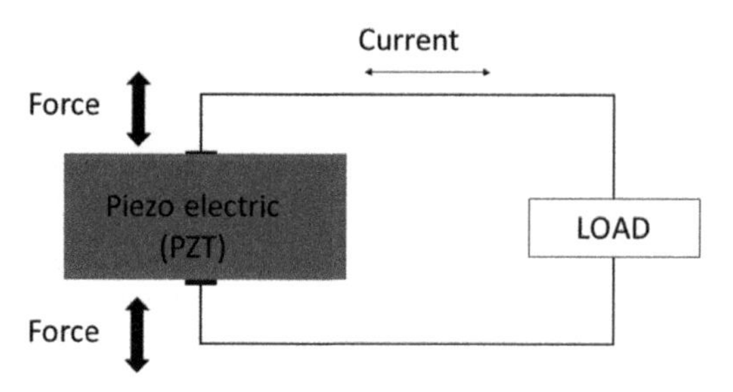

Figure 8.9 Working principle of the piezoelectric effect as used in energy harvesting.

8.5.2 Sustainable Structural Materials used in Piezoelectric Energy Harvesting

8.5.2.1 Cement and Cement Composites

Cement-based materials are one of the most abundant construction materials. Despite their energy intensive manufacturing processes, their sustainability lies in their long life span such that this energy cost is spread over a long period. The piezoelectric properties of hydrated hardened cement were reported by Sun et al. [98], who showed that some amount of electric current can be produced when a compressive force is applied. Piezoelectricity was attributed to the movement of water molecules and free ions under loading. The amount of electricity produced is however small; therefore, various admixtures have been explored for improving the properties of piezoelectric materials.

In one study, steel fibers with 8 μm diameter and 6 mm length were added to cement paste (0.18% by volume) [99]. The fibers had 47% polyvinyl alcohol binder to assist with the spreading of fibers within the cement during mixing. The results showed that the longitudinal piezoelectric coupling coefficient at 10 kHz increased to 2.5×10^{-11} m/V, compared to the cement paste without the admixture, while the piezoelectric voltage coefficient remained the same at 1.1×10^{-3} m^2/C.

8.5.2.2 Steel

Steel is a well-known structural material. It is a carbon-iron alloy with superior mechanical properties and good corrosion resistance. It is highly recyclable and reusable [100]. In piezoelectric energy harvesting, steel is often used as the carrying substrate [101].

Grzybek et al. [101] developed a prismatic cantilever beam. It comprised of a steel substrate which had two patches of macro fiber composite P2 (the piezoelectric material) glued to it on both sides. The substrate provides structural support to the piezoelectric material. Experiments were conducted for ten values of load resistance. A maximum current of 0.8 mA was generated for a 10 kΩ load, whereas the maximum power generated for a 60 kΩ load was about 15.2 mW.

8.5.2.3 Biopolymer-based Materials

Certain biopolymers such as keratins, collagens, cellulose and other fibrous materials have been reported for their soft mechanical properties and piezoelectric behavior [102]. Their piezoelectricity is derived from the anisotropic assembly of polar monomer units [103]. These materials are biocompatible and biodegradable, which makes them attractive sustainable solutions.

A piezoelectric nanogenerator was fabricated from pomelo fruit membrane bio-waste in [104]. The nanogenerator derived its piezoelectric properties from the complex interaction between cellulose and hemicellulose biopolymers and the pectin polysaccharide found in the fruit membrane. The nanogenerator was formed by cutting a $2 \times 2 \times 0.045$ cm^3 layer of the dried fruit membrane (without any special processing) and pasting conductive aluminium tape on either side of the membrane to act as electrodes. Copper wires were connected to the electrodes. The whole nanogenerator was encased in polypropylene for environmental protection. The energy harvesting performance of the nanogenerator was tested experimentally using finger tapping with an applied pressure of about 1.1 kPa. The nanogenerator generated about 15 V, 130 μA and 487.5 μW cm^{-2} of output voltage, current and power density respectively when connected to a full wave rectifier circuit and a 220 kΩ load. The device was also able to power multiple LEDs. The durability of the device was also tested. After 100 cycles of tensile testing with a gauge length of 40 mm, testing speed of 1 mm min^{-1} and a constant load of 400 gf, no observable deformation was observed.

Wood is one of the most abundant structural biopolymer composites available from the environment. It is biocompatible, biodegradable and renewable [105], making it an attractive material for sustainable applications. Wood is largely composed of cellulose, hemicellulose and lignin [106]. The presence of cellulose in wood gives it its piezoelectric properties, albeit with an inherently low piezoelectric constant [107].

A sustainable and biodegradable piezoelectric nanogenerator fabricated from delignified balsa wood has been demonstrated in a study [105]. Therein, lignin and hemicellulose were extracted from the wood to create extremely compressible wood sponges (130-fold increase in compressibility). The dimensions of the sponge in uncompressed state were 15 mm × 15 mm × 14 mm in the longitudinal, radial and tangential directions. The nanogenerator was able to provide an output of 0.69 V – 85 times than that generated by untreated wood. The mechanical robustness of the wood sponge was tested using repeated loading and unloading cycles under a constant stress of 13.3 kPa. After merely 50 cycles, a slight plastic deformation and degradation in output current were observed. To demonstrate scalability, 30 wood sponges were connected in parallel. This larger nanogenerator was able to produce 205 nA of output current that was able to drive an LED screen and an LCD screen by finger pressing or foot taping respectively. Furthermore, the end-of-life biodegradability was tested using cellulose degrading fungi. The results showed that the device could potentially reduce the amount of waste produced by energy harvesting devices.

8.5.3 Electromagnetic Energy Harvesting Theory

This method uses the interaction between a magnetic field and a conductor to generate electricity. This method can be used in a variety of applications, such as powering wireless sensors in industrial environments or powering small electronic devices. The major advantage of this method is that it is relatively easy to implement and can be used in a wide range of environments. However, the efficiency of this method is highly dependent on the strength of the magnetic field and the specific implementation. The working principle of electromagnetic energy harvesting is shown in Figure 8.10.

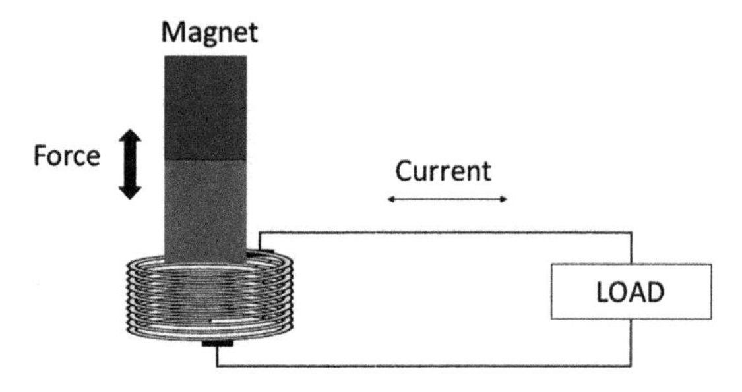

Figure 8.10 Working principle of electromagnetic energy harvesting.

From Faraday's law of electromagnetic induction, when a changing flux passes through a closed coil, electric current is induced. The associated induced voltage is given by [108]:

$$E(t) = -\frac{Nd\phi}{dt} \tag{11}$$

where N is the number of turns in the coil, ϕ is the magnetic flux and t is time. The negative sign is introduced due to Lenz's law.

8.5.4 Sustainable Structural Materials used in Electromagnetic Energy Harvesting

8.5.4.1 Steel

Gholikhani et al. explored the potential of electromagnetic energy harvesting from speed bumps on roads [109]. Speed bumps are considered as an optimum location for electromagnetic energy harvesting in road infrastructure since large deflections can be produced by vehicles being driven over the speed bumps, and potentially significant amounts of power can be generated (541–646 W peak [110]). Gholikhani created a device with a top plate to simulate a speed bump, cylindrical supports for structural integrity, and compression springs to regulate movement of the top plate under loading and restore the top plate to its original position. The cylindrical supports and compression springs were both made from stainless steel due its durability and good mechanical properties. Two versions of the protype were fabricated (as shown in Figure 8.11)—one with a cantilever mechanism and another with a rotation mechanism. In both the cases, these mechanisms facilitated the movement of a magnet through the flux of a current-carrying coil. The experiments yielded root mean square power outputs of 0.43 W and 0.04 W for the cantilever and the rotation mechanism respectively, and maximum peak power outputs of 2.8 W and 0.25 W respectively were measured for a 10 kN load.

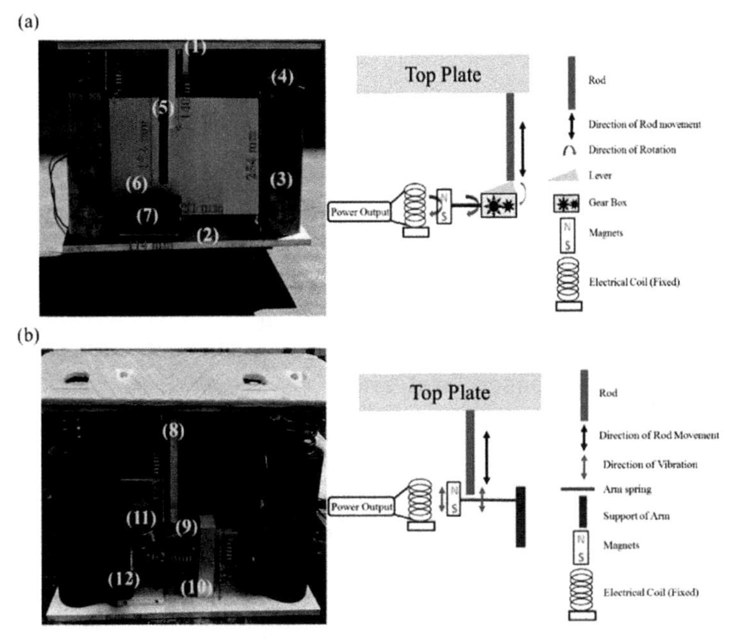

Figure 8.11 Electromagnetic energy harvesting using steel compression springs and cylindricalsupport for (a) cantilever mechanism and (b) rotational mechanism, with (1) top plate, (2) bottom plate, (3) cylindrical support, (4) compression springs, (5) a two-part rod, (6) a lever, (7) a box that includes magnets, coils, gears, and torsional springs, (8) a rod, (9) a spring arm, (10) arm support, (11) magnets, and (12) electrical coil. Reprinted from [109].

8.5.5 Electrostatic/Triboelectric Energy Harvesting Theory

Triboelectric effect is the production of charges when two materials with opposite polarities (tendencies to gain and lose electrons) come into contact and separate from each other (Figure 8.12). If the process of contact and separation is repeated cyclically, an alternating current is produced [111]. The efficiency of triboelectric energy harvesting depends heavily on the structural design of the charge generation layers, and the mode of operation [102]. There are four modes: contact separation mode (Figure 8.12(a)), sliding mode (Figure 8.12(b)), freestanding mode (Figure 8.12(c)), and single-electrode mode (Figure 8.12(d)) [112].

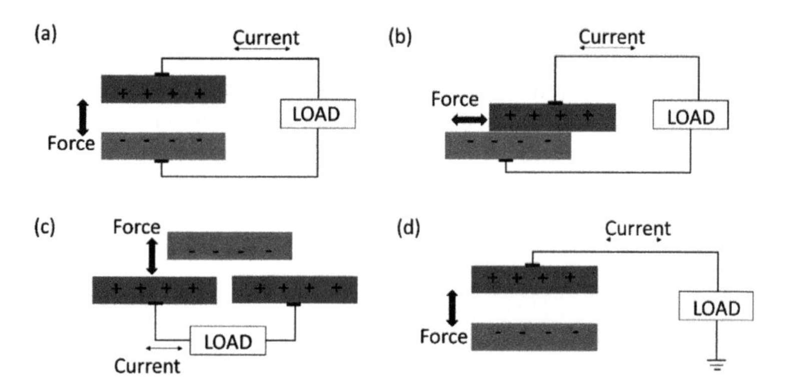

Figure 8.12 Triboelectric effect using (a) contact separation mode, (b) sliding mode, (c) freestanding mode, and (d) single electrode mode.

8.5.6 Sustainable Structural Materials used in Electrostatic/ Triboelectric Energy Harvesting

8.5.6.1 Biopolymer-based Materials

Bamboo is an attractive natural alternative to synthetic polymers because it is cheap, renewable, non-toxic, environmentally friendly, and fully biodegradable [113]. Zhang et al. developed three low-cost triboelectric nanogenerators [114]. The three nanogenerators were made from bamboo fiber, wood fiber, and cotton fiber (for the positive electrode) recycled from disposable facial towels. The negative electrodes were fabricated from polydimethylsiloxane (PDMS). From performance comparison under the same frequency and stroke condition (2 Hz and 2 cm respectively), it was found that the bamboo fiber based nanogenerator gave the best performance, with an output power density of 20.56 $\mu W/cm^2$ for a 50 MΩ load resistance. As a proof-of-concept demonstration, the prototype was used to power a commercial calculator and light 53 LEDs through finger tapping.

8.5.6.2 Graphene and Related Materials

Graphene's unique and exceptional properties of high transparency, excellent conductivity, elasticity, and mechanical flexibility make it well suited for applications in flexible electronics. Kim et al. [115] demonstrated triboelectric energy harvesting using a graphene-based triboelectric nanogenerator. They developed one-layer, two-layer, three-layer and four-layer flexible transparent graphene nanogenerators. For the oppositely charged contact, a polyethylene terephthalate polymer was used. Vertical compressive stress was applied to the devices, and their output voltage and current density were measured. The one-layer nanogenerator yielded a high output voltage of 5 V and a current density of 500 nA cm^{-2}. The prototypes were able to power a liquid crystal display (LCD), an LED, and an electroluminescence (EL) display unit through a rectifier circuit.

8.6 CONCLUSION

The world is undergoing a pivotal shift towards environmentally responsible engineering practices, focusing on safeguarding our environment and conserving resources for future generations. Within the realm of structural materials, this underscores the growing importance of sustainable structural materials. These materials harmonize structural functionality with eco-consciousness, promising significant benefits across various domains, including transportation, construction, energy generation, protective gear, and microelectronics. This chapter has particularly emphasized their role in energy harvesting applications, highlighting their potential to drive the development of more efficient and environmentally friendly systems.

Commencing with an introduction to structural materials and their sustainable variants, the chapter has traversed through the landscape of four major energy harvesting methodologies (solar, thermal, kinetic, and electrochemical). For each method, various sustainable structural materials that have been employed for energy harvesting were discussed. The discussion elucidated the intricate interplay between sustainable structural materials, their properties, and the associated energy capture mechanisms. Through an examination of fundamental theoretical frameworks and real-world applications, the chapter has highlighted the pivotal role of sustainable structural materials in charting a path towards a more sustainable future. These materials are poised to be utilized for both structural and energy harvesting applications, contributing to a greener and more sustainable world.

REFERENCES

[1] Structural Materials. https://tmi.utexas.edu/research/structural-materials (accessed Dec. 10, 2022).

[2] Markandan, K., Zhang, Z., Chin, J., Cheah, K.H. and Tang, H.B. 2019. Fabrication and preliminary testing of hydroxylammonium nitrate (HAN)-based ceramic microthruster for potential application of nanosatellites in constellation formation flying. Microsyst. Technol. 25(11): 4209–4217. doi:10.1007/s00542-019-04484-2

[3] Provis, J.L. 2015. Grand challenges in structural materials. Front. Mater. 2: 31. doi: 10.3389/fmats.2015.00031.

[4] Raabe, D., Tasan, C.C. and Olivetti, E.A. 2019. Strategies for improving the sustainability of structural metals. Nature. 575(7781): 64–74. doi: 10.1038/s41586-019-1702-5.

[5] Pelz, J.S., Ku, N., Meyers, M.A. and Vargas-Gonzalez, L.R. 2021. Additive manufacturing of structural ceramics: a historical perspective. J. Mater. Res. Technol. 15: 670–695. doi: 10.1016/j.jmrt.2021.07.155.

[6] Mishnaevsky, L., Branner, K., Petersen, H., Beauson, J., McGugan, M. and Sørensen, B. 2017. Materials for wind turbine blades: An overview. Materials (Basel). 10(11): 1285. doi: 10.3390/ma10111285.

[7] Wimmers, G. 2017. Wood: A construction material for tall buildings. Nat. Rev. Mater. 2(12): 17051. doi: 10.1038/natrevmats.2017.51.

[8] Fan, Z., Sun, H., Zhang, L., Zhao, X. and Hu, Y. 2022. Lightweight, high-strength wood prepared by deep eutectic solvent treatment as a green structural material. ACS Sustain. Chem. Eng. 10(29): 9600–9611. doi: 10.1021/acssuschemeng.2c02606.

[9] Oladazimi, A., Mansour, S., Hosseinijou, S. and Majdfaghihi, M. 2021. Sustainability identification of steel and concrete construction frames with respect to triple bottom line. Buildings. 11(11): 565. doi: 10.3390/buildings11110565.

[10] Yadav M. and Mathur, A. 2021. Bamboo as a sustainable material in the construction industry: An overview. Mater. Today Proc. 43: 2872–2876. doi: 10.1016/j.matpr.2021.01.125.

[11] Donkor P. and Obonyo, E. 2015. Earthen construction materials: Assessing the feasibility of improving strength and deformability of compressed earth blocks using polypropylene fibers. Mater. Des. 83: 813–819. doi: 10.1016/j.matdes.2015.06.017.

[12] Bahrami, A., Vall, A. and Khalaf, A. 2021. Comparison of cross-laminated timber and reinforced concrete floors with regard to load-bearing properties. Civ. Eng. Archit. 9(5): 1395–1408. doi: 10.13189/cea.2021.090513.

[13] Zhang, H., Yang, T. and Zhang, K. 2022. Low-corrosivity structural timber for a sustainable future. Matter. 5(7): 1992–1995. doi: 10.1016/j.matt.2022.06.023.

[14] Singh, V.P., Kumar, M., Srivastava, R.S. and Vaish, R. 2021. Thermoelectric energy harvesting using cement-based composites: a review," Mater. Today Energy 21: 100714. doi: 10.1016/j.mtener.2021.100714.

[15] Kiziroglou and M.E. Yeatman, E.M. 2012. Materials and techniques for energy harvesting. In: Functional Materials for Sustainable Energy Applications. Elsevier. 541–572. doi: 10.1533/9780857096371.4.539.

[16] US Energy Information Administration. Internations Energy Outlook 2021. 2022. https://www.eia.gov/outlooks/ieo/narrative/consumption/sub-topic-01.php (accessed Dec. 17, 2022).

[17] Pop-Vadean, A., Pop, P.P., Latinovic, T., Barz, C. and Lung, C. 2017. Harvesting energy an sustainable power source, replace batteries for powering WSN and devices on the IoT, IOP Conf. Ser. Mater. Sci. Eng. 200: 012043. doi: 10.1088/1757-899X/200/1/012043.

[18] Bowen, C.R., Taylor, J., LeBoulbar, E., Zabek, D., Chauhan, A. and Vaish, R. 2014. Pyroelectric materials and devices for energy harvesting applications. Energy Environ. Sci. 7(12): 3836–3856. doi: 10.1039/C4EE01759E.

[19] Hao, D., Lingfei, Q., Alaeldin, M.T., Ammar, A., Ali, A., Dabing, L., et al. 2022 Solar energy harvesting technologies for PV self-powered applications: A comprehensive review. Renew. Energy. 188: 678–697. doi: 10.1016/j.renene.2022.02.066.

[20] Liu, H., Fu, H., Sun, L., Lee, C. and Yeatman, E.M. 2021. Hybrid energy harvesting technology: From materials, structural design, system integration to applications. Renew. Sustain. Energy Rev. 137: 110473. doi: 10.1016/j.rser.2020.110473.

[21] Akinaga, H. 2020. Recent advances and future prospects in energy harvesting technologies. Jpn. J. Appl. Phys. 59(11): 110201. doi: 10.35848/1347-4065/abbfa0.

[22] Forman, C., Muritala, I.K., Pardemann, R. and Meyer, B. 2016. Estimating the global waste heat potential. Renew. Sustain. Energy Rev. 57: 1568–1579. doi: 10.1016/j.rser.2015.12.192.

[23] Kishore R. and Priya, S. 2018. A review on low-grade thermal energy harvesting: Materials, methods and devices. Materials (Basel). 11(8): 1433. doi: 10.3390/ma11081433.

[24] Wang, H., Jasim, A. and Chen, X. 2018. Energy harvesting technologies in roadway and bridge for different applications—A comprehensive review," Appl. Energy. 212: 1083–1094. doi: 10.1016/j.apenergy.2017.12.125.

[25] Wang, L., Zhenxuan, F., Youchao, Q., Chi, Z., Libo, Z., Zhuangde, J., et al. 2022. Overview of human kinetic energy harvesting and application. ACS Appl. Energy Mater. 5(6): 7091–7114. doi: 10.1021/acsaem.2c00703.

[26] Gawron, P., Wendt, T.M., Stiglmeier, L., Hangst, N. and Himmelsbach, U.B. 2021. A review on kinetic energy harvesting with focus on 3d printed electromagnetic vibration harvesters. Energies. 14(21): 6961. doi: 10.3390/en14216961.

[27] Zou, H.-X., Lin-Chuan, Z., Qiu-Hua, G., Lei, Z., Feng-Rui, L., Ting, T., et al. 2019. Mechanical modulations for enhancing energy harvesting: Principles, methods and applications. Appl. Energy 255: 113871. doi: 10.1016/j.apenergy.2019.113871.

[28] Shukla, A.K., Mitra, S., Dhakar, S., Maiti, A., Sharma, S. and Dey, K.K. 2023. Electrochemical energy harvesting using microbial active matter. ACS Appl. Bio Mater. 6(1): 117–125. doi: 10.1021/acsabm.2c00785.

[29] Wang, H., Park, J.-D. and Ren, Z.J. 2015. Practical energy harvesting for microbial fuel cells: A review. Environ. Sci. Technol. 49(6): 3267–3277. doi: 10.1021/es5047765.

[30] Roy, K., Devi, P. and Kumar, P. 2021. Magnetic-field induced sustainable electrochemical energy harvesting and storage devices: Recent progress, opportunities, and future perspectives. Nano Energy 87: 106119. doi: 10.1016/j.nanoen.2021.106119.

[31] Kim S. et al. 2016. Electrochemically driven mechanical energy harvesting. Nat. Commun. 7(1): 10146. doi: 10.1038/ncomms10146.

[32] Preimesberger, J.I., Kang, S. and Arnold, C.B. 2020. Figures of merit for piezoelectrochemical energy-harvesting systems. Joule. 4(9): 1893–1906. doi: 10.1016/j.joule.2020.07.019.

[33] Sim H.J. and Choi, C. 2022. Microbuckled mechano-electrochemical harvesting fiber for self-powered organ motion sensors. Nano Lett. 22(21): 8695–8703. doi: 10.1021/acs.nanolett.2c03296.

[34] Allsopp, B.L., Robin, O., Simon, R.J., Ian, B., Gavin, S., Peter, S., et al. 2020. Towards improved cover glasses for photovoltaic devices. Prog. Photovoltaics Res. Appl. 28(11): 1187–1206. doi: 10.1002/pip.3334.

[35] Sucupira L. and Castro-Gomes, J. 2021. Review of energy harvesting for buildings based on solar energy and thermal materials. Civil Eng. 2(4): 852–873. doi: 10.3390/civileng2040046.

[36] Copeland, A.W., Black, O.D. and Garrett, A.B. 1942. The Photovoltaic Effect. Chem. Rev. 31(1): 177–226.

[37] United States Department of Energy. (n.d.), Solar Energy Technologies Office, Solar Photovoltaic Technology Basics. https://www.energy.gov/eere/solar/solar-photovoltaic-technology-basics.

[38] Freitas S. and Brito, M.C. 2019. Solar façades for future cities. Renew. Energy Focus. 31: 73–79. doi: 10.1016/j.ref.2019.09.002.

[39] A. Jóźwik. 2022. Application of glass structures in architectural shaping of all-glass pavilions, extensions, and links. Buildings. 12(8): 1254. doi: 10.3390/buildings12081254.

[40] Landi, D., Germani, M. and Marconi, M. 2019. Analyzing the environmental sustainability of glass bottles reuse in an Italian wine consortium. Procedia CIRP. 80: 399–404. doi: 10.1016/j.procir.2019.01.054.

[41] Burrows and K. Fthenakis, V. 2015. Solar energy materials and solar cells glass needs for a growing photovoltaics industry. Sol. Energy Mater. Sol. Cells. 132: 455–459. doi: 10.1016/j.solmat.2014.09.028.

[42] Barraud, E. 2013. Stained glass solar windows for the swiss tech convention center. Chimia (Aarau). 67(3): 181–182, 2013, doi: 10.2533/chimia.2013.181.

[43] Cai Y. and Guo, Z. 2019. Spectral investigation of solar energy absorption and light transmittance in a water-filled prismatic glass louver. Sol. Energy. 179: 164–173. doi: 10.1016/j.solener.2018.12.066.

[44] Powell, D.M., Winkler, M.T., Choi, H.J., Simmons, C.B., Needleman, D.B. and Buonassisi, T. 2012. Crystalline silicon photovoltaics: A cost analysis framework for determining technology pathways to reach baseload electricity costs. Energy Environ. Sci. 5(3): 5874–5883. doi: 10.1039/C2EE03489A.

[45] Solutions I. and Solar, F.O.R. Solaronix solar cells. 1–8.

[46] Navaratnam, S., Selvaranjan, K., Jayasooriya, D., Rajeev, P. and Sanjayan, J. 2023. Applications of natural and synthetic fiber reinforced polymer in infrastructure: A suitability assessment. J. Build. Eng. 66: 105835. doi: 10.1016/j.jobe.2023.105835.

[47] Han, J., Wan, C., Fang, H. and Bai, Y. 2020. Development of self-floating fibre reinforced polymer composite structures for photovoltaic energy harvesting. Compos. Struct. 253: 112788. doi: 10.1016/j.compstruct.2020.112788.

[48] Fang, H., Bai, Y., Liu, W., Qi, Y. and Wang, J. 2019. Connections and structural applications of fibre reinforced polymer composites for civil infrastructure in aggressive environments. Compos. Part B Eng. 164: 129–143. doi: 10.1016/j.compositesb.2018.11.047.

[49] Yang, J., Banerjee, A. and Guha, S. 2003. Amorphous silicon based photovoltaics—from earth to the 'final frontier.' Sol. Energy Mater. Sol. Cells. 78(1–4): 597–612. doi: 10.1016/S0927-0248(02)00453-1.

[50] Sreejith, S., Ajayan, J., Kollem, S. and Sivasankari, B. 2022. A comprehensive review on thin film amorphous silicon solar cells. Silicon 14(14): 8277–8293. doi: 10.1007/s12633-021-01644-w.

[51] Dinwoodie T.L. and Shugar, D.S. Optimizing roof-integrated photovoltaics: A case study of the PowerGuard roofing tile. *In*: Proceedings of 1994 IEEE 1st World Conference on Photovoltaic Energy Conversion—WCPEC (A Joint Conference of PVSC, PVSEC and PSEC). 1: 1004–1007. doi: 10.1109/WCPEC.1994.520130.

[52] Solar Energy Technologies Office. Cadmium Telluride" https://www.energy.gov/eere/solar/cadmium-telluride (accessed Feb. 17, 2023).

[53] Birkmire R.W. and McCandless, B.E. 2010. CdTe thin film technology: Leading thin film PV into the future. Curr. Opin. Solid State Mater. Sci. 14(6): 139–142. doi: 10.1016/j.cossms.2010.08.002.

[54] A Comprehensive Guide to Solar Energy Systems. Elsevier, 2018. doi: 10.1016/C2016-0-01527-9.

[55] Breeze, P., Solar Power Generation. Elsevier, 2016. doi: 10.1016/C2014-0-04849-6.

[56] Anctil A. and Fthenakis, V. 2013. Critical metals in strategic photovoltaic technologies: abundance versus recyclability. Prog. Photovoltaics Res. Appl. 21(6): 1253–1259. doi: 10.1002/pip.2308.

[57] Alrashidi, H., Issa, W., Sellami, N., Ghosh, A., Mallick, T.K. and Sundaram, S. 2020. Performance assessment of cadmium telluride-based semi-transparent glazing for power saving in façade buildings. Energy Build. 215: 109585. doi: 10.1016/j.enbuild.2019.109585.

[58] Partoens B. and Peeters, F.M. 2006. From graphene to graphite: Electronic structure around the K point. Phys. Rev. B. 74(7): 075404. doi: 10.1103/PhysRevB.74.075404.

[59] Falkovsky, L.A. 2008. Optical properties of graphene. J. Phys. Conf. Ser. 129: 012004. doi: 10.1088/1742-6596/129/1/012004.

[60] Bonaccorso, F., Sun, Z., Hasan, T. and Ferrari, A.C. 2010. Graphene photonics and optoelectronics. Nat. Photonics. 4(9): 611–622. doi: 10.1038/nphoton.2010.186.

[61] Li, X., Hongwei, Z., Kunlin, W., Anyuan, C., Jinquan, W., Chunyan, L., et al. 2010. Graphene-on-silicon schottky junction solar cells. Adv. Mater. 22(25): 2743–2748. doi: 10.1002/adma.200904383.

[62] Lee, C., Wei, X., Kysar, J.W. and Hone, J. 2008. Measurement of the elastic properties and intrinsic strength of monolayer graphene. Science. 321(5887): 385–388. doi: 10.1126/science.1157996.

[63] Yoon, J., et al. 2017. Superflexible, high-efficiency perovskite solar cells utilizing graphene electrodes: Towards future foldable power sources. Energy Environ. Sci. 10(1): 337–345. doi: 10.1039/C6EE02650H.

[64] Honig, J.M. 2005. Irreversible thermodynamics and basic transport theory in solids. *In*: Encyclopedia of Condensed Matter Physics. Elsevier. 34–43. doi: 10.1016/B0-12-369401-9/00712-9.

[65] Daghigh R. and Khaledian, Y. 2018. Effective design, theoretical and experimental assessment of a solar thermoelectric cooling-heating system. Sol. Energy. 162: 561–572. doi: 10.1016/j.solener.2018.01.012.

[66] Rowe, D.M. 2018. Thermoelectrics Handbook: Macro to Nano. CRC Press. [Online]. Available: https://books.google.ae/books?id=0iwERQe5IKQC

[67] Boukai, A.I., Bunimovich, Y., Tahir-Kheli, J., Yu, J.-K., Goddard III, W.A. and Heath, J.R. 2008. Silicon nanowires as efficient thermoelectric materials. Nature. 451(7175): 168–171. doi: 10.1038/nature06458.

[68] Li, J., Xinyue, Z., Zhiwei, C., Siqi, L., Wen, L., Jiahong, S., et al. 2018. Low-symmetry rhombohedral gete thermoelectrics, Joule. 2(5): 976–987. doi: 10.1016/j.joule.2018.02.016.

[69] Hong, M, Zhi-Gang, C., Lei, Y., Yi-Chao, Z., Dargusch, M.S., Wang, H., et al. 2018. Realizing zT of 2.3 in $Ge_{1-x-y}Sb_xIn_yTe$ via reducing the phase-transition temperature and introducing resonant energy doping. Adv. Mater. 30(11): 1705942. doi: 10.1002/adma.201705942.

[70] Shuai, J., Tan, X.J., Guo, J.T., Xu, A., Gellé, R., Gautier, J.-F., et al. 2019. Enhanced thermoelectric performance through crystal field engineering in transition metal–doped GeTe. Mater. Today Phys. 9: 100094. doi: 10.1016/j.mtphys.2019.100094.

[71] Hinterleitner, B., Knapp, I., Poneder, M., Shi, Y., Müller, H., Eguchi, G., et al. 2019. Thermoelectric performance of a metastable thin-film Heusler alloy. Nature. 576(7785): 85–90. doi: 10.1038/s41586-019-1751-9.

[72] Hu, P., Tian-Ran, W., Pengfei, Q., Yan, C., Jiong, Y., Xun, S., et al. 2019. Largely enhanced seebeck coefficient and thermoelectric performance by the distortion of electronic density of states in Ge 2 Sb 2 Te 5. ACS Appl. Mater. Interfaces. 11(37): 34046–34052. doi: 10.1021/acsami.9b12854.

[73] AlJaberi, F.Y., Hussein, A.A., Ali, A.M. and Al-khateeb, R.T. 2022. A review on the enhancement of asphalt cement using different additives. 020032. doi: 10.1063/5.0107845.

[74] Kowalski, K.J., Król, J,. Radziszewski, P., Casado, R., Blanco, V., Pérez, D., et al. 2016. Eco-friendly materials for a new concept of asphalt pavement. 2016. Transp. Res. Procedia. 14: 3582–3591. doi: 10.1016/j.trpro.2016.05.426.

[75] Zhu, X., Yu, Y. and Li, F. 2019. A review on thermoelectric energy harvesting from asphalt pavement: Configuration, performance and future. Constr. Build. Mater. 228: 116818. 2019, doi: 10.1016/j.conbuildmat.2019.116818.

[76] Adwan, I., Abdalrhman, M., Zubair, A.M., Iswandaru, W., Nuryazmin, A.Z., Naeem, A.M., et al. 2021. Asphalt pavement temperature prediction models: A review. Appl. Sci. 11(9): 3794. doi: 10.3390/app11093794.

[77] Chang, S.Y., Bahar, S.K.A., Husain, A.A.M. and Zhao, J. 2015. Advances in Civil Engineering and Building Materials IV: Selected papers from the 2014 4th International Conference on Civil Engineering and Building Materials (CEBM 2014), 15-16 November 2014, Hong Kong. CRC Press. [Online]. Available: https://books.google.ae/books?id=GSSsCQAAQBAJ

[78] Malhotra, V.M. 1999. Making concrete 'Greener' with Fly Ash. Concr. Int. -DETROIT- TA - TT - 21(5): 61–66.

[79] Müller, H.S., Breiner, R., Moffatt, J.S. and Haist, M. 2014. Design and properties of sustainable concrete. Procedia. Eng. 95: 290–304. doi: 10.1016/j.proeng.2014.12.189.

[80] Ji, T., Zhang, X. and Li, W. 2016. Enhanced thermoelectric effect of cement composite by addition of metallic oxide nanopowders for energy harvesting in buildings. Constr. Build. Mater. 115: 576–581. doi: 10.1016/j.conbuildmat.2016.04.035.

[81] Asim, N., Badiei, M., Samsudin, N.A., Mohammad, M., Razali, H., Soltani, S., et al. 2022. Application of graphene-based materials in developing sustainable infrastructure: An overview," Compos. Part B Eng. 245: 110188. doi: 10.1016/j.compositesb.2022.110188.

[82] Krystek, M., Pakulski, D., Patroniak, V., Górski, M., Szojda, L., Ciesielski, A., et al. 2019. High-performance graphene-based cementitious composites. Adv. Sci. 6(9): 1801195. doi: 10.1002/advs.201801195.

[83] Beloin-Saint-Pierre D. and Hischier, R. 2021. Towards a more environmentally sustainable production of graphene-based materials. Int. J. Life Cycle Assess. 26(2): 327–343. doi: 10.1007/s11367-020-01864-z.

[84] Munuera, J., Britnell, L., Santoro, C., Cuéllar-Franca, R. and Casiraghi, C. 2021. A review on sustainable production of graphene and related life cycle assessment. 2D Materials 9(1): 012002. https://doi.org/10.1088/2053-1583/ac3f23.

[85] Ghosh, S., Harish, S., Rocky, K.A., Ohtaki, M. and Saha, B.B. 2019. Graphene enhanced thermoelectric properties of cement based composites for building energy harvesting. Energy Build. 202: 109419. doi: 10.1016/j.enbuild.2019.109419.

[86] Lin, Y. et al. 2019. An integral p-n connected all-graphene fiber boosting wearable thermoelectric energy harvesting. Compos. Commun. 16: 79–83. doi: 10.1016/j.coco.2019.09.002.

[87] Yu, C., Kim, H., Youn, J.R. and Song, Y.S. 2021. Enhancement of structural stability of graphene aerogel for thermal energy harvesting. ACS Appl. Energy Mater. 4(10): 11666–11674. doi: 10.1021/acsaem.1c02390.

[88] Gomez-Gualdrón, D.A., Burgos, J.C., Yu, J. and Balbuena, P.B. 2011. Carbon nanotubes. 175–245. doi: 10.1016/B978-0-12-416020-0.00005-X.

[89] Janas, D. 2020. From bio to nano: A review of sustainable methods of synthesis of carbon nanotubes. Sustainability. 12(10): 4115. doi: 10.3390/su12104115.

[90] Maliheh, D., Vareli, I., Liebscher, M., Tzounis, L., Sgarzi, M., Paipetis, A., et al. 2021. Thermoelectric energy harvesting from single-walled carbon nanotube alkali-activated nanocomposites produced from industrial waste materials. Nanomaterials. 11(5): 1095. doi: 10.3390/nano11051095.

[91] Lingam, D., Parikh, A.R., Huang, J., Jain, A. and Minary-Jolandan, M. 2013. Nano/microscale pyroelectric energy harvesting: challenges and opportunities. Int. J. Smart Nano Mater. 4(4): 229–245. doi: 10.1080/19475411.2013.872207.

[92] Cuadras, A., Gasulla, M., and Ferrari, V. 2010. Thermal energy harvesting through pyroelectricity. Sensors Actuators A Phys. 158(1): 132–139. doi: 10.1016/j.sna.2009.12.018.

[93] Hossain S.S. and Roy, P.K. 2020. Sustainable ceramics derived from solid wastes: A review. J. Asian Ceram. Soc. 8(4): 984–1009. doi: 10.1080/21870764.2020.1815348.

[94] Qingping, W., Bowen, C.R., Lewis, R., Chen, J., Lei, W., Zhang, H., et al. 2019. Hexagonal boron nitride nanosheets doped pyroelectric ceramic composite for high-performance thermal energy harvesting. Nano Energy. 60: 144–152. doi: 10.1016/j.nanoen.2019.03.037.

[95] Qingping, W., He, S., Bowen, C.R., Xiao, X., Oh, J.A.S., Sun, J., et al. 2022. Porous pyroelectric ceramic with carbon nanotubes for high-performance thermal to electrical energy conversion. Nano Energy. 102: 107703. doi: 10.1016/j.nanoen.2022.107703.

[96] Takeno, A., Okui, N., Kitoh, T., Muraoka, M., Umemoto, S. and Sakai, T. 1991. Preparation and piezoelectricity of β form poly(vinylidene fluoride) thin film by vapour deposition. Thin Solid Films 202(2): 205–211. doi: 10.1016/0040-6090(91)90090-K.

[97] Safaei, M., Sodano, H.A. and Anton, S.R. 2019. A review of energy harvesting using piezoelectric materials: state-of-the-art a decade later (2008–2018). Smart Mater. Struct. 28(11): 113001. doi: 10.1088/1361-665X/ab36e4.

[98] Sun, M., Li, Z. and Song, X. 2004. Piezoelectric effect of hardened cement paste. Cem. Concr. Compos. 26(6): 717–720. doi: 10.1016/S0958-9465(03)00104-5.

[99] Wen S. and Chung, D.D.L. 2002. Piezoelectric cement-based materials with large coupling and voltage coefficients. Cem. Concr. Res. 32(3): 335–339. doi: 10.1016/S0008-8846(01)00682-2.

[100] Broniewicz F. and Broniewicz, M. 2020. Sustainability of steel office buildings. Energies. 13(14): 3723. doi: 10.3390/en13143723.

[101] Grzybek D. and Sioma, A. 2021. Vision analysis of the Influence of piezoelectric energy harvesting on vibration damping of a cantilever beam. Energies. 14(21): 7168. doi: 10.3390/en14217168.

[102] Annamalai, P.K., Nanjundan, A.K., Dubal, D.P. and Baek, J. 2021. An overview of cellulose-based nanogenerators. Adv. Mater. Technol. 6(3): 2001164. doi: 10.1002/admt.202001164.

[103] Matveev, N.N., Evsikova, N.Y., Kamalova, N.S. and Korotkikh, N.I. 2013. Role of cellulose crystallites in the polarization of a biopolymer composite: Wood in a nonuniform temperature field. Bull. Russ. Acad. Sci. Phys. 77(8): 1076–1077. doi: 10.3103/S1062873813080261.

[104] Bairagi, S., Ghosh, S. and Ali, S.W. 2020. A fully sustainable, self-poled, bio-waste based piezoelectric nanogenerator: electricity generation from pomelo fruit membrane. Sci. Rep. 10(1): 12121. doi: 10.1038/s41598-020-68751-3.

[105] Sun, J., Guo, H., Ribera, J., Wu, C., Tu, K., Binelli, M., et al. 2020. Sustainable and biodegradable wood sponge piezoelectric nanogenerator for sensing and energy

harvesting applications. ACS Nano. 14(11): 14665–14674. doi: 10.1021/acsnano.0c05493.

[106] Li, J., Liu, Y., Wu, M., Yao, K., Gao, Z., Gao, Y., et al. 2023. Thin, soft, 3D printing enabled crosstalk minimized triboelectric nanogenerator arrays for tactile sensing. Fundam. Res. 3(1): 111–117. doi: 10.1016/j.fmre.2022.01.021.

[107] Li, X., Gao, L., Wang, M., Dong, L., He, P., Xie, Y., et al. 2023. Recent development and emerging applications of robust biomimetic superhydrophobic wood. J. Mater. Chem. A. doi: 10.1039/D2TA09828H.

[108] Muscat, A., Bhattacharya, S. and Zhu, Y. 2022. Electromagnetic vibrational energy harvesters: A review. Sensors. 22(15): 5555. doi: 10.3390/s22155555.

[109] Gholikhani, M., Tahami, S.A., Khalili, M. and Dessouky, S. 2019. Electromagnetic energy harvesting technology: key to sustainability in transportation systems. Sustainability. 11(18): 4906. doi: 10.3390/su11184906.

[110] Wang, L., Todaria, P., Pandey, A., O'Connor, J., Chernow, B. and Zuo, L. 2016. An electromagnetic speed bump energy harvester and its interactions with vehicles. 2016. IEEE/ASME Trans. Mechatronics. 21(4): 1985–1994. doi: 10.1109/TMECH.2016.2546179.

[111] Ibrahim, A., Ramini, A. and Towfighian, S. 2020. Triboelectric energy harvester with large bandwidth under harmonic and random excitations. Energy Reports. 6: 2490–2502. doi: 10.1016/j.egyr.2020.09.007.

[112] Kim, D.W., Lee, J.H., Kim, J.K. and Jeong, U. 2020. Material aspects of triboelectric energy generation and sensors. NPG Asia Mater. 12(1): 6. doi: 10.1038/s41427-019-0176-0.

[113] Roslan, S.A.H., Rasid, Z.A. and Hassan, M.Z. 2018. Bamboo reinforced polymer composite—A comprehensive review. IOP Conf. Ser. Mater. Sci. Eng. 344: 012008. doi: 10.1088/1757-899X/344/1/012008.

[114] Zhang, P., Deng, L., Zhang, H., Li, P. and Zhang, W. 2022. High performance triboelectric nanogenerator based on bamboo fibers with trench structure for self-powered sensing. Sustain. Energy Technol. Assessments. 53: 102489. doi: 10.1016/j.seta.2022.102489.

[115] Kim, S., Gupta, M.K., Lee, K.Y., Sohn, A., Kim, T.Y., Shin, K.S., et al. 2014. Transparent flexible graphene triboelectric nanogenerators. Adv. Mater. 26(23): 3918–3925. doi: 10.1002/adma.201400172.

Chapter 9

Load Distribution and Energy Absorption Characteristics in Stretch and Bending Dominated Lattice Structures Processed by Additive Manufacturing

Abdirahman Yasin Ibrahim, Vagish Ganason, Nahren Rajandran, Ang Chun Kit, Elango Natarajan and Kalaimani Markandan*

Faculty of Engineering, Technology and Built Environment, UCSI University, 56000, Malaysia

9.1 INTRODUCTION

Low-density cellular structures, found abundantly in nature and harnessed by human ingenuity, encompass a diverse range of fascinating phenomena. Among these remarkable formations are the cytoskeletons of living cells, the intricate frameworks of cork, wood, trabecular bone, and coral [1, 2]. It is worth noting that the concept of 'cellular' structures was formulated by the pioneering mind of Hooke, inspired by his observations of the intricate architecture of cork. Building upon these natural wonders, the synthetic polymer foam industry has flourished,

*For Correspondence: kalaimani@ucsiuniversity.edu.my

offering invaluable solutions in the realms of cushioning, packaging, and energy absorption [3].

Recent advances have pushed the boundaries of lightweight construction, resulting in the development of polymeric, metallic and ceramic foams that strive to achieve the elusive combination of strength and rigidity. These innovations have been notably applied in the fabrication of sandwich structures [1,4], allowing the creation of lightweight yet robust materials. Undeniably, the mechanical properties inherent to these cellular formations remain of paramount significance.

Conceptually, cellular materials can be understood as complex composites, elegantly composed of solid matter interwoven with voids [5]. These solid elements intertwine, forming a network of interconnected struts or plates that yield a porous framework, permitting the graceful passage of fluids or enclosing intimate spaces within the cellular matrix. Critical attributes defining these materials include the composition of the solid constituent(s), the relative density parameter (denoted as ρ, reflecting the proportion of occupied space), the dimensions of individual cells and their accompanying wall thickness, as well as the intricate connectivity and geometric regularity displayed by the cellular walls and edges. By adjusting the lattice structural parameters, such as the connectivity of cells or the dimensions of cell size and struts, it is possible to significantly modify the physical behaviour of these structures. This allows for properties that cannot be achieved by the original constituent materials [6,7].

Cellular solids can be categorized into two- and three-dimensional lattice structures. These structures are commonly referred to as lattice materials due to their microarchitecture, which enables them to be treated as monolithic materials possessing their own effective properties [8]. Lattice structures offer numerous advantageous characteristics that position them as a promising solution across diverse applications. These characteristics include their high specific stiffness and strength, rendering them suitable for lightweight structures. Additionally, their expansive surface area makes them well-suited for heat exchange purposes. Lattice structures also excel as energy absorbers, capable of undergoing significant deformation at relatively low stress levels. Furthermore, the multitude of internal pores in lattice structure enables them to serve as effective acoustic insulators [1,7,8].

The fundamental principles governing cellular properties are relevant to all types of cellular solids such as ceramics, metals and polymers. These principles are primarily shaped by three key factors [7,9,10]: (a) the specific characteristics of the solid material(s) constituting the cellular solid, (b) arrangement and shape of the cell edges and faces, including their connectivity, and (c) the relative density, representing the foam density compared to the density of the solid material [9]. These design variables are briefly summarized in Figure 9.1.

Three-dimensional structures composed of members resembling struts can be categorized based on their primary deformation mechanism, which can be either strut bending or stretching [11]. To assess this, Maxwell's stability criterion is employed to evaluate the level of connectivity among struts and the corresponding degrees of freedom within a unit cell. This criterion allows for the classification of cell configurations into two distinct categories: (a) those dominated by bending, exhibiting exceptional energy absorption capacity, and (b) those dominated by stretching, demonstrating remarkable stiffness and initial collapse strength [12–15]. The terms 'stretching-dominated' and 'bending-dominated' describe the primary mode of deformation observed when a material is subjected to external stress. In response to an imposed displacement, the structural members undergo either bending or stretching to accommodate the applied force [16].

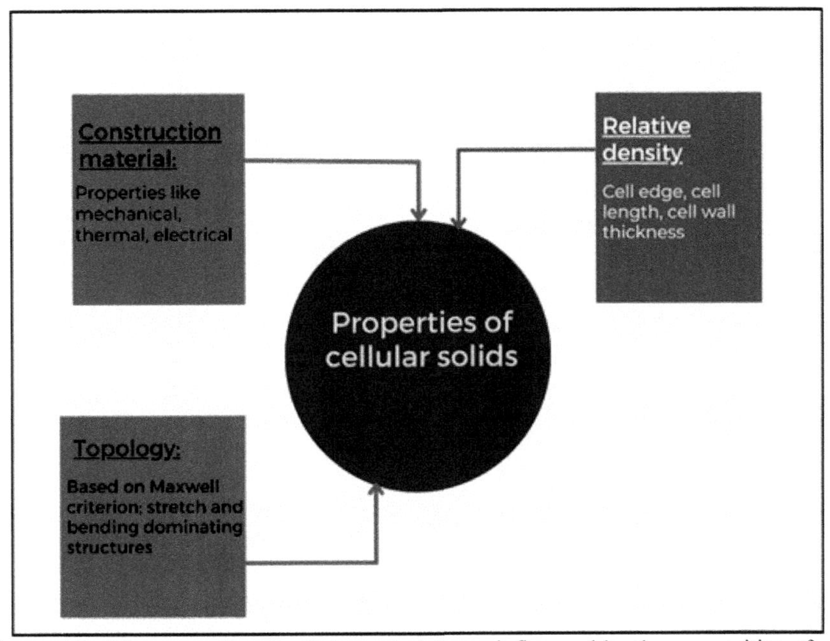

Figure 9.1 The properties of cellular materials are influenced by the composition of cell walls, cell topology, and relative density, which serve as design variables.

9.2 TYPES OF STRUCTURES

9.2.1 Bending-dominated Lattice Structures

Bending-dominated lattices are cellular solids created by dispersing polymers, alloys, ceramics or glasses in an emulsifying agent. The term 'emulsifying agent' encompasses various techniques used to efficiently trap gas, resembling the process employed by yeast during bread production [12]. Figure 9.2(A) depicts a representative example of a lattice-structured material belonging to the class dominated by bending. This particular structure consists of interconnected struts joined at junctions. The distinctive feature of this class is the relatively low level of connectivity observed at these junctions, indicated by the limited number of struts converging at each point [17]. Bending-dominated lattice structures are composed of solid struts that enclose empty spaces filled with either gas or fluid, as depicted in Figure 9.2(B) [9, 18]. In the broader context, cellular solids, including these bending-dominated lattice structures, are distinguished by a fundamental parameter: their relative density, more commonly referred to as 'solid fraction' [12, 19], which can be defined as the fraction of solid matter present in the cubes [20]. Figure 9.2(C) and (D) show Kelvin cell lattice structure—types of bending-dominated structures—designed using CAD software and printed.

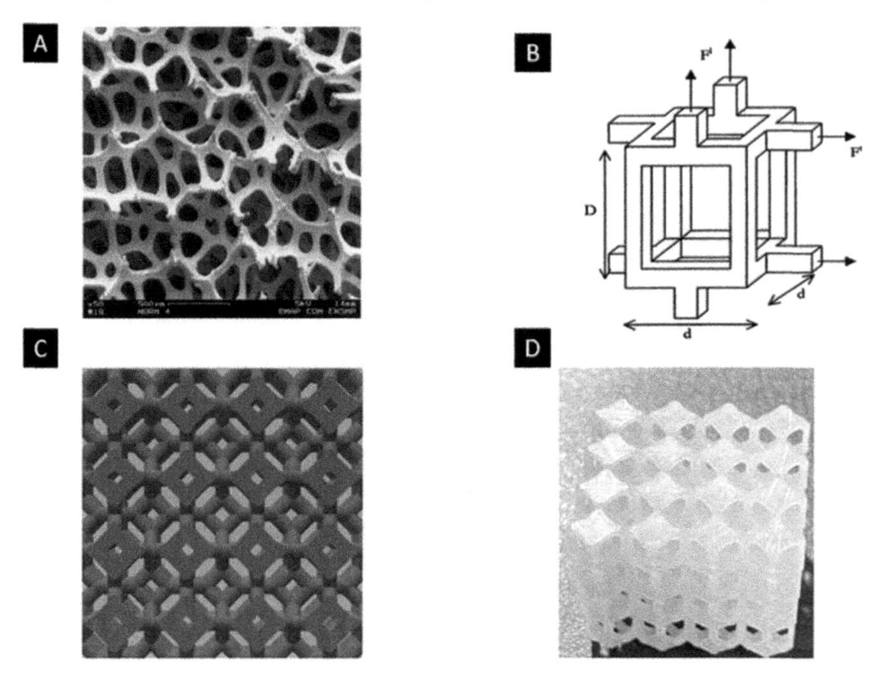

Figure 9.2 (A) Cellular structure of nickel foam [7, 17] (B) idealized bending-dominated lattice [7] (C) Kelvin cell designed using SolidWorks (D) 3D printed kelvin cell lattice.

During the thriving era of construction in the 1800s, the development of design theories for structures was a significant concern. A pioneering Figure. in this domain was Maxwell, who made notable contributions to the field of structural rigidity mathematics. In 1864, he presented a seminal paper introducing a fundamental theorem that characterizes the stability of rigid truss structures, a principle that extends to encompass cellular materials and lattice structures. Maxwell's stability criterion, derived from his insightful findings, involves the calculation of a metric denoted as M for lattice-like structures with b struts and j joints. In the context of three-dimensional structures, this metric is computed using the formula:

$$M = b - 3j + 6.$$

According to Maxwell's criterion, if the computed value of M is less than zero, the structure is categorized as a bending-dominated structure [7, 21].

The equation below shows the relative density characterization of bending-dominated structures.

$$\frac{\tilde{\rho}}{\rho_s} = \left(\frac{t}{L}\right)^2$$

Here, $\tilde{\rho}$ represents the density of the cell, ρ_s is the density of the solid content in the cell, L denotes the size of the cell, and t is the thickness of the cell edges ($t < L$). The ratio of the relative density of the cell to the density of the solid it is made of is called solid fraction $(\tilde{\rho}/\rho_s)$ [21].

9.2.2 Stretching-dominated Lattice Structures

Traditional foams are not very stiff because the way their cells are arranged allows them to bend easily. Thus, it would be helpful to investigate different designs where the edges of the cells can stretch. The concept of micro-truss lattice structures is derived from this line of thinking. To gain a deeper understanding of these structures, it is necessary to refer back to Maxwell's stability criterion. The application of Maxwell's criterion to cellular materials provides valuable insights despite its apparent simplicity. It elucidates the design principles of lattice materials and offers substantial understanding in this context. If M is greater than zero, the structure is a stretching-dominated structure.

$$M = b - 3j + 6.$$

Figure 9.3(A) illustrates an example of a micro-truss lattice structure, representing a specific instance of stretching-dominated structures. In this particular structure, the computed value of M is 18, indicating its characteristic of having no mechanism but numerous potential self-stress states. It is worth noting that this structure is just one among many with M values greater than zero, all of which belong to the category of stretching-dominated structures. Figure 9.3(B) presents the unit cell of the aforementioned structure, with an M value of 18, denoting its unique characteristics [7].

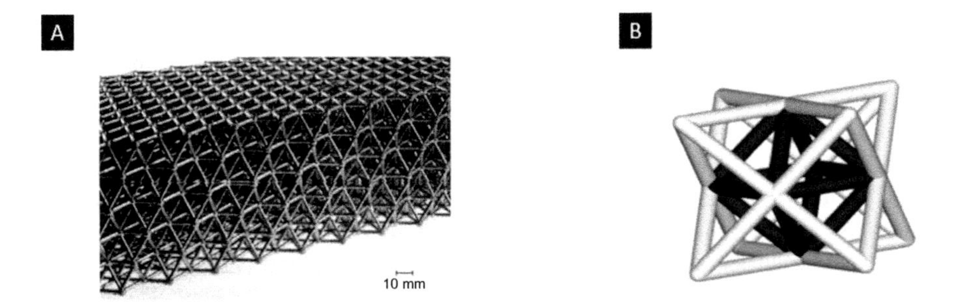

Figure 9.3 (A) Micro-truss structure exhibits an M value greater than zero; (B) unit cell for structures with $M > 0$ [7].

In summary, according to Maxwell's criterion for 3D lattice structures, if M is less than 0, the frame is classified as a mechanism or bending-dominated lattice structure. In this scenario, the structure lacks both stiffness and strength, making it susceptible to collapsing under any applied load [22,23]. This happens as a result of the fact that the frame lacks an adequate number of struts to balance external force while also balancing the moment generated at the nodes. Hence, bending stresses arise in the struts, leading to a behaviour where bending phenomena plays a dominant role [6]. Once the joints of the frame are locked, rotation is prohibited as observed in Figure 9.4(A), and the bars of the frame get deformed when the structure is subjected to a load [7,22]. On the other hand, if the value of M is greater than zero, the frame no longer functions as a mechanism. When the frame is subjected to a load, its members experience either tension or compression, meaning that no bending occurs at nodes and, as a result, the frame transforms into a stretching-dominated structure as shown in Figure 9.4(B) [6,18].

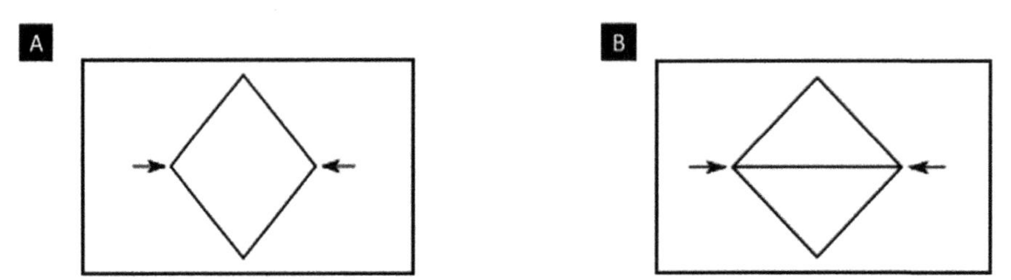

Figure 9.4 (A) Bending-dominated structures [14], (B) stretching-dominated structures [14].

9.3 MECHANICAL PROPERTIES OF BENDING-DOMINATED AND STRETCHING-DOMINATED LATTICE STRUCTURES

The mechanical properties of lattice structures, like elastic modulus and yield strength, slightly differ from those of continuous bulk materials. Lattice structures converge to specific values as the number of unit cells increases. Unlike bulk materials, lattice structures can have auxetic properties, with a negative Poisson's ratio [6]. This means that they expand in the transverse direction when stretched longitudinally, whereas bulk materials typically contract in the transverse direction [6,24]. The Poisson's ratio for auxetic lattice structures can range from –1 to 0.5, depending on the lattice geometry and auxetic level [25]. Bulk materials, on the other hand, have a positive Poisson's ratio between 0 and 0.5 [26]. Auxetic lattice structures can also exhibit other unique characteristics like negative stiffness, compressibility, or thermal expansion coefficient, as well as high stiffness of reduced mass [6,16].

9.3.1 Mechanical Properties of Bending-dominated Lattice Structures

The compressive stress-strain graph of bending-dominated lattices, as shown in Figure 9.5(A), depicts the behavior of the lattice structure. The structure is linearly elasticated to its elastic limit, where the cell-edges yield, buckle or fracture. The structure continues to fail under roughly constant stress ('plateau stress'

$\tilde{\sigma}pl$) until opposing sides of the cells collide ('densification strain' $\tilde{\varepsilon}d$), at which point the stress rises dramatically [22]. The bending deflection (δ) of the cell edges in the open-celled structure can be deduced from the remote compressive stress. The force (F) exerted on the cell edges is proportional to the square of the characteristic length (L) of the cells. This force causes the cell edges to bend, resulting in a bending deflection. Based on the figure's open-celled structure, the following equation can be derived [18]:

$$\delta = \frac{FL^3}{E_s I}$$

where:

- δ represents the bending deflection of the cell edges
- F is the force exerted on the cell edges, which is proportional to L^2
- L is the characteristic length of the cells
- E_s is the elastic modulus of the material from which the structure is constructed
- I is the moment of inertia of the cell edges.

The structure's modulus ($\tilde{E}$) can be obtained by compiling the previously mentioned results. The solid modulus of the material from which the structure is constructed is represented by E_s. The second moment of area of the cell edge with a square cross-section, $t \times t$, is denoted by $I = t^4/12$. The compressive strain experienced by the entire cell is given by $\varepsilon = 2\delta/L$. Based on these considerations, the correlation for the structure's modulus can be expressed as [7]:

$$\frac{\tilde{E}}{Es} \propto \left(\frac{\tilde{\rho}}{\rho s}\right)^2$$

Seeing that $\tilde{E} = Es$ for $\tilde{\rho} = \rho s$, we anticipate that the constant of proportionality will be close to one. Experimentation and numerical simulation both validate this hypothesis [22].

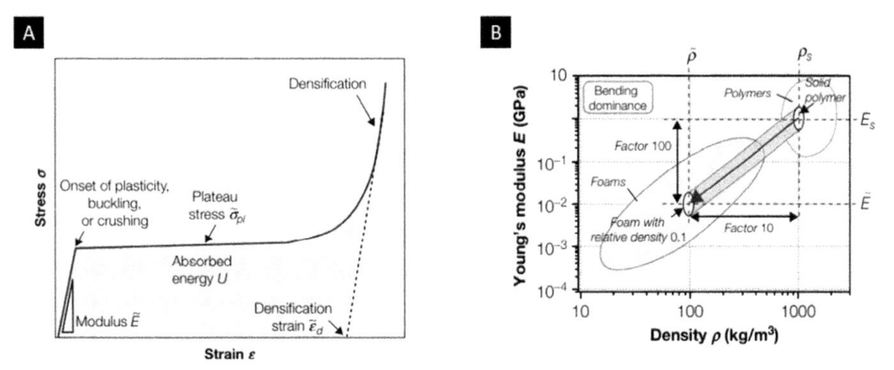

Figure 9.5 (A) Stress vs. strain graph of bending-dominated lattices [22]; (B) relationship between solid fraction and Young's modulus [7].

Figure 9.5(B) illustrates the relationship between relative density (solid fraction), denoted as ρ, and Young's modulus, represented as E. The graph demonstrates a quadratic dependence between these two variables. As a result, even a small decrease in relative density leads to a significant decrease in the modulus. This means that bending-dominated structures exhibit lower values of modulus as well as density. The quadratic relationship highlights the sensitivity of the modulus to changes in relative density, emphasizing the impact of structural characteristics on the mechanical properties of the lattice system [7].

Both the collapse load and the structure's plateau stress can be effectively modeled using a similar methodology. Figure 9.6(A) provides a visual representation of the consequences of a force applied to a cell wall exceeding the completely plastic moment. This indicates a critical point in the behavior of the structure, where the cell wall undergoes irreversible plastic deformation and potential failure. In the equation below, the relationship between moment, yield strength and thickness is expressed [22]:

$$M_f = \frac{\sigma_s t^3}{4}$$

whereas σ_s is the yield strength of the solid the lattice is composed of. This moment is linked to the remote stress as $M \propto FL \propto \sigma L^3$. Putting these findings together yields the failure strength $\tilde{\sigma} pl$, resulting in the following equation:

$$\frac{\sigma_{pl}}{\sigma_{f,s}} = C\left(\frac{\tilde{\rho}}{\rho_s}\right)^{1.5}$$

The proportionality constant, denoted as C, in the context of the discussed phenomena, has been determined through a combination of experimental observations and numerical computations. The findings from these complementary approaches have converged, yielding a consistent value of C as 0.3 [22].

Elastomeric foams experience collapse through a mechanism characterized by elastic buckling rather than yielding, as depicted in Figure 9.6(B). This implies that the foam structure undergoes reversible deformation due to compressive forces, causing the cell walls to buckle. On the other hand, fragile foams collapse through cell-wall fracture, as illustrated in Figure 9.6(C). In this case, the foam structure fails due to irreversible fracture or breakage of the cell walls. These distinct collapse mechanisms highlight the different behaviors exhibited by elastomeric and fragile foams under compressive loading conditions [1,7,22].

Buckling collapse happens when the stress transcends the $\tilde{\sigma} el$, explained by the equation below [22]:

$$\frac{\tilde{\sigma}_{el}}{E_s} \approx 0.05\left(\frac{\tilde{\rho}}{\rho_s}\right)^2$$

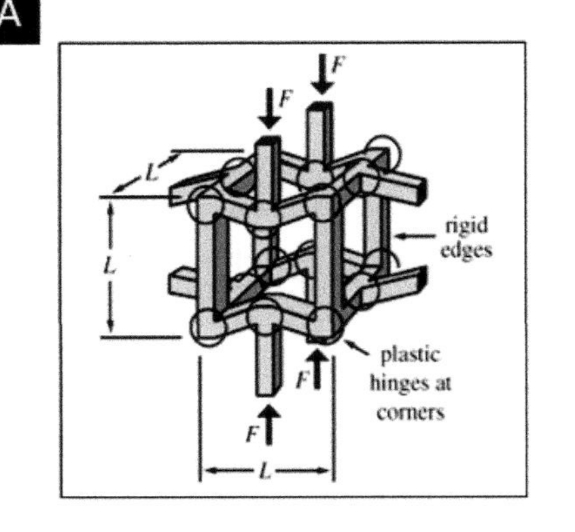

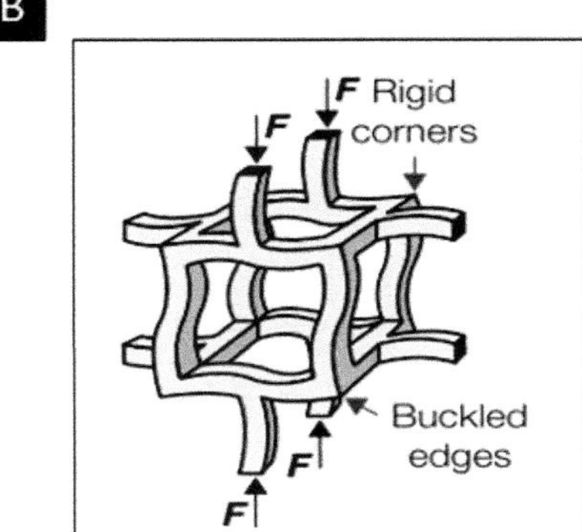

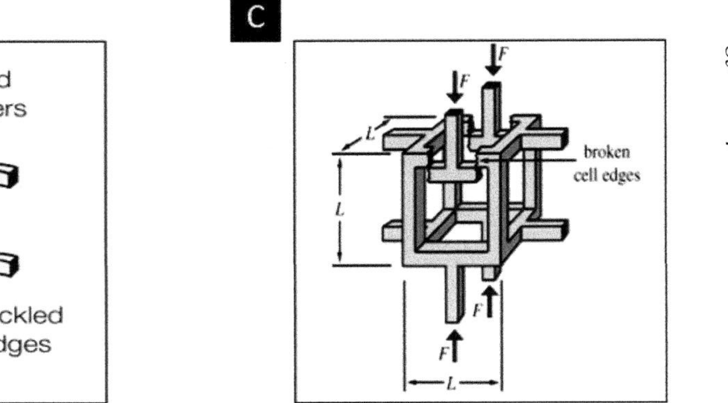

Figure 9.6 (A) Bending-dominated lattices composed of ductile substances collapse by the plastic bending of the cell edges [7]; (B) cell edge buckling compresses elastomeric foam [22]; (C) brittle foam collapse by cell fracture [22].

Collapse by cell fracture occurs when the stress goes beyond $\tilde{\sigma}cr$, where $\tilde{\sigma}cr$ is the material's flexural strength for cell walls, explained by equation below [22]:

$$\frac{\tilde{\sigma}cr}{\sigma_{cr,s}} \approx 0.05\left(\frac{\tilde{\rho}}{\rho_s}\right)^2$$

When the stress undergoes a rapid increase, densification takes place as a direct geometric consequence. This phenomenon arises because the opposing sides of the cells are pressed together, preventing them from further bending or buckling. The occurrence of densification can be precisely quantified by a crucial strain value known as $\tilde{\varepsilon}d$, which represents the densification strain [18,22]. The relationship between stress, densification and strain can be described by the following equation [22]:

$$\tilde{\varepsilon}_d = 1 - 1.4\left(\frac{\tilde{\rho}}{\rho_s}\right)$$

An estimate of the amount of useful energy a foam can absorb per unit volume is expressed as [22]:

$$\tilde{u} = \tilde{\sigma}_{Pl}\tilde{\varepsilon}_d$$

where σpl is the plateau stress, which is equal to the foam's lowest yield, buckling, or fracturing strength.

Table 9.1 presents an illustrative example of bending-dominated structures, accompanied with their respective properties. The table provides a comprehensive overview of these structures, highlighting key characteristics and parameters that define their mechanical behavior.

9.3.2 Mechanical Properties of Stretching-dominated Lattice Structures

As previously mentioned, based on Maxwell's stability criterion, stretching-dominated structures exhibit a value of M greater than 0. When considering the tensile loading of the material, the initial response of the structure is the elastic stretching of its struts. In the case of simple tension loading, regardless of the direction of the applied force, approximately one-third of the bars in the structure experience tension. Consequently, the following equation is applicable to describe this behavior [22]:

$$\frac{\tilde{E}}{E_s} = \frac{1}{3}\left(\frac{\tilde{\rho}}{\rho_s}\right)$$

Figure 9.7(A) displays a density-modulus curve that exhibits a linear relationship rather than a quadratic one. This linear relationship implies that for a similar density, the structure possesses significantly higher stiffness. Consequently, when the cell edges yield, the structure experiences collapse, leading to collapse

Table 9.1 Examples of bending-dominated structures, and their properties.

Lattice Type	Material	Solid Fraction	Young's Modulus, $\tilde{E}$ (GPA)	Flexural Strength $\tilde{\sigma}pl$ (MPa)	Lattice Image	References
Bending-dominated	HMW polyethylene foam	0.16	0.63	0.63		[22, 27]
	Alporas foam	0.05	-	0.63		[1, 28]
	Pyramidal core	0.05	-	8.19	Pyramidal core	[1]

stress [22]. When one or more sets of trusses within the structure collapse, either due to plastic deformation, buckling, or fracture, it signifies that the structure has exceeded its elastic limit. The strength of the structure is determined by the mechanism that exhibits the lowest collapse load. Employing a similar rationale as previously discussed, in the case of truss plasticity, the collapse stress can be quantified using the equation presented below. This equation provides valuable insights into the behavior of the structure under collapse conditions and aids in the assessment of its overall strength and performance [7,29].

$$\frac{\tilde{\sigma}_{P_1}}{\tilde{\sigma}_{y,s}} = \frac{1}{3}\left(\frac{\tilde{\rho}}{\rho_s}\right)$$

If the struts within the structure are thin, they may buckle before experiencing collapse. In such cases, the determination of 'buckling strength' follows a similar logical approach, leading to the correlation depicted below:

$$\frac{\tilde{\sigma}_{el}}{E_s} \propto \left(\frac{\tilde{\rho}}{\rho_s}\right)^2$$

The only distinction between this correlation and the previously mentioned one for the collapse of bending-dominated structures by buckling is the specific constant of proportionality, which is 0.2 in this instance:

$$\frac{\tilde{\sigma}_{el}}{E_s} = 0.2x\left(\frac{\tilde{\rho}}{\rho_s}\right)^2$$

During the fracture failure of struts, a lattice composed of ceramics or another brittle material is prone to collapse. In line with the principles of stretch dominance, the struts experiencing tension are expected to fail first. Building upon the preceding discussion, we anticipate that the collapse stress, denoted as $\tilde{\sigma}cr$, scales according to the relationship outlined in the correlation below:

$$\frac{\tilde{\sigma}_{Cr}}{\sigma_{cr,s}} \propto \left(\frac{\tilde{\rho}}{\rho_S}\right)$$

In the given equation, $\sigma_{cr,s}$ represents the tensile fracture strength of the strut's material. It is worth noting that the proportionality constant in this case is less precisely defined. This is due to the stochastic nature of brittle fracture, which is influenced by the presence and distribution of flaws within the struts. The occurrence of failure in the first truss does not necessarily guarantee the failure of the entire structure. The width of the distribution of fracture characteristics plays a pivotal role in determining whether the failure of a single truss will propagate throughout the entire lattice. The extent of this distribution width impacts the structural reliability and the overall failure behavior of the lattice system [22].

In the context of stretch domination, the stress-strain curve illustrated in Figure 9.7(B) demonstrates the distinct behavior of a structure. This type of structure

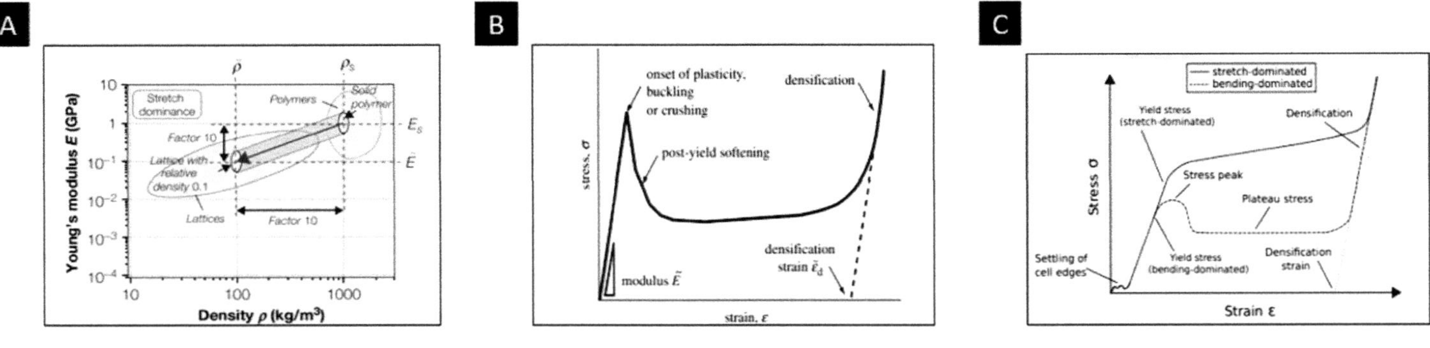

Figure 9.7 (A) Stretching-dominated lattices have moduli that are significantly higher than those of bending-dominated lattices with the same density [22]; (B) compressive stress-strain graph of stretching-dominated lattices [7]; (C) comparison of the stress-strain curves of bending- and stretching-dominated lattice structures [6].

exhibits a notable characteristic known as post-yield softening, which involves a decrease in stress and strain after reaching the yield point. Despite this softening behavior, the structure initially possesses high strength and stiffness, enabling it to withstand substantial loads without experiencing permanent deformation [22].

Based on these findings, it can be concluded that a stretch-dominated lattice exhibits a significantly higher modulus and early collapse strength compared to a bending-dominated cellular material with a similar solid fraction. These characteristics make stretch-dominated cellular solids an optimal choice for lightweight structural applications [30]. However, it is important to note that the deformation mechanisms in stretch-dominated structures involve 'hard' modes such as tension and compression, rather than 'soft' bending modes. This makes them less suitable for energy absorption applications that require a stress-strain curve with a long, flat plateau. In stretch-dominated structures, the initial yield is accompanied with plastic buckling or brittle collapse of the struts, leading to post-yield softening. This post-yield regime continues until the densification strain is reached, at which point the stress sharply increases. The sharp increase in stress beyond the densification strain signifies the end of the post-yield regime. These observations highlight the unique mechanical behavior exhibited by stretch-dominated cellular solids and provide insights into their limitations and strengths for various applications [7, 31]. Figure 9.7(C) compares the stress-strain curve of bending- and stretching-dominated lattice structures.

Table 9.2 shows different types of stretching-dominated lattice structures and their mechanical properties.

9.4 VISCOELASTIC PROPERTIES OF CELLULAR SOLIDS

Most materials exhibit viscoelastic behavior, meaning that they possess a combination of characteristics found in both elastic solids and viscous liquids. By definition, viscoelastic behavior entails a confluence of elastic and viscous characteristics, wherein the application of stress initiates an immediate elastic strain, succeeded by a time-dependent viscous strain [33]. Polymers deviate from ideal adherence to Hooke's Law, even within the linear elastic range, where the stress-strain relationship is not perfectly linear. In Figure 9.8(A), the depicted curvature is intentionally exaggerated for illustrative purposes [34]. Despite the expectation that stress in a material would simply follow the strain curve back to zero upon release, a phenomenon known as hysteresis becomes apparent. Hysteresis indicates that the behavior of a material depends on its prior history. In the context of polymers, Hooke's Law suggests that subjecting a specific sample to a particular strain would result in a corresponding stress (or vice versa). However, this relationship is influenced by whether the sample is being stretched or allowed to relax. When the sample returns to its initial length during relaxation, the observed strain is typically lower than what was observed during stretching [34]. This behavior is illustrated in the exaggerated curvature of the curve presented in Figure 9.8(B).

Table 9.2 Examples of stretching-dominated structures, and their properties.

Lattice Type	Material	Solid Fraction	Young's Modulus, $\tilde{E}$ (GPA)	Flexural Strength $\tilde{\sigma}$ (MPa)	Lattice Image	Reference
Stretching-dominated	HMW polyethene foam	0.16	0.05	1.8	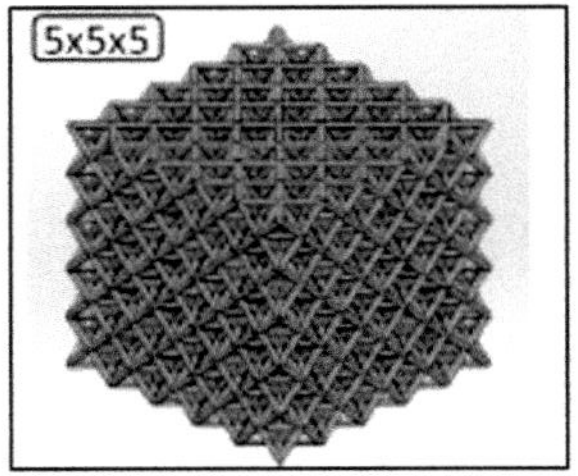 5x5x5	[22, 32]
	Square honeycomb	0.05	-	10.9	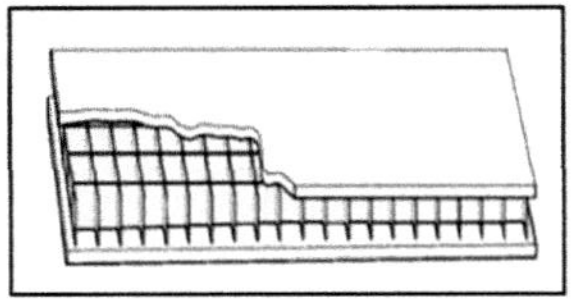	[1]
	Diamond core	0.05	-	3.33	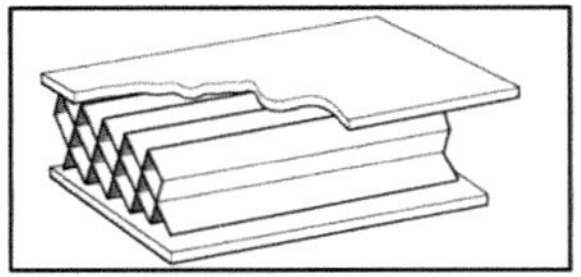	[1]
	Corrugated core	0.05	-	4.23	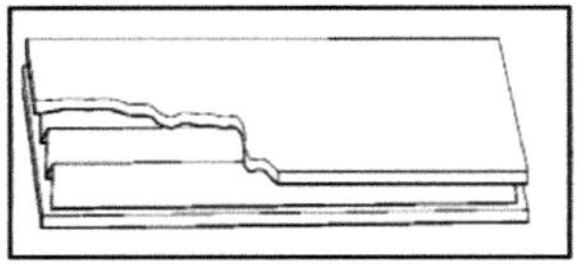	[1]

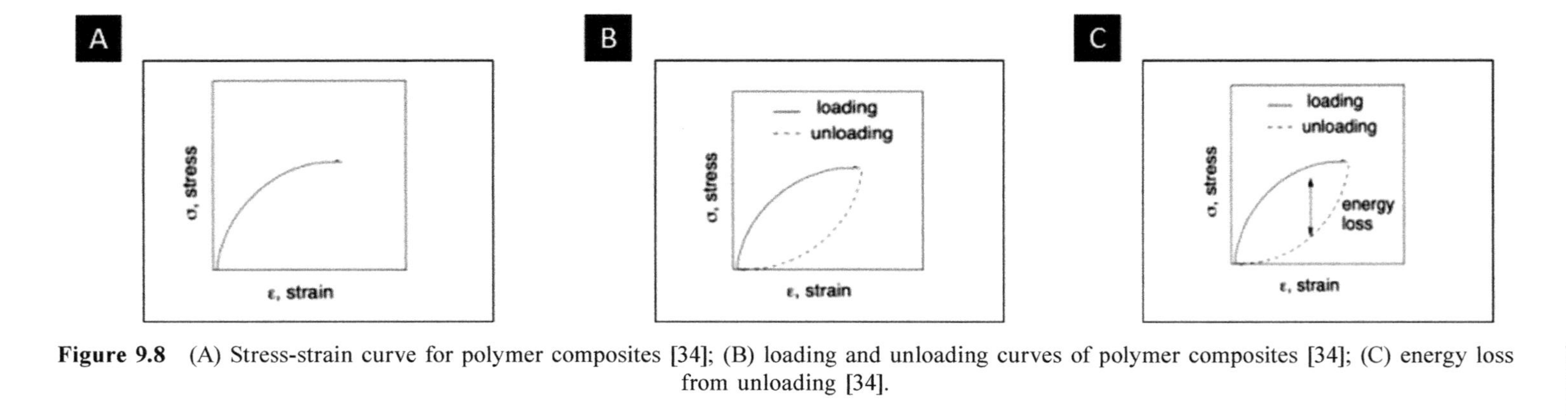

Figure 9.8 (A) Stress-strain curve for polymer composites [34]; (B) loading and unloading curves of polymer composites [34]; (C) energy loss from unloading [34].

The dissimilarity between the loading curve (when stress was initially applied) and the unloading curve (when the stress was removed) as shown in Figure 9.8(C) signifies energy dissipation. When a force is applied to displace a sample or a section of it over a certain distance, the force exerted is not equal to the initial force applied when the sample returns to its original position. Consequently, a certain amount of energy is lost during this process [34].

The storage modulus E', serves as a fundamental metric gauging the magnitude of energy input necessary to elicit deformation in a given sample, thereby encapsulating its inherent resistance to distortion. The storage modulus quantifies the amount of energy required to deform a sample. The disparity between the loading and unloading curves is referred to as loss modulus, denoted as E''. It represents the energy dissipated during the cyclic strain process. The primary mechanism contributing to energy loss in the polymer experiment is related to the movement of polymer chains. The resistance to deformation in polymers arises from the entanglement of chains, including physical crosslinks and other intermolecular interactions as chains encounter each other while undergoing conformational changes to accommodate the new shape of the material. When the stress is removed, the material reverts to its equilibrium shape. However, there is no requirement for the chains to follow the exact same conformational pathway to return to their original conformations [34].

9.4.1 Storage Moduli

The storage modulus, often referred to as the material's energy storage capacity, is a fundamental property that reflects the stiffness and elasticity of a material. It quantifies the material's ability to store and return energy when subjected to external forces or deformations [35]. In a research study conducted by Aw et al., the dynamic storage modulus (E') of polymers and polymer composites was investigated under varying printing patterns and infill densities. The data obtained from the study, as depicted in Figure 9.9 (a) and 9(b), revealed a noticeable decrease in the E' curve as the temperature increased from 30 to 100 degrees Celsius. This reduction in E' can be attributed to the enhanced molecular mobility of polymer chains, caused by the elevated temperature [36]. The incorporation of ZnO fillers in the composites induces a stiffening effect, resulting in higher values of the dynamic storage modulus (E'), compared to pure ABS and CABS materials at equivalent temperatures [36].

The use of fillers that exhibit strong interfacial adhesion with the matrix contributes to an increase in the storage modulus (E') [37]. However, inadequate stress transmission occurs between the fillers and the matrix when there is a greater distance between filament strands. Consequently, the stiffening impact of fillers is diminished, resulting in a lower E' value. As the infill density increases, tighter raster and deposited fibers are generated, leading to a denser structure and an enhancement in the E' of the printed pieces [38]. Notably, rectilinear samples, by virtue of their less rigid structure, exhibit slightly lower storage modulus values than line samples at temperatures below the glass transition temperature, Tg [37].

A

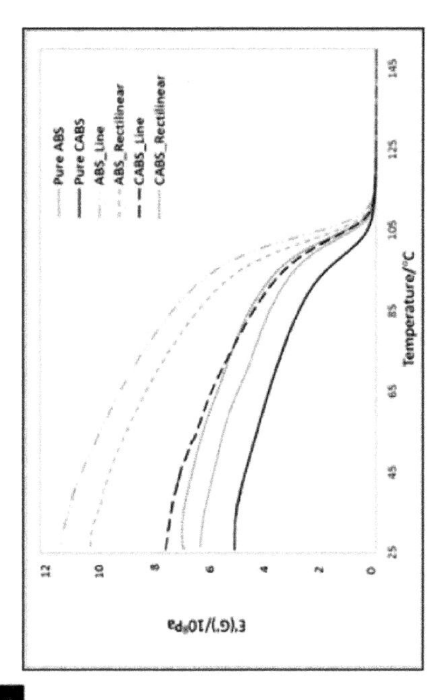

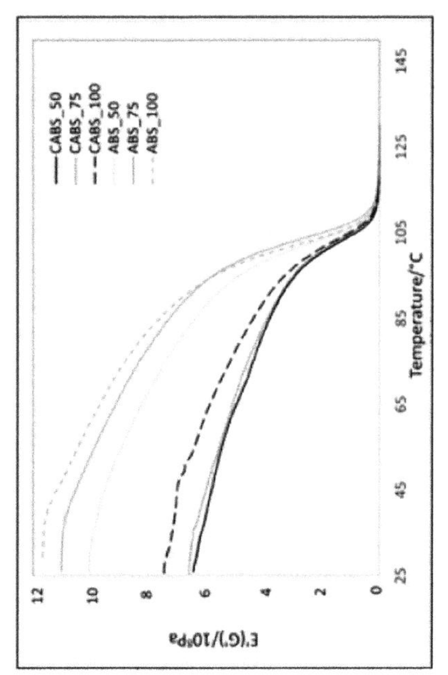

Figure 9.9 (A, B) Influence of infill density and printing pattern on storage moduli [36].

In a separate investigation conducted by Xu et al., the impact of heat treatment or annealing on the rheological, microstructural, protein structure and 3D printing characteristics of egg yolk was examined. The study involved subjecting egg yolk to different heat treatment durations (ranging from 2 to 12 minutes) at temperatures of 72, 76, 80 and 84°C. The results demonstrated that with increasing heating time, the storage modulus (E'), loss modulus (E''), and viscosity exhibit an upward trend. Figure 9.10 presents the viscoelastic properties of the heat-induced egg yolk pastes, with the storage modulus (E') reflecting solid-like behavior, while the loss modulus (E'') presents liquid-like characteristics as the frequency increases. The findings indicated that the frequency dependency of egg yolks decreased with longer heating times and higher temperatures [39]. This phenomenon can be attributed to the enhanced gelation degree, where the physical gel transforms into a covalently crosslinked gel network [40]. Furthermore, the storage modulus (E') and the loss modulus (E'') of the egg yolks increased as the heating time was extended. Notably, after being heated for specific durations (e.g., 8 minutes at 76°C, 6 minutes at 80°C, and 6 minutes at 84°C), the egg yolks exhibited solid-like behavior (E') or liquid-like behavior (E'') (Figure 9.10). This solid-like behavior after heating facilitated better resistance to collapse during and after the 3D printing process, allowing the egg yolk to retain its shape. These findings highlight that heat-treated egg yolk behaves more like a solid and retains its shape after 3D printing, exhibiting improved printability and structural integrity [39].

9.4.2 Loss Moduli

The dissipation of energy in a material is quantified by its loss modulus, which exhibits an inverse relationship with its storage modulus. The loss modulus represents the material's capacity to dissipate energy [41]. Figures 11(A) and 9.11(B) present the loss modulus, E", of composites at different printing patterns and infill densities, as investigated by Aw et al. The research findings indicate that rectilinear pattern samples, which are characterized by lower stiffness, exhibit higher values of loss modulus compared to line pattern samples. This can be attributed to the fact that materials with lower stiffness demonstrate greater dissipation of energy, leading to an elevated loss modulus. Comparing ABS and CABS composites to pure ABS and CABS, it was observed that the peaks of the loss modulus and tan δ shifted towards higher temperatures. This shift suggests that the inclusion of fillers in the composites enhanced their thermal stability. Furthermore, as the polymer chain mobility increased beyond the peak point, the loss modulus experienced a decline [36].

The glass transition temperature (Tg) of a material can be determined using both the peak of the loss modulus and the tan δ (loss tangent or damping factor), and these values may vary. tan δ represents the ratio of the loss modulus to the storage modulus and is used to assess the glass transition temperature (Tg) of the material. In the said investigation, tan δ was used to evaluate the Tg of the material. The addition of fillers resulted in an increase in the Tg of pure ABS and CABS, raising it from a range of 94–95°C to 101–103°C. The changes in infill density or printing pattern did not significantly impact the Tg of the material. When the tan δ value increases, the material exhibits a higher capability to dissipate energy. Conversely, when tan δ decreases, the material demonstrates a greater ability to store energy when subjected to a load [37].

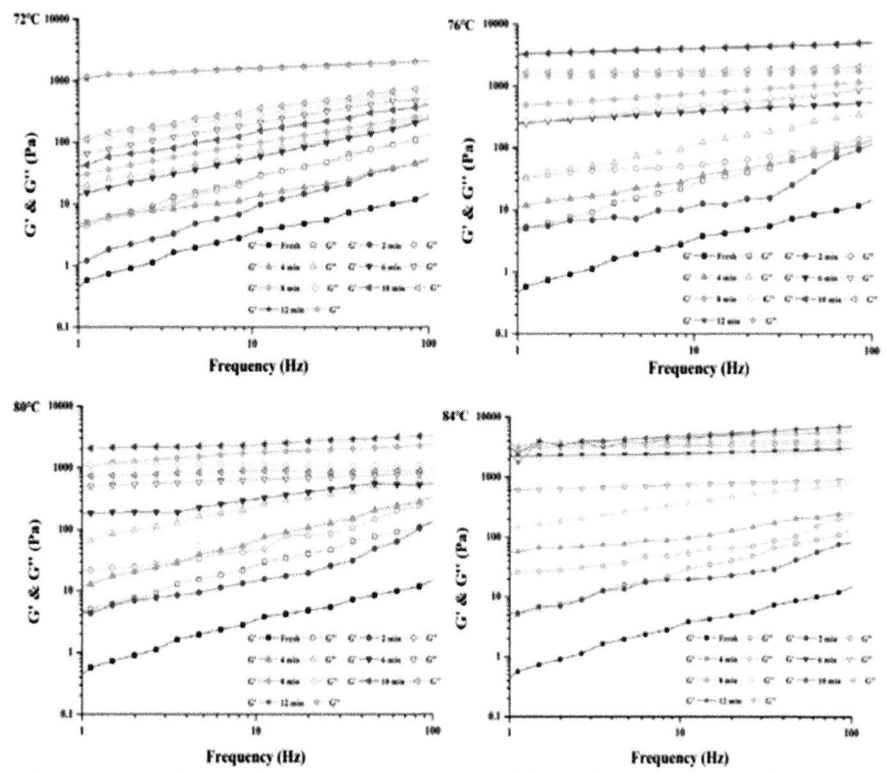

Figure 9.10 Influence of heating on Storage Modulus (E') and Loss Modulus (E'') of egg yolk [39].

In a separate investigation conducted by Masood et al., the dynamic mechanical properties of ABS material fabricated through fused deposition modelling were examined. Figure 9.12 depicts the relationship between the maximum loss modulus and the built style utilized. The experimental data revealed that an increase in temperature led to an elevation in the maximum loss modulus. The loss modulus represents the material's ability to dissipate energy and is inversely proportional to the storage modulus, which represents the material's capacity to store energy. Moreover, the experimental findings demonstrated that the solid regular built style exhibited the highest maximum loss modulus, highlighting its superior performance compared to other built styles. The double dense style, with its intermediate values, fell between the solid regular and sparse styles. This can be attributed to the presence of cross hatch patterns in the intermediate layers of the double dense style, which imparted greater strength compared to the unidirectional patterns observed in the middle layers of the sparse style. These results provide valuable

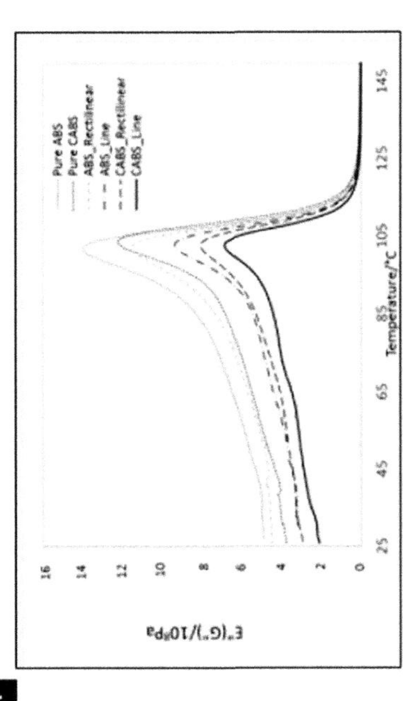

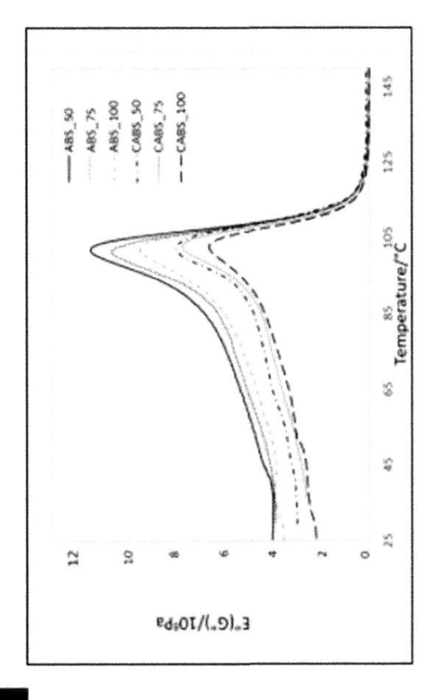

Figure 9.11 (A, B) Influence of infill density and printing pattern on loss moduli [36].

insights into the influence of built styles on the dynamic mechanical properties of ABS material during the fused deposition modelling process [37].

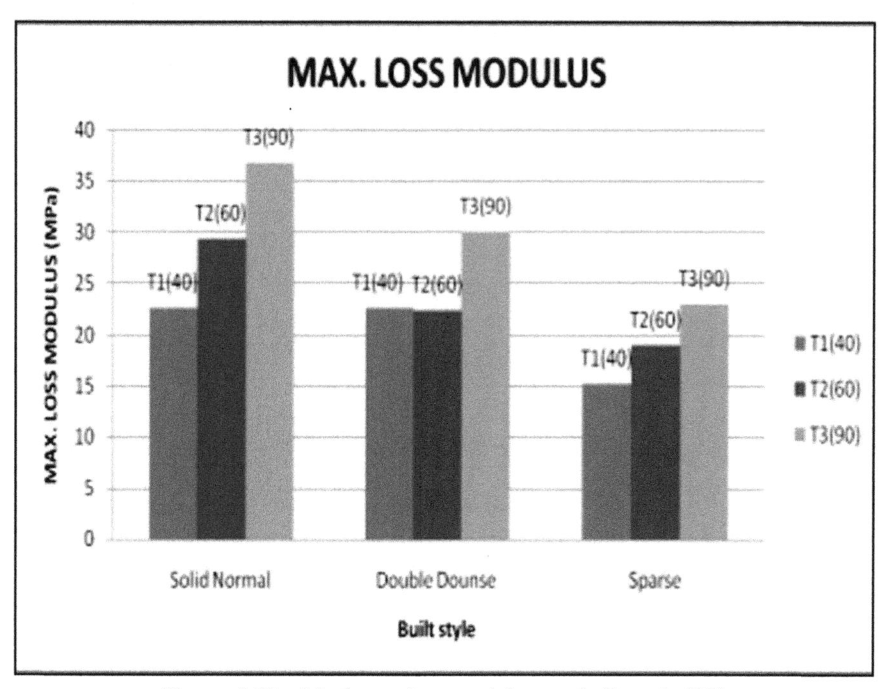

Figure 9.12 Maximum loss modulus vs. built style [37].

9.4.3 Effect of Density on Mechanical Properties

Based on a comprehensive review of the literature findings presented in Table 9.3, it is evident that the Young's modulus values exhibit variations depending on the lattice structure. Specifically, plate-based lattice exhibits the highest Young's modulus, whereas shell-based lattice exhibits the lowest Young's modulus. This observed discrepancy in Young's modulus values can be attributed to the plate-based lattice's ability to effectively distribute and carry load capacity due to the presence of its plates. The integration and configuration of the plate-based structure likely contribute to its superior mechanical performance. Additionally, the literature findings suggest that the relative density of the lattice structures, influenced by manufacturing variations, plays a significant role in determining their mechanical properties, including Young's modulus and ultimate strength. Higher relative density results in improved mechanical properties such as higher Young's modulus values and ultimate strength in the lattice structures. These results underscore the importance of considering and evaluating various polymer processing procedures, particularly density control, to achieve optimal mechanical efficiency in 3D-printed polymer lattices. By carefully managing the density during the manufacturing process, it becomes possible to enhance the mechanical properties and overall performance of these lattice structures [42].

Table 9. 3 Effect of different densities on the mechanical properties of 3D printed lattices.

Lattice Structure	Relative Density	Young's Modulus (MPa)	Ultimate Strength (MPa)	Lattice	References
ABS/Shell-based lattice	0.253	59.5	—		[42]
ABS/Truss-based lattice	0.255	70	—		[42]
ABS/Plate-based lattice	0.257	168	—		[42]
BCC − 0.5ρr	0.5	222 ± 38	27 ± 3.6		[43]
BCC − 0.4ρr	0.4	157 ± 10	11 ± 1.3	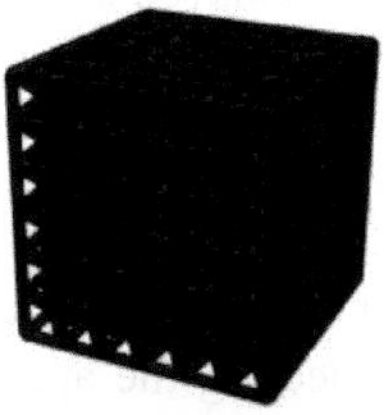	[43]
BCC − 0.3ρr	0.3	71 ± 8	5 ± 0.1	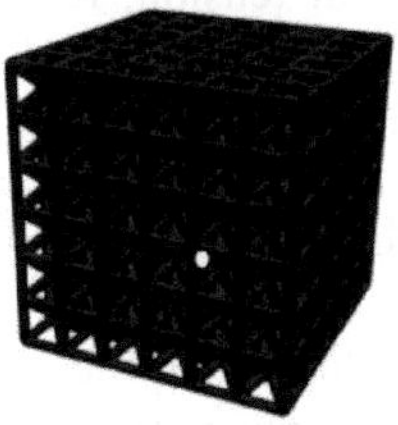	[43]

9.4.4 Effect of Thermal Curing on Mechanical Properties

In a research work by N. Jayanth, heat treatment in the form of annealing was employed to enhance the mechanical properties of 3D-printed poly lactic acid (PLA) composites. The study investigated three different annealing temperatures: 90°C, 100°C and 120°C, each applied for three different durations: 60 minutes, 120 minutes and 240 minutes. The findings of the study indicated that annealing had a significant positive impact on the mechanical properties of the PLA composite. Notably, the tensile strength of the specimens increased as a result of annealing, which can be attributed to the reduction in stress caused by the temperature differential experienced during the 3D printing process. Furthermore, the annealed specimens exhibited a considerably lower heat distortion temperature compared to the unannealed specimens. This observation suggests that the annealing process affected the thermal behavior of the 3D-printed parts. Based on the tensile and heat distortion tests conducted, the results suggested that annealing the 3D-printed PLA parts for a duration of 4 hours at approximately 100°C yielded optimal results, surpassing those of the unannealed samples. These findings emphasize the significance of annealing as a post-processing technique to enhance the mechanical properties and thermal behavior of 3D-printed PLA composites [44].

However, the specific temperature and duration of heat treatment vary depending on the type of resin utilized and the desired outcome. The main objective of heat treatment is to reduce the internal tension that develops within the 3D-printed polymer composite during the manufacturing process, thereby improving its mechanical properties. The presence of inadequate crystallization and voids between the layers of polymer resin contributes to the formation of internal tension. By reducing the porosity within the polymer composite structure, heat treatment promotes the creation of a final material that is denser and stronger [45]. For instance, a study conducted using Polylactic Acid (PLA) in FDM printing demonstrated significant improvements in tensile strength through post-printing heat treatment. Specifically, tensile strength enhancements of 17% and 26% were achieved by subjecting the printed samples to heat treatment at 80°C and 65°C for a specific period of time [46]. In another study focused on PLA specimens, the interlayer tensile strength increased by 11%, 13% and 17%, following annealing at 90°C for 30 minutes, 240 minutes and 480 minutes respectively [47]. These research findings highlight the efficacy of post-printing heat treatment in enhancing the mechanical properties of 3D-printed objects. By carefully selecting the appropriate temperature and duration, it is possible to mitigate internal tension, reduce porosity, and improve the overall strength and performance of the final 3D-printed material.

9.5 APPLICATIONS OF CELLULAR SOLIDS

The use of cellular metallic materials is steadily growing, as they offer a diverse range of applications. However, the feasibility of employing a suitable porous metal or metal foam to address a specific problem is contingent upon several factors, which can be summarized by the following keywords [48]:

- **Morphology:** This refers to the desired type of porosity, such as open or closed, the required amount of porosity, the desired size of pores, and the total internal surface area of the cellular material required.
- **Metallurgy:** This encompasses the specific metal or alloy composition or the desired microstructural state that the cellular material needs to exhibit to meet the application requirements.
- **Processing:** This pertains to the available techniques for shaping the foam or cellular solid, as well as the feasibility of manufacturing composites by integrating the foam with conventional sheets or profiles.
- **Economy:** Consideration of cost-related factors is crucial, including the economic feasibility of large-volume production and the overall suitability of the cellular material for cost-effective manufacturing processes.

By carefully evaluating these factors, researchers and engineers can determine the most appropriate porous metal or metal foam solution for a given application, considering the morphology, metallurgical requirements, processing capabilities, and economic considerations [48].

9.5.1 Automotive Industry

The demand for improved automobile safety has resulted in increased vehicle weight, posing challenges for fuel efficiency. Additionally, there is a need for compact cars without compromising passenger space. This requires innovative solutions such as compact engines and optimized structures. However, this approach presents new issues including heat dissipation, crash safety, and acoustic emissions. Resolving these challenges necessitates a holistic approach, integrating weight reduction, safety standards, passenger comfort, heat management, crash safety, and acoustic performance. The automotive industry strives to address these demands through advanced technologies and materials to enhance overall performance and efficiency [48].

Metal foams, particularly aluminium foams, offer potential solutions for various challenges in the automotive industry. Figure 9.13 provides an overview of the three main fields of application for metal foams. The inner circles represent distinct fields, while the outer boxes highlight the foam properties responsible for their advantages in each field. Ideally, an application would involve a lightweight panel that can absorb energy during crashes while also providing sound or heat absorption functionality (intersection of the three circles in Figure 9.13). However, finding such multifunctional applications is challenging, and a two-fold application is often considered satisfactory (for example, a structural lightweight panel that also serves as a sound absorber). Metal foams thus offer versatility and the potential to address multiple requirements simultaneously [48].

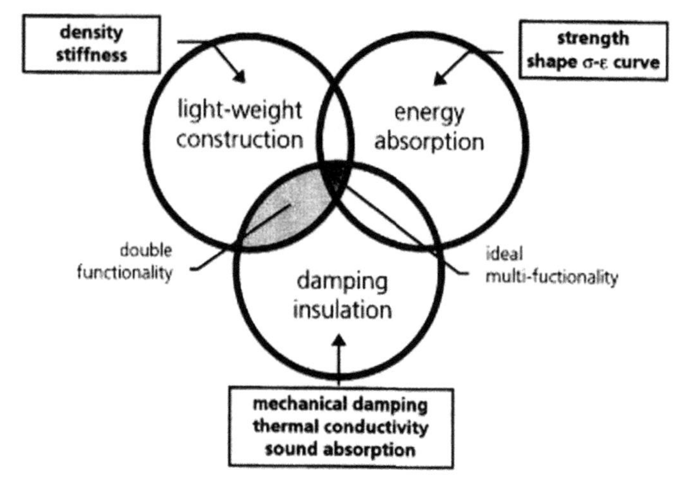

Figure 9.13 Primary application areas of structural metal foams in the automotive industry [48].

9.5.2 Aerospace Industry

Foamed metals offer similar lightweight construction benefits in aerospace and automotive sectors. In aerospace applications, replacing expensive honeycomb structures with foamed aluminium sheets or metal foam sandwich panels can improve performance while reducing costs. Foams provide enhanced buckling and crippling resistance, along with isotropic mechanical properties and the ability to create composite structures without adhesive bonding [48,49]. This allows for improved fire behavior and structural integrity. Boeing has explored the use of large titanium foam sandwich parts made by the gas entrapment technique as shown in Figure 9.14, and aluminium foam core sandwiches for helicopter tail booms offering advantages such as the ability to fabricate curved and 3D-shaped components compared to flat honeycomb structures. Helicopter manufacturers are increasingly considering aluminium foam parts as replacements for existing honeycomb components [48].

Additional applications involve the use of structural metal foams in turbines, where their improved stiffness and enhanced damping properties are advantageous. Porous metals are also utilized for seals between different stages of engines. In this case, the turbine blade shapes the porous material to form a nearly gas-tight seal during its initial operation. Aluminium foam has been investigated for its potential applications in space technology. It has been evaluated as an energy-absorbing crash element for space vehicle landing pads and as reinforcement for load-bearing structures in satellites. This is done to replace materials that pose challenges in the harsh environmental conditions of space, including temperature changes and vacuum. For space applications, there has also been consideration of using highly reactive but lightweight alloy foams such as Li-Mg foams. These alloys, which are typically not suitable due to their high reactivity, may find utility in a vacuum environment [48].

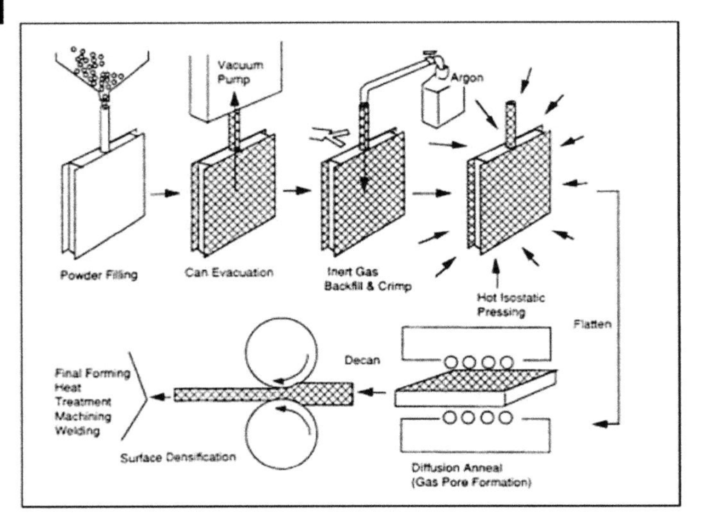

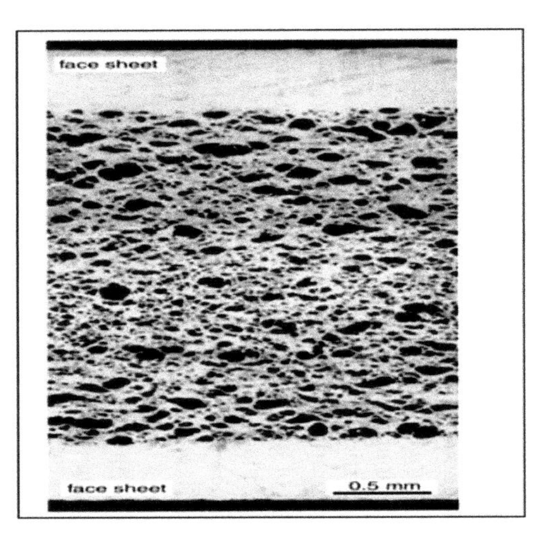

Figure 9.14 (A) Gas entrapment technique [48, 50]; (B) TiAl6V4 sandwich structure with a porous core made by the gas entrapment technique [48, 49].

9.5.3 Biomedical Industries

The biomedical and healthcare sector has embraced the use of cellular structures due to their advantageous properties such as high strength-to-weight ratio and increased surface area [51–58]. In this field, due to their biocompatibility, cobalt-chromium and titanium alloys are commonly used for dental implants and prostheses. One particularly beneficial application of such materials is the promotion of osseointegration, and improved implant fixation compared to porous coatings. The presence of a porous layer on the external surface of the implant, as depicted in Figure 9.15, plays a crucial role in facilitating tissue ingrowth, leading to reduced recovery times. By understanding the relationship between the modulus and the density of cellular structures, medical implants can be manufactured with the desired properties of modulus, biocompatibility and porosity. Additionally, these structures find utility in biological instruments, bone reconstruction, and various biomedical implants [48,59–62].

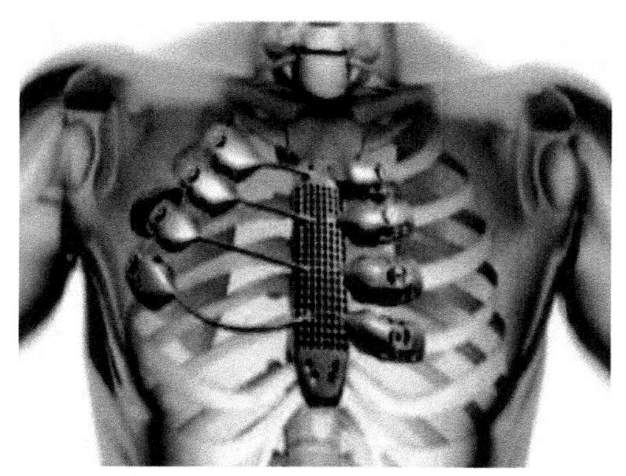

Figure 9.15 3D-printed lattice structure implant for medical applications [53].

9.6 CONCLUSION

Additive manufacturing (AM) technology offers unprecedented possibilities in terms of geometric and material flexibility. One notable application is the fabrication of lattice structures which possess superior properties compared to solid materials and conventional structures. Lattice structures produced through AM exhibit remarkable architectural, mechanical and functional versatility. They blur the distinction between material and structure, allowing for the integration of multiple functions within a single physical component. This opens up practical solutions for a wide range of applications. The principal objective of this piece was to comprehensively investigate the mechanical viscoelastic properties of lattice structures, with the ultimate goal of enhancing their applicability in the domain of additive manufacturing design.

In lattice structures, two predominant deformation modes can be observed: stretching and bending. Understanding the difference between stretching-dominated and bending-dominated lattice structures is crucial for their effective design and application. In stretching-dominated lattice structures, the primary mode of deformation is stretching or the axial elongation of the lattice members. These structures typically have a high stiffness along the axial direction, making them suitable for applications where resistance to tensile loads is crucial. Due to their predominantly linear deformation behavior, stretching-dominated lattice structures exhibit a relatively high modulus of elasticity. However, their ability to absorb energy is relatively limited, as the stretching deformation does not allow for significant energy dissipation or redistribution.

On the other hand, bending-dominated lattice structures experience deformation primarily through bending or the flexural deflection of lattice members. These structures exhibit a lower stiffness along the axial direction compared to stretching-dominated counterparts. However, they offer enhanced energy absorption capabilities due to the inherent flexural deformation mechanism. The bending-dominated lattice structures can dissipate a substantial amount of energy through the bending and recovery of their members, making them suitable for applications that require impact resistance or shock absorption.

Viscoelastic properties play a significant role in understanding the mechanical behavior of cellular solids, which are characterized by their cellular structure and interconnected voids. These materials exhibit both solid-like and liquid-like responses, making their viscoelastic properties crucial for a wide range of applications.

One important aspect is the storage modulus (E'), which represents the material's ability to store elastic energy when subjected to a mechanical load. In cellular solids, the storage modulus is influenced by factors such as density and material composition. As the density increases, the storage modulus generally rises, indicating an increase in the material's stiffness and resistance to deformation. This relationship between density and storage modulus is crucial in industries where mechanical strength is vital, as higher-density cellular solids tend to exhibit improved load-bearing capabilities.

The loss modulus (E") is another key viscoelastic property, that characterizes a material's ability to dissipate energy during cyclic loading. In cellular solids, the loss modulus is influenced by factors such as density, structural arrangement and cell wall properties. Higher-density cellular solids typically exhibit higher loss moduli, indicating greater energy dissipation and damping capabilities. This property is advantageous in industries where shock absorption or vibration damping is desired.

Thermal curing or post-curing processes can also significantly impact the viscoelastic properties of cellular solids. These processes involve subjecting the material to elevated temperatures, which can induce chemical reactions or enhance intermolecular bonding. Thermal curing can lead to an increase in the storage modulus, resulting in improved mechanical properties such as stiffness and strength. Additionally, it can affect the loss modulus by altering the material's internal friction and energy dissipation characteristics.

Cellular solids find diverse applications across different industries, including automotive, aerospace and biomedical sectors. In the automotive industry, cellular solids are utilized for lightweight structural components, providing improved fuel efficiency and reduced emissions. They are employed in applications such as interior panels, seat cushions, and impact-absorbing materials for enhanced occupant safety. In the aerospace industry, cellular solids are employed in lightweight structural components, contributing to weight reduction and increased fuel efficiency. They find use in aircraft interiors, sandwich structures, and thermal insulation materials. In the biomedical field, cellular solids are used for tissue engineering scaffolds, implantable devices, and drug delivery systems. The unique porous structure of cellular solids allows for cell ingrowth, nutrient diffusion, and controlled release of therapeutic agents, making them valuable in regenerative medicine and biomedical applications.

REFERENCES

[1] Fleck, N.a. 2004. An overview of the mechanical properties of foams and periodic lattice materials. Cell. Met. Polym. 2004: 1–4.

[2] Neubauer, J. 2001. Highlighted Topics. J. Appl. Physiol. 90: 1593–1599.

[3] Awson, I.A.N.L. 2016. Crafting the microworld: how Robert Hooke constructed knowledge about small things. Notes Rec.70: 23–44. doi: 10.1098/rsnr.2015.0057.

[4] Markandan, K., Lim, R., Kumar Kanaujia, P., Seetoh, I., bin Md. Rosdi, M.R., Tey, Z.H., et al. 2020. Additive manufacturing of composite materials and functionally graded structures using selective heat melting technique. J. Mater. Sci. Technol. 47: 243–252. https://doi.org/10.1016/j.jmst.2019.12.016.

[5] Lal Lazar, P.J., Subramanian, J., Natarajan, E., Markandan, K. and Ramesh, S. 2023. Anisotropic structure-property relations of FDM printed short glass fiber reinforced polyamide TPMS structures under quasi-static compression. J. Mater. Res. Technol. 24: 9562–9579. https://doi.org/10.1016/j.jmrt.2023.05.167.

[6] Maconachie, T., Leary, M., Lozanovski, B., Zhang, X., Qian, M., Faruque, O., et al. 2019. SLM lattice structures: Properties, performance, applications and challenges. Mater. Des. 183: 108137. https://doi.org/10.1016/j.matdes.2019.108137.

[7] Ashby, M.F. 2006. The properties of foams and lattices. Philos. Trans. R. Soc. A Math. Phys. Eng. Sci. 364: 15–30. https://doi.org/10.1098/rsta.2005.1678.

[8] Case Western Reserve University, IEEE Instrumentation and Measurement Society, Institute of Systems Control and Instrumentation Engineers, Institute of Electrical and Electronics Engineers, ISFA 2016 : International Symposium on Flexible Automation : August 1–3, 2016, Cleveland, Ohio, (2016) 1–3.

[9] Ashby, I. and Gibson, M.F. 1988. Cellular Solids: Structure and Properties. Cambridge University Press.

[10] Natarajan, E., Freitas, L.I., Santhosh, M.S., Markandan, K., Majeed Al-Talib, A.A. and Hassan, C.S. 2022. Experimental and numerical analysis on suitability of S-Glass-Carbon fiber reinforced polymer composites for submarine hull. Def. Technol. 19: 1–11. https://doi.org/10.1016/j.dt.2022.06.003.

[11] Kaur, M., Yun, T.G., Han, S.M., Thomas, E.L. and Kim, W.S. 2017. 3D printed stretching-dominated micro-trusses. Mater. Des. 134: 272–280. https://doi.org/10.1016/j.matdes.2017.08.061.

[12] Fleck, N.A., Ashby, M.F. and Deshpande, V.S. 2002. The topology of cellular structures. 81–89. https://doi.org/10.1007/978-94-015-9930-6_7.

[13] Li, C., Lei, H., Zhang, Z., Zhang, X., Zhou, H., Wang, P., et al. 2020. Architecture design of periodic truss-lattice cells for additive manufacturing. Addit. Manuf. 34: 101172. https://doi.org/10.1016/j.addma.2020.101172.

[14] Souza, J., Großmann, A. and Mittelstedt, C. 2018. Micromechanical analysis of the effective properties of lattice structures in additive manufacturing. Addit. Manuf. 23: 53–69. https://doi.org/10.1016/j.addma.2018.07.007.

[15] Leary, M., Mazur, M., Elambasseril, J., McMillan, M., Chirent, T., Sun, Y., et al. 2016. Selective laser melting (SLM) of $AlSi1_2Mg$ lattice structures, Mater. Des. 98: 344–357. https://doi.org/10.1016/j.matdes.2016.02.127.

[16] Rayneau-Kirkhope, D. 2018. Stiff auxetics: Hierarchy as a route to stiff, strong lattice based auxetic meta-materials. Sci. Rep. 8: 1–10. https://doi.org/10.1038/s41598-018-30822-x.

[17] Badiche, X., Forest, S., Guibert, T., Bienvenu, Y., Bartout, J.D., Ienny, P., et al. 2000. Mechanical properties and non-homogeneous deformation of open-cell nickel foams: Application of the mechanics of cellular solids and of porous materials. Mater. Sci. Eng. A. 289: 276–288. https://doi.org/10.1016/S0921-5093(00)00898-4.

[18] Deshpande, M.F.A. and Deshpande, N.A.F.V.S. 2000. Foam topology bending versus stretching dominated architectures. Int. J. Non. Linear. Mech. 106: 144–154. https://doi.org/10.1016/j.ijnonlinmec.2018.08.006.

[19] Deshpande, V.S., Fleck, N.A., Ashby, M.F. 2001. Food protein-polysaccharide conjugates obtained via Maillard reaction: Review. 49: 1747–1769.

[20] Markandan, K. and Lai, C.Q. 2020. Enhanced mechanical properties of 3D printed graphene-polymer composite lattices at very low graphene concentrations. Compos. Part A Appl. Sci. Manuf. 129: 105726. https://doi.org/10.1016/j.compositesa.2019.105726.

[21] Bhate, D., Penick, C.A., Ferry, L.A. and Lee, C. 2019. Classification and selection of cellular materials in mechanical design: Engineering and biomimetic approaches. Designs. 3: 1–31. https://doi.org/10.3390/designs3010019.

[22] Ashby, M., 2011. Materials Selection in Mechanical Design, 4th Ed. Elsevier.

[23] Wagner, M.A., Lumpe, T.S., Chen, T. and Shea, K. 2019. Programmable, active lattice structures: Unifying stretch-dominated and bending-dominated topologies. Extrem. Mech. Lett. 29: 100461. https://doi.org/10.1016/j.eml.2019.100461.

[24] Desmoulins, A., Zelhofer, A.J. and Kochmann, D.M. 2016. Auxeticity in truss networks and the role of bending versus stretching deformation, Smart Mater. Struct. 25: 0–21. https://doi.org/10.1088/0964-1726/25/5/054003.

[25] Babaee, S., Shim, J., Weaver, J.C., Chen, E.R., Patel, N. and Bertoldi, K. 2013. 3D soft metamaterials with negative poisson's ratio, Adv. Mater. 25: 5044–5049. https://doi.org/10.1002/adma.201301986.

[26] Gercek, H. 2007. Poisson's ratio values for rocks, Int. J. Rock Mech. Min. Sci. 44: 1–13. https://doi.org/10.1016/j.ijrmms.2006.04.011.

[27] Liu, J., Qin, S., Wang, G., Zhang, H., Zhou, H. and Gao, Y. 2020. Batch foaming of ultra-high molecular weight polyethylene with supercritical carbon dioxide: Influence of temperature and pressure, Polym. Test. 106974. https://doi.org/10.1016/j.polymertesting.2020.106974.

[28] Sadighi, S.T.R.H.M. 2021. Dynamic crushing behavior of closed—cell aluminum foams based on different space—filling unit cells, Arch. Civ. Mech. Eng. 1: 1–15. https://doi.org/10.1007/s43452-021-00251-1.

[29] Fleck, N. 2016. Metal foams: A design guide metal foams: A design guide, 3069: 264.

[30] Markandan, K. and Quan, C. 2023. Fabrication, properties and applications of polymer composites additively manufactured with filler alignment control: A review. Compos. Part B. 256: 110661. https://doi.org/10.1016/j.compositesb.2023.110661.

[31] Ashby, M.F., Medalist, R.E.M. 1983. The mechanical properties of cellular solids. Compos. Struct. 153: 866–875. https://doi.org/10.1016/j.compstruct.2016.07.018.

[32] Bolan, M., Dean, M., Bardelcik, A. 2023. The energy absorption behavior of 3D-printed polymeric octet-truss lattice structures of varying strut length and radius. Polymers. 15(3): 713. https://doi.org/10.3390/polym15030713.

[33] Choi, H.R., Jung, K., Koo, J.C., Nam, J. and Lee, Y. 2008. Chapter 25 Micro-Annelid-Like Robot Actuated by Artificial Muscles Based on Dielectric Elastomers. 8–11.

[34] Schaller, C. 2023. Polymer Chemistry. LibreTexts. 99–132.

[35] Hernandez, D.D. 2015. Factors affecting dimensional precision of consumer 3D printing. Int. J. Aviat. Aeronaut. Aerosp. 2(4). https://doi.org/10.15394/ijaaa.2015.1085.

[36] Aw, Y.Y., Yeoh, C.K., Idris, M.A., Teh, P.L., Hamzah, K.A. and Sazali, S.A. 2018. Effect of printing parameters on tensile, dynamic mechanical, and thermoelectric properties of FDM 3D printed CABS/ZnO composites. Materials (Basel). 11. https://doi.org/10.3390/ma11040466.

[37] Arivazhagan, A. and Masood, S.H. 2012. Dynamic mechanical properties of ABS material processed by fused deposition modelling. Int. J. Eng. Res. Appl. 2: 2009–2014. https://doi.org/10.3844/ajeassp.2014.307.315.

[38] Debnath, N., Panwar, V., Bag, S., Saha, M. and Pal, K. 2015. Effect of carbon black and nanoclay on mechanical and thermal properties of ABS-PANI/ABS-PPy blends. J. Appl. Polym. Sci. 132: 16–19. https://doi.org/10.1002/app.42577.

[39] Xu, L., Gu, L., Su, Y., Chang, C., Wang, J., Dong, S., et al. 2020. Impact of thermal treatment on the rheological, microstructural, protein structures and extrusion 3D printing characteristics of egg yolk. Food Hydrocoll. 100: 105399. https://doi.org/10.1016/j.foodhyd.2019.105399.

[40] Kaewmanee, T., Benjakul, S., Visessanguan, W. 2012. Effect of acetic acid and commercial protease pretreatment on salting and characteristics of salted duck egg. Food Bioprocess Technol. 5: 1502–1510. https://doi.org/10.1007/s11947-011-0510-1.

[41] Liang, J.-Z. 2010. Predictions of storage modulus of glass bead-filled low-density-polyethylene composites. Mater. Sci. 1: 343–349. https://doi.org/10.4236/msa.2010.16050.

[42] Abusabir, A., Khan, M.A., Asif, M. and Khan, K.A. 2022. Effect of architected structural members on the viscoelastic response of 3D printed simple cubic lattice structures. Polymers (Basel). 14. https://doi.org/10.3390/polym14030618.

[43] Egan, P.F., Khatri, N.R., Parab, M.A. and Arefin, A.M.E. 2022. Mechanics of 3D-printed polymer lattices with varied design and processing strategies. Polymers (Basel). 14. https://doi.org/10.3390/polym14245515.

[44] Jayanth, N., Jaswanthraj, K., Sandeep, S., Mallaya, N.H. and Siddharth, S.R. 2021. Effect of heat treatment on mechanical properties of 3D printed PLA. J. Mech. Behav. Biomed. Mater. 123: 104764. https://doi.org/10.1016/j.jmbbm.2021.104764.

[45] Guduru, K.K. and Srinivasu, G. 2020. Effect of post treatment on tensile properties of carbon reinforced PLA composite by 3D printing. Mater. Today Proc. 33: 5403–5407. https://doi.org/10.1016/j.matpr.2020.03.128.

[46] Srithep, L.S., Nealey, Y. and Turng, P. 2013. Effects of annealing time and temperature on the crystallinity and heat resistance behavior of injection-molded poly(lactic acid). Polym. Eng. Sci. 53: 580–588. https://doi.org/10.1002/pen.

[47] Bhandari, S., Lopez-Anido, R.A. and Gardner, D.J. 2019. Enhancing the interlayer tensile strength of 3D printed short carbon fiber reinforced PETG and PLA composites via annealing. Addit. Manuf. 30: 100922. https://doi.org/10.1016/j.addma.2019.100922.

[48] Banhart, J. 2001. Manufacture, characterisation and application of cellular metals and metal foams. Prog. Mater. Sci. 46: 559–632. https://doi.org/10.1016/S0079-6425(00)00002-5.

[49] Schwartz, D.D. Shih, D.S., Lederich, D.S., Martin, R.J., Porous, R.L. 1998. Cellular Materials for Structural Applications, Mater. Res. Soc. 521: 225. https://doi.org/10.1002/pc.24494.

[50] Kennedy, A. 2012. Porous metals and metal foams made from powders. Powder Metall. https://doi.org/10.5772/33060.

[51] Sutradhar, A., Park, J., Carrau, D. and Miller, M.J. 2014. Experimental validation of 3D printed patient-specific implants using digital image correlation and finite element analysis. Comput. Biol. Med. 52: 8–17. https://doi.org/10.1016/j.compbiomed.2014.06.002.

[52] Sutradhar, A., Paulino, G.H., Miller, M.J. and Nguyen, T.H. 2010. Topological optimization for designing patient-specific large craniofacial segmental bone replacements. Proc. Natl. Acad. Sci. U.S.A. 107: 13222–13227. https://doi.org/10.1073/pnas.1001208107.

[53] Nazir, A., Abate, K.M., Kumar, A. and Jeng, J.Y. 2019. A state-of-the-art review on types, design, optimization, and additive manufacturing of cellular structures. Int. J. Adv. Manuf. Technol. 104: 3489–3510. https://doi.org/10.1007/s00170-019-04085-3.

[54] Arafat, M.T., Gibson, I. and Li, X. 2014. State of the art and future direction of additive manufactured scaffolds-based bone tissue engineering. Rapid Prototyp. J. 20: 13–26. https://doi.org/10.1108/RPJ-03-2012-0023.

[55] Hutmacher, D.W. 2000. Scaffolds in tissue engineering bone and cartilage. pp. 175–189. *In*: D.F. Williams (ed). The Biomaterials: Silver Jubilee Compendium. Elsevier. https://doi.org/10.1016/B978-008045154-1.50021-6.

[56] Giannitelli, S.M., Accoto, D., Trombetta, M. and Rainer, A. 2014. Current trends in the design of scaffolds for computer-aided tissue engineering, Acta Biomater. 10: 580–594. https://doi.org/10.1016/j.actbio.2013.10.024.

[57] Armillotta, A. and Pelzer, R. 2008. Modeling of porous structures for rapid prototyping of tissue engineering scaffolds. Int. J. Adv. Manuf. Technol. 39: 501–511. https://doi.org/10.1007/s00170-007-1247-x.

[58] Dias, M.R., Guedes, J.M., Flanagan, C.L., Hollister, S.J. and Fernandes, P.R. 2014. Optimization of scaffold design for bone tissue engineering: A computational and experimental study, Med. Eng. Phys. 36: 448–457. https://doi.org/10.1016/j.medengphy.2014.02.010.

[59] Dabrowski, B., Swieszkowski, W., Godlinski, D. and Kurzydlowski, K.J. 2010. Highly porous titanium scaffolds for orthopaedic applications, J. Biomed. Mater. Res.—Part B Appl. Biomater. 95: 53–61. https://doi.org/10.1002/jbm.b.31682.

[60] Cooper, L.F. 2000. A role for surface topography in creating and maintaining bone at titanium endosseous implants. J. Prosthet. Dent. 84(5): 522–534. doi: 10.1067/mpr.2000.111966. PMID: 11105008.

[61] Hansson, S., and Norton, M. 1999. The relation between surface roughness and interfacial shear strength for bone-anchored implants. A mathematical model. J. Biomech. 32: 829–836. https://doi.org/10.1016/S0021-9290(99)00058-5.

[62] Colombo, P. and Degischer, H.P. 2021. Highly porous metals and ceramics. Adv. Eng. Mater. 14: 1051. https://doi.org/10.1002/adem.201200347.

Chapter **10**

Additive Manufacturing of PLA-based UV-curable Resin with Plant Extracts as Fillers

Elango Natarajan*[1,2], Kevin Kumar[1], Kalaimani Markandan[1], Saravanakumar Nesappan[2] and Anto Dilip Albert Selvaraj[2]

*[1]Faculty of Engineering, Technology and Built Environment, UCSI University, Kuala Lumpur, Malaysia

[2]Department of Mechanical Engineering, PSG Institute of Technology and Applied Research, Coimbatore, Tamil Nadu, India

10.1 INTRODUCTION

Recent developments in 3D printing have allowed rapid prototyping of complex structures in a time efficient manner with reduced material wastage. Among the various 3D printing techniques that have been reported in literature till date, stereolithography stands out due to several advantages such as surface smoothness, high resolution, good fabrication speed and the possibility of obtaining composites with isotropic properties. However, mechanical properties of stereolithographically 3D printed structures exhibit fairly poor mechanical properties which can be improved by increasing the specific stiffness and strength of the material through addition of reinforcement particles such as iron and copper [1, 2] tungsten [3], aluminium [4], alumina [5–8], barium and calcium titanate [9], diamond microparticles [10], carbon fibres [11] and glass beads [12].

*For Correspondence: elango@ucsiuniversity.edu.my

In the study presented in this chapter, a biodegradable (bio-based) polymer, i.e. polylactic acid (PLA), was used as the matrix material which can be considered as an effective eco-friendly alternative to conventional plastics [13]. The bio-based PLA polymer exhibits good mechanical properties such as high stiffness and eco-friendly characteristics such as biodegradability, hydrophilicity, biocompatibility and non-toxicity – all of which are highly desirable for use in various applications such as agriculture, medical packaging, as well as in the automobile industry [1–4]. However, the significance of PLA polymer for large-scale manufacturing can be restrained by limitations such as high brittleness and poor thermal stability.

In this study, contrary to the aforementioned common fillers, 3D printed PLA polymer was reinforced with a bio-filler—Cissus quadrangularis L. (CQ). Cissus quadrangularis Linn. Wall. Ex. Wight (family: Vitaceae) is an edible plant found in the hotter parts of India, Sri Lanka, East Africa, Malaysia and Thailand. It has been used in traditional medicine since ancient times and is commonly known as 'bone setter'. The stout fleshy quadrangular stem of this plant has been traditionally used for the treatment of bone fractures, skin infections, gastritis, constipation, eye diseases, piles, anaemia, asthma, irregular menstruation, burns and wounds [14, 15]. Several previous in vivo experiments have demonstrated that CQ promotes ALP activity and enhances collagen synthesis in the fracture healing process [16].

10.2 MATERIALS AND METHODS

Dried coarse powder of CQ (1 kg) was extracted using 80% v/v aqueous ethanol, through cold maceration at room temperature for 72 hours. After the extraction was complete, the extract was filtered, and concentrated to a dry state in a rotary evaporator under reduced pressure and controlled temperature (40–50°C). The dark brown semisolid mass thus obtained was then stored in a vacuum desiccator until further use. A photosensitive liquid mixture of methacrylate oligomers and monomers was added to the CQ extract and mixed using a sonicator for an hour to ensure homogeneity of the resin. The CQ concentration was varied from 0.2 wt.% to 2 wt.%. CQ-polymer composite (10 mm × 10 mm × 10 mm) cubes were 3D printed using a commercial digital light processing (DLP) 3D printer (Photon Mono X, Anycubic Co. with sliced-layer thickness of 50 μm). For post-print thermal treatment, the CQ-polymer composites were placed in an oven (Memmert Universal, Schwabach, Germany) and cured at 120°C for an hour. The temperature in the oven was stable before the samples were placed in.

The axial stress withstanding ability of the composites was evaluated by cutting the samples according to ASTM D412 Type D standards. Tensile testing was carried out using a universal testing machine (Instron 4855, Norwood, MA, USA) equipped with a 100 kN load cell at 50mm diameter Unless otherwise stated, all the samples were tested three times and the average results were computed.

10.3 RESULTS AND DISCUSSION

10.3.1 Microstructural Characterisation

Figure 10.1 shows the scanning electron microscopy (SEM) images of 3D printed PLA composites reinforced with the CQ extract in concentrations varying from 0.02 wt.% to 1.5 wt.%. It can be seen that at lower concentrations, the CQ extract was easily dispersed within the PLA matrix, without significant agglomeration. Besides, there was good interfacial adhesion between PLA and the CQ extract, which was ascertained by the bonding between the PLA matrix and the CQ extract reducing fibre pull-out. On the other hand, at higher concentrations (above 1 wt.% CQ extract), the interfacial bonding between the PLA matrix and the CQ extract reduced significantly. For example, a higher number of pores/voids were visible on the surface of the polymer. Besides, it was difficult to achieve a homogenous dispersion of CQ at higher concentrations since there was higher

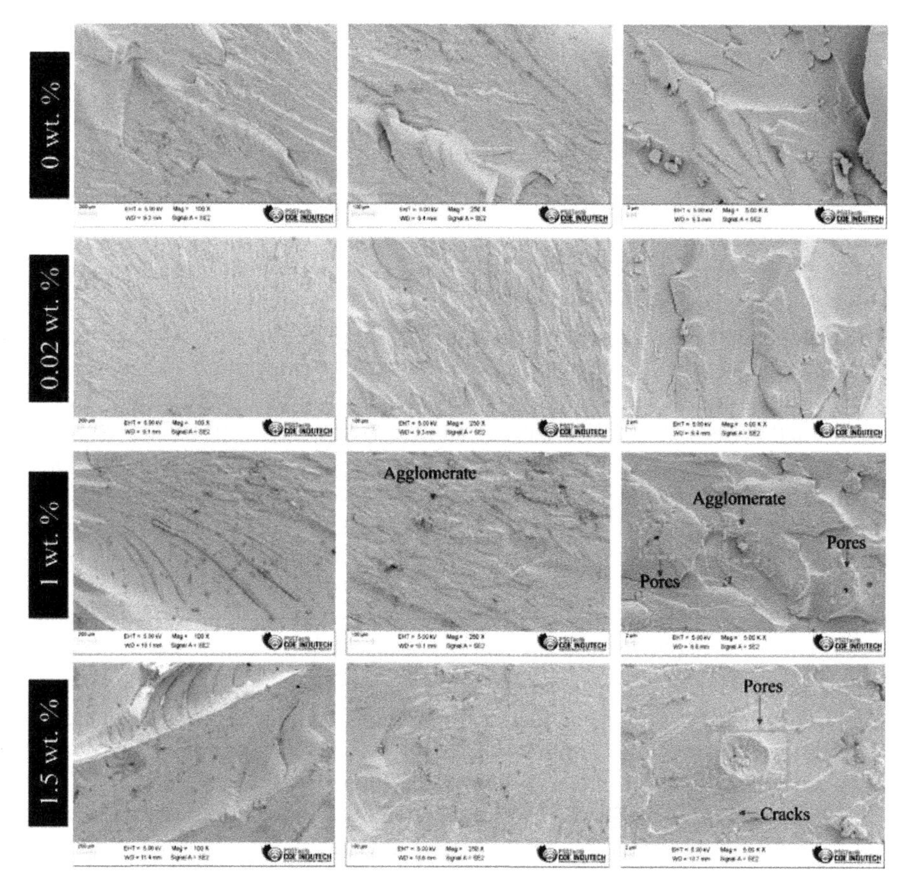

Figure 10.1 Scanning electron microscopy (SEM) images of 3D printed PLA composites reinforced with varying concentrations of CQ extract: 0.02 wt.% to 1.5 wt.%.

agglomeration in these composites. It is also interesting to note that the surface of a neat polymer and PLA reinforced with 0.02 wt.% CQ extract was smooth. However, when reinforced with CQ extract at higher concentrations (above 1 wt.%), the presence of cracks on the surface was significant. The cracks reduce the interfacial adhesion and bonding between the fibre and the matrix, which in turn prevents interlocking strength of the composites.

10.3.2 Mechanical Characterisation

Table 10.1 summarises the mechanical properties of composites fabricated in the present study. It can noted that thermal treatment enhances the mechanical properties of all the 3D printed composites reinforced with CQ extract. For example, the tensile strength and modulus of PLA composite reinforced with 0.2 wt.% CQ was enhanced by 10.26% and 10.2% respectively upon thermal treatment. A similar finding was reported for PLA reinforced with 1, 1.5 and 2 wt.% CQ. The enhancement in mechanical properties upon thermal treatment has been explained in previous studies [17, 18]. During the polymerisation process in 3D printing, some monomers will be confined within the solid polymer before the formation of a macromolecular chain (i.e., incomplete polymerisation). When the polymers are heated (post-print treatment) above their glass transition temperature for a specific duration (i.e., 120°C for an hour in the present study), the polymer chains become more flexible to allow the remaining monomers to be included in the macromolecular chain. This in turn leads to a more complete polymerisation process which enhances the mechanical properties of the composite.

Table 10.1 Mechanical properties of 3D printed CQ-reinforced PLA composites.

Wt.% CQ	Ultimate Tensile Strength [MPa]		Modulus [MPa]		Tensile Stress at Yield (Offset 0.2%) [MPa]	
	Before thermal treatment	After thermal treatment	Before thermal treatment	After thermal treatment	Before thermal treatment	After thermal treatment
0	34.11	32.88	2093.36	2019.60	22.00	21.55
0.2	43.68	48.16	2640.06	2909.33	29.88	34.06
1.0	25.44	34.16	1691.03	2061.52	16.10	22.87
1.5	26.44	34.48	1896.41	2084.87	19.00	23.58
2.0	11.24	12.46	318.54	352.09	5.12	6.25

Figure 10.2 shows the stress-strain curves of the 3D printed PLA composites reinforced with varying concentrations of CQ extract. Similar to the findings summarised in Table 10.1, it can be seen that addition of CQ extract at lower concentrations enhances the mechanical properties, whereas at higher concentrations (i.e., >0.2 wt.%), the tensile strength and modulus are significantly lower compared to the neat polymer. For example, addition of 0.2 wt.% CQ extract after thermal treatment increases the tensile strength and modulus of the neat PLA polymer by 46.5% and 44% respectively. The addition of CQ extract at lower concentrations provides for continuous stress transfer between the fibre and the polymer matrix

(enhanced load transfer efficiency from the matrix to the fibres), which in turn increases the tensile strength and modulus of the composite. On the other hand, the addition of 2 wt.% CQ extract reduces the tensile strength and modulus of neat PLA polymer by 62.1% and 82.6% respectively. This is because there are possibilities of filler agglomeration due to poor dispersion of CQ in PLA at higher filler concentrations. This in turn generates a week bonding between the fibre and the matrix (poor interfacial adhesion), as well as discontinuous stress transfer between them, thus deteriorating the mechanical properties of the composites [19].

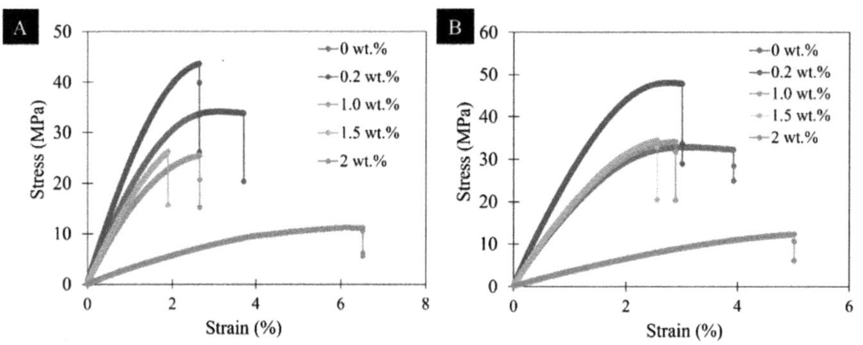

Figure 10.2 Stress-strain curve of 3D printed CQ-reinforced PLA composites (a) before and (b) after thermal treatment.

10.4 CONCLUSION

To conclude, in the study discussed in this chapter, PLA composites reinforced with varying concentrations of Cissus quadrangularis L. (CQ)—a perennial herb of the Vitaceae family—were successfully 3D printed via stereolithography. The experimental findings revealed that thermal treatment enhances the mechanical properties of PLA composites reinforced with CQ extract. Addition of CQ extract at lower concentrations provides for continuous stress transfer between the fibre and the polymer matrix, enhancing the load transfer efficiency from the matrix to the fibres, ultimately increasing the tensile strength and modulus of the composite. However, beyond 0.2 wt.% of CQ extract, the fibres agglomerate with the PLA resin due to poor dispersion, which results in week bonding and poor interfacial adhesion between the fibre and the matrix, in turn degrading the mechanical properties of the composites significantly. Since the composites can play a vital role in the field of nanotechnology or regenerative biomaterials, the homogeneity of CQ in PLA at higher concentrations, along with failure mechanics should be investigated in detail and is left to future studies.

REFERENCES

[1] An, J., Teoh, J.E.M., Suntornnond, R. and Chua, C.K. 2015. Design and 3D Printing of Scaffolds and Tissues. Engineering. 1(2): 261–268. doi: 10.15302/J-ENG-2015061.

[2] Liu, T., Guessasma, S., Zhu, J., Zhang, W., Nouri, H. and Belhabib, S. 2018. Microstructural defects induced by stereolithography and related compressive behaviour of polymers. J. Mater Process Technol. 251: 37–46. doi: 10.1016/j.jmatprotec.2017.08.014.

[3] Sniderman, B., Baum, P. and Rajan, V. 2016. 3D opportunity for life: Additive manufacturing takes humanitarian action. Deloitte Review. 19.

[4] Keleş, Ö., Blevins, C.W. and Bowman, K.J. 2017. Effect of build orientation on the mechanical reliability of 3D printed ABS. Rapid Prototyp. J. 23(2): 320–328. doi: 10.1108/RPJ-09-2015-0122.

[5] Mertz, L. 2013. Dream it, design it, print it in 3-D: What can 3-D printing do for you? IEEE Pulse. 4(6): 15–21. doi: 10.1109/MPUL.2013.2279616.

[6] Finnes T. and Letcher, T. 2015. High definition 3D printing-comparing SLA and FDM printing technologies. [Online]. Available: http://openprairie.sdstate.edu/jurhttp://openprairie.sdstate.edu/jur/vol13/iss1/3

[7] Hoy, M.B. 2013. 3D printing: Making things at the library. Med. Ref. Serv. Q. 32(1): 94–99. doi: 10.1080/02763869.2013.749139.

[8] Cui, X., Boland, T., Lima, D. D.D. and Lotz, M.K. 2012. Thermal inkjet printing in tissue engineering and regenerative medicine. Recent Pat. Drug Deliv. Formul. 6(2): 149–155. doi: 10.2174/187221112800672949.

[9] Banks, J. 2013. Adding value in additive manufacturing: Researchers in the United Kingdom and Europe look to 3D printing for customization. IEEE Pulse. 4(6): 22–26. doi: 10.1109/MPUL.2013.2279617.

[10] Ursan, I., Chiu, L. and Pierce, A. 2013. Three-dimensional drug printing: A structured review. J.Am. Pharma. Assoc. 53(2): 136–144. doi: 10.1331/JAPhA.2013.12217.

[11] Bartlett, S. 2013. Printing organs on demand. Lancet. Respir. Med. 1(9): 684. doi: 10.1016/S2213-2600(13)70239-X.

[12] Mertz, L. 2013. New world of 3-D printing offers 'completely new ways of thinking': Q&A with author, engineer, and 3-D printing expert Hod Lipson. IEEE Pulse. 4(6): 12–14. doi: 10.1109/MPUL.2013.2279615.

[13] Noor, N., Shapira, A., Edri, R., Gal, I., Wertheim, L. and Dvir, T. 2019. 3D printing of personalized thick and perfusable cardiac patches and hearts. Ad. Sci. 6(11): 1900344. doi: 10.1002/advs.201900344.

[14] Asolkar, L. 1992. Second supplement to glossary of Indian medicinal plants with active principles. Part-1 (A-K). CSIR, New Delhi, 414.

[15] Kirtikar, K.R., Das Basu, B., Blatter, E., Mhaskar, K.S. and Caius, J.F. 2000. Kirtikar and Basu's Illustrated Indian Medicinal Plants: their usage in Ayurveda and Unani Medicines. 9. Delhi: Sri Satguru Publications.

[16] Shirwaikar, A., Khan, S. and Malini, S. 2003. Antiosteoporotic effect of ethanol extract of Cissus quadrangularis Linn. on ovariectomized rat. J. Ethnopharmacol. 89(2–3): 245–250. 2003, doi: 10.1016/j.jep.2003.08.004.

[17] Markandan, K. and Quan, C. 2023. Fabrication, properties and applications of polymer composites additively manufactured with filler alignment control: A review. Composites Part B. 256: 110661. doi: 10.1016/j.compositesb.2023.110661.

[18] Markandan, K. and Lai, C.Q. 2020. Enhanced mechanical properties of 3D printed graphene-polymer composite lattices at very low graphene concentrations. Compos. Part A: Appl. Sci. Manuf. 129: 105726. doi: 10.1016/j.compositesa. 2019.105726.

[19] Kumar A.A.J., and Prakash, M. 2020. Mechanical and morphological characterization of basalt/Cissus quadrangularis hybrid fiber reinforced polylactic acid composites. Proceedings of the Institution of Mechanical Engineers. Part C: J. Mech. Eng. Sci. 234(14): 2895–2907. doi: 10.1177/0954406220911072.

Chapter **11**

CO_2 Absorbing Composite Materials: Fundamentals and Properties

Harshini Pakalapati
Institute of Biotechnology, RWTH Aachen University,
Worringerwerg 3, 52074, Aachen, Germany
Email: harshini.pakalapati@gmail.com

11.1 INTRODUCTION

The increasing concentration of carbon dioxide (CO_2) in the atmosphere has become a significant concern for global warming and climate change. Thus, there is an urgent need to develop effective strategies to reduce CO_2 emissions and to capture atmospheric CO_2. Several strategies have been proposed to mitigate CO_2 emissions, such as precombustion, post combustion, and oxyfuel combustion technologies, which capture CO_2 before, during or after fuel combustion [1]. Among these technologies, absorption, adsorption, membrane separation, and cryogenic separation have been extensively studied for CO_2 capture and separation. However, these technologies have their own advantages as well as disadvantages such as high energy consumption, high cost, low selectivity and limited durability [2]. To address these challenges, researchers have turned to the development of composite materials that can enhance the efficiency, selectivity, and stability of CO_2 capture and separation.

Composite materials have emerged as a promising class of CO_2 absorbing materials due to their unique properties and versatility. They are made up of two or more distinct materials, each with their own unique properties, that are combined to form a new material with enhanced properties. The resulting material can be tailored to meet specific requirements and can exhibit properties not found in any of the constituent materials alone. Carbon dioxide absorbing composite materials offer several advantages over traditional CO_2 capture technologies, including high selectivity, high capacity and low energy requirements [3]. These materials can be synthesized using a variety of methods, including sol-gel, electrospinning, and chemical vapor deposition, and can be tuned to optimize their properties for specific applications. Various materials like fillers and membranes have been used such as metal-organic frameworks (MOFs), zeolites, silica, polymers, and carbon-based materials, which can improve the adsorption, permeability, and selectivity of the host material [4]. For example, MOFs have been shown to have high CO_2 adsorption capacity and selectivity, while zeolites have good thermal and chemical stability. Membrane technologies, such as polymer membranes, have been used for gas separation, due to their high selectivity and ease of operation.

This chapter provides the fundamentals and properties of carbon dioxide absorbing composite materials, along with their synthesis and characterization. Furthermore, different methods used to prepare these composite materials, their potential applications and future prospects will be presented in this chapter.

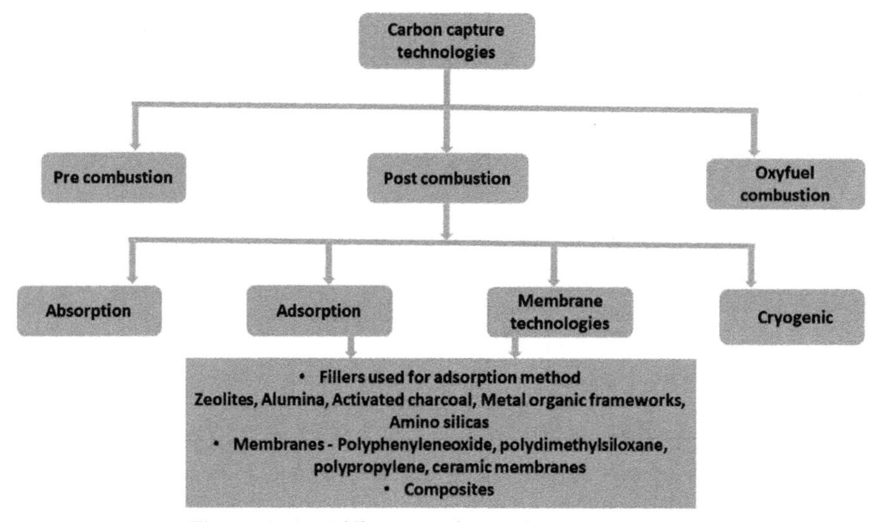

Figure 11.1 Different methods of carbon capture.

11.2 SYNTHESIS OF COMPOSITE MATERIALS

CO_2 absorbing composite materials are typically synthesized by combining a carbon dioxide absorbing material like polymers, either organic and inorganic, and can include materials such as zeolites, amine functionalized materials and activated carbon. There are different methods for synthesizing CO_2 absorbing composite materials, which include solution casting, polymerization, electrospinning, polymer blending and chemical vapor deposition.

Solvent casting is a widely used technique for synthesizing composites [5]. In this technique, a polymer matrix is dissolved in a suitable solvent, and CO_2 absorbing fillers are added to the solution. The mixture is then cast into a desired shape or onto a substrate, using methods such as spin coating or dip coating. The solvent is then evaporated or removed through a curing process, resulting in the formation of a composite material. This method allows for the incorporation of various types of fillers, including zeolites, metal-organic frameworks (MOFs), or carbon-based materials.

In-situ polymerization technique involves the simultaneous polymerization of monomers and the incorporation of CO_2 absorbing fillers or materials [6]. The monomers and fillers are mixed together and undergo a polymerization reaction, resulting in the formation of a composite material. The advantage of this method is the intimate mixing of the fillers within the polymer matrix, leading to enhanced interaction between the fillers and the polymer. This technique is particularly suitable for the synthesis of polymer-based composites with embedded CO_2 absorbing materials [7]. Electrospinning is a versatile technique for fabricating polymer fibers with nanoscale diameters. It is used for synthesizing CO_2 absorbing composite materials by incorporating fillers or materials into the polymer solution before electrospinning [8]. Later, the solution is subjected to a high-voltage electric field which draws out the polymer fibers and the CO_2 absorbing fillers. The resulting composite fibers exhibit a high surface area and a well-dispersed distribution of the fillers, which enhances the CO_2 absorption properties of the material. Polymer blending involves the physical mixing of a polymer matrix with CO_2 absorbing fillers or materials. The fillers can be in the form of particles, fibers or other suitable morphologies [9]. Blending techniques such as melt blending or solution blending are employed to ensure the proper dispersion of fillers within the polymer matrix. The blended mixture is then processed using techniques such as extrusion or compression molding to form the composite material. The advantage of this method is the flexibility in choosing different polymers and fillers to achieve the desired CO_2 absorption properties.

Another commonly used method, Chemical Vapor Deposition (CVD), is a vapor phase method used to deposit thin films of carbon-based materials like graphene or carbon nanotubes onto a substrate – a pre-existing polymer matrix [10]. The process involves introducing gaseous precursors into a reaction chamber wherein they react and deposit onto the matrix. The resulting composite material exhibits enhanced CO_2 absorption properties due to the presence of carbon-based fillers [11]. Other techniques like template-assisted methods and impregnation are also used to synthesize CO_2 absorbing composites. Template-assisted methods involve the use of templates, i.e., sacrificial structures to create a desired composite morphology [12]. These methods are particularly useful for synthesizing CO_2 absorbing materials with well-defined structures and controlled porosity. One example is the synthesis of zeolite-based composites, where a template with a specific structure is used to guide the growth of zeolite crystals.

After synthesis, the template can be removed through calcination or other means, leaving behind a composite material with tailored CO_2 absorption properties. Impregnation methods involve the application of CO_2 absorbing materials onto the surface or their impregnation into the bulk of the substrate or matrix [13]. This technique allows for the modification of existing materials to enhance their CO_2 absorption properties. It is particularly useful when working with materials that already possess desirable mechanical or structural characteristics but require additional CO_2 absorption functionality. Table 11.1 summarizes the advantages and disadvantages of these methods. All these synthesis techniques offer different approaches for incorporating CO_2 absorbing fillers or materials into composite systems. The choice of method depends on various factors, including the desired composite structure, the compatibility of the materials, and the targeted application. Sometimes, a combination of techniques is also used to achieve the desired properties and performance of the composite material.

Table 11.1 Advantages and disadvantages of synthesis methods for CO_2 absorbing composites.

Synthesis Method	Advantages	Disadvantages
Solution Casting	Versatile method suitable for a wide range of polymer-based composites	Limited control over filler dispersion in the polymer matrix
In-situ Polymerization	Enhanced interaction between the filler and the polymer matrix	Difficulty in controlling the polymerization reaction
Electrospinning	Produces fibers with a high surface area and well-dispersed fillers	Complex setup and process
Polymer Blending	Flexibility in choosing different polymers and fillers	May result in phase separation or poor filler dispersion
Chemical Vapor Deposition	Enables precise control over the deposition of carbon-based fillers	Limited to the deposition of carbon-based materials
Template-assisted Methods	Allows for the synthesis of composites with well-defined structures and controlled porosity	Requires additional steps for template removal
	Can modify existing materials to enhance CO_2 absorption properties	Adhesion issues between the coating and the substrate

11.3 COMPOSITE MATERIALS USED FOR CO_2 ABSORPTION

Composite materials have gained a significant importance in the field of CO_2 absorption due to their superior properties and performance. Various types of composite materials are utilized for CO_2 absorption, with distinct advantages and mechanisms for capturing and storing CO_2. These materials play a crucial role in mitigating CO_2 emissions and addressing environmental challenges related to climate change.

11.3.1 Different Composites for CO_2 Absorption

11.3.1.1 Polymer-based Composites

Polymer-based composites combine polymers with CO_2 absorbing additives or fillers to enhance their CO_2 absorption capacity. These composites benefit from the versatility and tunability of polymers, allowing for the incorporation of functional groups or porous structures that facilitate CO_2 capture [14]. Examples include polymers such as polyethyleneimine (PEI) blended with amine-functionalized silica particles or porous polymers impregnated with CO_2 adsorbents [15].

11.3.1.2 Metal-Organic Frameworks (MOFs)

MOFs are crystalline materials composed of metal ions or clusters coordinated with organic ligands. They exhibit high surface areas, tunable porosity, and specific surface functionalities, making them excellent candidates for CO_2 absorption [16]. MOFs can be incorporated into a composite matrix, such as polymers or carbon-based materials, for enhanced CO_2 adsorption properties. The combination of MOFs with other materials provides synergistic effects and improved stability [17].

11.3.1.3 Zeolites

Zeolites are microporous aluminosilicate materials with well-defined channels and cavities. They have high CO_2 adsorption capacities due to their selective adsorption properties. Zeolites can be integrated into polymer matrices or combined with other materials to form composite structures [18]. This allows for the enhancement of CO_2 capture efficiency and the modulation of adsorption-desorption properties.

11.3.1.4 Carbon-based Composites

Carbon-based composites, including activated carbons, carbon nanotubes, and graphene-based materials, exhibit high surface areas and serve as tailored structures that facilitate CO_2 absorption [19]. Carbon-based materials can be combined with polymers or other matrices to form composite materials with improved CO_2 adsorption properties. The presence of carbon-based components provides a high adsorption capacity and facilitates the interaction and storage of CO_2 molecules [20].

11.3.1.5 Other Emerging Materials

Advancements in materials science continue to introduce novel composite materials for CO_2 absorption. These include hybrid materials, such as metal-organic polymers, covalent organic frameworks, and nanoparticle-based composites, which offer unique properties and potential for enhanced CO_2 capture [21,22].

Table 11.2 presents common composite materials. The selection of the appropriate composite material depends on factors such as the desired CO_2 absorption capacity, stability, regeneration capability, and application requirements. By tailoring the composition, structure, and surface properties of these composite materials, researchers can optimize their CO_2 absorption performance and contribute to the development of efficient CO_2 capture technologies [23].

Table 11.2 Composites of CO_2 composite materials CO_2 Absorbing Composites: Synthesis, Properties, and Challenges [15, 17, 18, 20, 22].

	Composite Material	Synthesis Method	Properties	Limitations
1.	Polymer-based Composites	Solution casting, in-situ polymerization, electrospinning	High CO_2 adsorption capacity, tunable porosity, chemical stability	Limited mechanical strength and durability
2.	Metal-Organic Frameworks	Solvothermal synthesis, microwave-assisted synthesis, electrochemical synthesis	High surface area, tunable porosity, selective CO_2 adsorption	Limited stability under certain environmental conditions
3.	Zeolites	Hydrothermal synthesis, ion exchange, template-assisted synthesis	High CO_2 adsorption capacity, well-defined channels, thermal stability	Limited regeneration capability, potential structural degradation
4.	Carbon-based Composites	Chemical Vapor Deposition (CVD), template-assisted methods	High surface area, tailored structures, enhanced CO_2 adsorption	Challenges in controlling composite structure and scalability
5.	Other Emerging Materials	Various synthesis methods depending on the material	Unique properties, potential for enhanced CO_2 capture	Limited research and development, scalability challenges

Additionally, graphene-based composites, which utilize graphene as a building block, are fabricated to enhance the CO_2 adsorption capacity and structural stability of composites. Ongoing research in the field of CO_2 absorption has led to the exploration of various emerging materials such as hybrid composites, metal-oxide-based composites, and novel nanomaterials. These materials offer unique properties and show promise in enhancing CO_2 absorption efficiency. Many factors are considered during the selection of composites for this particular application of CO_2 absorption, like adsorption capacity, selectivity, stability and durability, reusability and, finally, its scalability and cost effectiveness to ensure its viability for large-scale CO_2 capture applications.

11.3.2 Properties and Characterization of CO_2 Absorbing Composite Materials

Understanding the properties of CO_2 absorbing composite materials is essential for their successful design, development and application. By gaining insights into these properties, researchers and engineers can design composites to further optimize the strategies for CO_2 absorbing composites. The key properties that influence the performance and effectiveness of these materials in capturing and storing carbon dioxide have been discussed in the following sections.

11.3.2.1 CO_2 Absorption Capacity

CO_2 absorption capacity is a crucial property of composite materials designed for CO_2 capture. It refers to the amount of CO_2 that can be absorbed or adsorbed by the composite per unit mass or volume. This property depends on factors such as the nature of the filler, its surface area, porosity, and the interactions between the filler and the polymer matrix [24]. Understanding and optimizing CO_2 absorption capacity is essential for developing composite materials with high capture efficiency.

11.3.1.2 Selectivity

Selectivity is another important property that determines the ability of a composite material to preferentially absorb CO_2 over other gases. It depends on the interactions between the filler and CO_2 molecules, as well as the presence of other gases in the environment. A high selectivity ensures that the composite material can effectively capture CO_2 while minimizing the uptake of unwanted gases, enhancing the overall performance and efficiency of CO_2 absorption [25] [26].

11.3.1.3 Mechanical Strength

The mechanical strength of CO_2 absorbing composites is crucial for their practical applications. These materials should possess sufficient mechanical integrity and durability to withstand the operation conditions and handling during their use [27]. The strength and stiffness of the composite are influenced by factors such as the choice of matrix material, filler dispersion, and bonding between the filler and the matrix [28]. Optimizing the mechanical properties ensures the stability and reliability of the composite in various environments.

11.3.1.4 Thermal Stability

Thermal stability is an important property for CO_2 absorbing composites, especially in applications that involve high-temperature environments. These materials should exhibit sufficient thermal stability to maintain their structural integrity and CO_2 absorption properties at elevated temperatures. The choice of filler, matrix, and their compatibility play a significant role in determining the thermal stability of the composite material [29, 30].

11.3.2.5 Regeneration Capability

Regeneration capability refers to the composite material's ability to release the captured CO_2 for subsequent storage or utilization. It is an important property for applications where the captured CO_2 needs to be separated from the composite material and stored or processed further. Understanding the regeneration kinetics and optimizing the composite design can enhance the efficiency and effectiveness of CO_2 release [31,32].

In conclusion, the properties of CO_2 absorbing composite materials, including CO_2 absorption capacity, selectivity, mechanical strength, thermal stability, and regeneration capability, are key factors that determine their performance and applicability. By comprehensively characterizing and optimizing these properties for enhanced CO_2 absorption capabilities, we can be tailor them for specific applications in carbon capture, storage, and utilization.

To assess the properties and performance of CO_2 absorbing composite materials, various characterization techniques are employed. These may include techniques such as scanning electron microscopy (SEM) and transmission electron microscopy (TEM) to examine the composite's morphology and structure [33–35]. Moreover, Brunauer-Emmett-Teller (BET) analysis can be used to determine the surface area and porosity of composites. Furthermore, thermogravimetric analysis (TGA) and differential scanning calorimetry (DSC) provide information on the thermal stability and the decomposition behavior of these composites [36]. Gas adsorption measurements, such as volumetric or gravimetric methods, are conducted to evaluate the CO_2 adsorption capacity and selectivity of composite materials. These characterization techniques enable researchers to understand the structure-property relationships and optimize the performance of CO_2 absorbing composite materials.

11.4 CHALLENGES AND FUTURE PERSPECTIVES

Despite the significant progress in the development and application of CO_2 absorbing composite materials, several challenges remain to be addressed. One key challenge is the optimization of CO_2 absorption capacity and selectivity. While many composite materials show promising CO_2 absorption properties, further improvements are needed to enhance their efficiency and effectiveness in capturing and retaining CO_2. This involves exploring novel fillers, optimizing filler-matrix interactions, and improving the composite design to maximize CO_2 uptake. Another challenge would be the scalability and cost-effectiveness of CO_2 absorbing composite materials. For widespread adoption and implementation, it is crucial to develop synthesis methods that can be easily scaled up to industrial production levels. Additionally, the cost of materials and manufacturing processes need to be optimized to ensure the economic viability of CO_2 absorbing composites.

The future prospects for CO_2 absorbing composite materials are highly promising. As research continues, the development of new materials and the optimization of existing ones will yield composite materials with superior CO_2

absorption capabilities. Additionally, the integration of these materials into various industrial applications, such as carbon capture and storage, renewable energy systems, and catalysis, holds immense potential for mitigating greenhouse gas emissions and promoting sustainable practices. Furthermore, advancements in the understanding of composite material properties and their behavior under different environmental conditions will enable the design of more efficient and durable CO_2 absorbing composites. This includes exploring the long-term stability of these materials, their resistance to degradation, and their recyclability or reusability.

To conclude, while challenges exist in optimizing CO_2 absorption capacity and cost-effectiveness, the future of CO_2 absorbing composite materials is promising. With continued research and development efforts, these materials have the potential to play a vital role in combatting climate change, promoting sustainable technologies, and contributing to a greener future.

REFERENCES

[1] Jiang, D.E., Madhrin, S.M. and Dai, S. 2020. Mater. Carbon Capture. John Wiley & Sons.

[2] Gizer, S.G., Polat, O., Ram, M.K. and Sahiner, N. 2022. Recent developments in CO_2 capture, utilization, related materials, and challenges. Int. J. Energy Res. 46: 16241–16263.

[3] Wongchitphimon, S., Lee, S.S., Chauah, C.Y., Wang, R. and BAE, T.H. 2020. Composite materials for carbon capture. Mater. Carbon Capture. 237–266.

[4] Sumida, K., Rogow, D.L., Mason, J.A., Mcdonald, T.M., Bloch, E.D., Herm, Z.R., et al. 2012. Carbon dioxide capture in metal–organic frameworks. Chem. Rev. 112: 724–781.

[5] Nanni, F., Travagalia, P. and Valentini, M. 2009. Effect of carbon nanofibres dispersion on the microwave absorbing properties of CNF/epoxy composites. Compos. Sci. Technol. 69: 485–490.

[6] Xu, L., Huang, Z., Zhang, X., Cheng, D., Chen, X., Du, H., et al. 2023. Reinforcing styrene-butadiene rubber by silica/carbon black by-product composite through an in-situ polymerization process. Polym. Compos. 44: 663–672.

[7] Wu, K., Ting, T., Wang, G., Ho, W. and Shih, C. 2008. Effect of carbon black content on electrical and microwave absorbing properties of polyaniline/carbon black nanocomposites. Polym. Degrad. Stab. 93: 483–488.

[8] Chen, J., Wang, Y., Liu, Y., Tan, Y., Zhang, J., Liu, P., et al. 2023. Fabrication of macroporous magnetic carbon fibers via the cooperative etching-electrospinning technology toward ultra-light microwave absorption. Carbon. 208: 82–91.

[9] Akarzadeh, E., Shockravi, A. and Vatanpour, V. 2021. High performance compatible thiazole-based polymeric blend cellulose acetate membrane as selective CO_2 absorbent and molecular sieve. Carbohydr. Polym. 252: 117215.

[10] Zhu, W., Yao, Y., Zhang, Y., Jiang, H., Wang, Z., Chen, W., et al. 2020. Preparation of an amine-modified cellulose nanocrystal aerogel by chemical vapor deposition and its application in CO_2 capture. Ind. Eng. Chem. Res. 59: 16660–16668.

[11] Delhaes, P. 2002. Chemical vapor deposition and infiltration processes of carbon materials. Carbon. 40: 641–657.

[12] Singh, A.K., Shishkin, A., Koppel, T. and Gupta, N. 2018. A review of porous lightweight composite materials for electromagnetic interference shielding. Composites, Part B: Eng. 149: 188–197.

[13] Donnini, J., Corinaldesi, V. and Nanni, A. 2016. Mechanical properties of FRCM using carbon fabrics with different coating treatments. Composites, Part B: Eng. 88: 220–228.

[14] Nguyen, D., Murialdo, M., Hornbostel, K., Pang, S., Ye, C., Smith, W., et al. 2019. 3D printed polymer composites for CO_2 capture. Ind. Eng. Chem. Res. 58: 22015–22020.

[15] Zhao, Y., Ding, H. and Zhong, Q. 2013. Synthesis and characterization of MOF-aminated graphite oxide composites for CO_2 capture. Appl. Surf. Sci. 284: 138–144.

[16] Kearns, E.R., Gillespie, R. and D'alessandro, D.M. 2021. 3D printing of metal–organic framework composite materials for clean energy and environmental applications. J. Mater. Chem. A. 9: 27252–27270.

[17] Han, S., Yao, A., Ding, Y., Leng, Q., Teng, F., Zhao, L., et al. 2022. A dual-template imprinted polymer based on amino-functionalized zirconium-based metal–organic framework for delivery of doxorubicin and phycocyanin with synergistic anticancer effect. Eur. Polym. J. 170: 111161.

[18] Thakkar, H., Lawson, S., Rownaghi, A.A. and Rezaei, F. 2018. Development of 3D-printed polymer-zeolite composite monoliths for gas separation. Chem. Eng. J. 348: 109–116.

[19] Nguyen, D., Murialdo, M., Hornostel, K., Pang S., Ye, C., Smith, W., et al. 2019. 3D printed polymer composites for CO_2 capture. Ind. Eng. Chem. Res. 58: 22015–22020.

[20] Wang, J., Pu, Q., Ning, P. and Lu, S. 2021. Activated carbon-based composites for capturing CO_2: A review. Greenhouse Gases: Sci. Technol. 11: 377–393.

[21] Lin, H., Yang, Y., Hsu, Y.C., Zhang, J., Welton, C., Afolabi, I., et al. 2023. Metal-Organic Frameworks for Water Harvesting and Concurrent Carbon Capture: A Review for Hygroscopic Materials. Adv. Mater. 2209073.

[22] Wu, J., Liu, X., Zhang, R., Zhang, J., Si, H. and Wu, Z. 2023. A novel (CaO/CeO_2)@CeO_2 composite adsorbent based on microinjection titration-calcination strategy for CO_2 adsorption. Chem. Eng. J. 454: 140485.

[23] Liu, E., Lu, X. and Wang, D. 2023. A systematic review of carbon capture, utilization and storage: status, progress and challenges. Energies. 16: 2865.

[24] Kong, Y., Jin, L. and Qiu, J. 2013. Synthesis, characterization, and CO_2 capture study of micro-nano carbonaceous composites. Sci. Total Environ. 463: 192–198.

[25] Politakos, N., Barbarin, I., Cordero-Lanzac, T., Gonzalez, A., Zangi, R. and Tomovska, R. 2020. Reduced graphene oxide/polymer monolithic materials for selective CO_2 capture. Polymers. 12: 936.

[26] Sandru, M., Haukebo, S.H. and Hagg, M.-B. 2010. Composite hollow fiber membranes for CO_2 capture. J. Membr. Sci. 346: 172–186.

[27] Kazi, S.S., Aranda, A., Difelice, L., Meyer, J., Murillo, R. and Grasa, G. 2017. Development of cost effective and high performance composite for CO_2 capture in Ca–Cu looping process. Energy Procedia. 114: 211–219.

[28] Abdel-gawwad, H.A., Rashad, A.M., Mohammed, M.S. and Tawfik, T.A. 2021. The potential application of cement kiln dust-red clay brick waste-silica fume composites as unfired building bricks with outstanding properties and high ability to CO_2-capture. Journal Build. Eng. 42: 102479.

[29] Wang, W., Xiao, J., Wei, X., Ding, J., Wang, X. and Song, C. 2014. Development of a new clay supported polyethylenimine composite for CO_2 capture. Applied Energy. 113: 334–341.

[30] Papa, E., Medri, V., Paillard, C., Contri, B., Murri, A.N., Vaccari, A., et al. 2019. Geopolymer-hydrotalcite composites for CO_2 capture. J. Cleaner Prod. 237: 117738.

[31] Qian, D., Lei, C., Hao, G.-P., Li, W.-C. and Lu, A.-H. 2012. Synthesis of hierarchical porous carbon monoliths with incorporated metal–organic frameworks for enhancing volumetric based CO_2 capture capability. ACS Appl. Mater. Interface. 4: 6125–6132.

[32] Wang, Q., Luo, J., Zhong, Z. and Borgna, A. 2011. CO_2 capture by solid adsorbents and their applications: current status and new trends. Energy Environ. Sci. 4: 42–55.

[33] Zhao, Y., Ding, H. and Zhaong, Q. 2013. Synthesis and characterization of MOF-aminated graphite oxide composites for CO_2 capture. Appl. Surf. Sci. 284: 138–144.

[34] Minellil, M., Papa, E., Medri, V., Miccio, F., Benito, P., Doghieri, F., et al. 2018. Characterization of novel geopolymer–zeolite composites as solid adsorbents for CO_2 capture. Chem. Eng. J. 341: 505–515.

[35] Mohamed, M.G., Samy, M.M., Mansoure, T.H., Li, C.-J., Li, W.-C., Chen, J.-H., et al. 2022. Microporous carbon and carbon/metal composite materials derived from bio-benzoxazine-linked precursor for CO_2 capture and energy storage applications. Int. J. Mol. Sci. 23: 347.

[36] Valverde, J., Perejon, A. and Perez-maqueda, L.A. 2012. Enhancement of fast CO_2 capture by a nano-SiO_2/CaO composite at Ca-looping conditions. Environ. Sci. Tech. 46: 6401–6408.

Chapter **12**

A Review on Biological Pathways for Key Indoor Pollutants

Shmitha Arikrishnan
Ramboll Singapore Pte Ltd, 100 Amoy St, 069920, Singapore
Email: shmitha.ak@gmail.com

12.1 INTRODUCTION

Indoor air quality is an important issue that is addressed by several of the Sustainable Development Goals (SDGs) adopted by the United Nations. These goals aim to promote sustainable development and the well-being of people around the world. For example, SDG 3 on "good health and well-being" highlights that poor indoor air quality can harm people's health, especially those with respiratory or cardiovascular conditions. By promoting good indoor air quality, SDG 3 aims to improve people's health and well-being. In addition, SDG 11 on sustainable cities and communities emphasizes indoor air quality as the key component of the built environment. By promoting sustainable urban planning and design, this goal aims to improve indoor air quality in homes, offices and other buildings. Most of the air pollution is generated by the inefficient burning of fossil fuels. Coal, oil and gasoline are examples of fuels used to produce energy for power supply or transportation. The quantity of carbon monoxide (CO) emitted at high concentrations reflects the amount of fossil fuels burnt. Other harmful pollutants, such as nitrogen oxides, are released into the atmosphere as a result. The capacity of the heart to pump adequate oxygen is reduced by breathing contaminated air created by the combustion of natural gas and fossil fuels. As a result, a variety of respiratory and cardiovascular disorders may develop.

To this end, recent studies have focused on green buildings as one of the ways to enhance the air quality around us, particularly for indoor air quality and the environment. Some important considerations to improve indoor air quality include adopting sound design, construction, commissioning, maintenance, and operating techniques that promote indoor environment quality (IEQ). Airborne bacteria, mould, and fungus, as well as radon, can be avoided using a building envelope design that successfully manages moisture sources from the outside as well as inside the building, along with HVAC system designs that are good at controlling interior humidity [1].

When creating a green building, it is important to choose materials that do not produce pollutants or at least emit them sparingly. The following sections will discuss in detail the factors affecting indoor air quality as well as the biological pathway of particulate matter, volatile organic compounds (VOCs), and carbon dioxide (CO_2).

12.2 FACTORS AFFECTING INDOOR AIR QUALITY

There are several factors that can affect indoor air quality IAQ, including ventilation, temperature, humidity, and the presence of pollutants. Common indoor air pollutants include volatile organic compounds (VOCs), particulate matter, carbon monoxide, radon and mould.

- VOCs are chemicals that are released by various building materials and household products, such as paints, cleaning agents, and furniture. These chemicals can cause irritation of the eyes, nose and throat, as well as headaches, dizziness and fatigue. Long-term exposure to VOCs has been linked to cancer and other serious health problems.
- Particulate matter refers to tiny particles that are suspended in the air, such as dust, pollen and smoke. These particles can cause respiratory problems, such as asthma and bronchitis, and can exacerbate existing respiratory conditions as well. They can also cause eye and skin irritation, and other health problems.
- Radon is a radioactive gas that is naturally present in the soil and rocks and can seep into buildings through cracks and gaps in the foundation. Long-term radon exposure has been linked to lung cancer.
- Mould is a type of fungus that can grow in damp areas of buildings, such as bathrooms and basements. Exposure to mould can cause respiratory problems, as well as eye and skin irritation.

In addition to these specific pollutants, there are also general factors that can affect IAQ. One of the most important of these is ventilation. Proper ventilation is essential for maintaining good IAQ, as it helps in removing pollutants and bringing fresh air into the building. Poor ventilation can lead to a build-up of pollutants, which can cause health problems. Temperature and humidity can also affect IAQ.

High temperatures and humidity can promote the growth of mould and other biological contaminants, while low temperatures and humidity can cause dryness and irritation of the eyes, nose and throat.

According to the report 'Household Air Pollution and Health,' in 2020, household air pollution was thought to be the cause of 3.2 million annual deaths, including approximately 237,000 deaths of children under the age of five [2]. Furthermore, exposure to household air pollution increases the risk of developing non-communicable diseases such as lung cancer, chronic obstructive pulmonary disease (COPD), ischemic heart disease, and stroke [3].

The COVID pandemic has provided the possibility of working from home to avoid commuting. This has made human beings in the digital era more sedentary than before. Hence, an understanding of the impact that indoor air pollutants have on the occupants is critical. Studies have shown that indoor air pollutants affect the productivity and the well-being of occupants. However, there is little understanding of how the pollutants' biological pathway affects the key organs of an occupant being exposed to the pollution. Therefore, this review focuses on key indoor pollutants, such as particulate matter (PM), volatile organic compounds (VOCs), and carbon dioxide (CO_2), that are commonly studied to create a better understanding of the biological pathway.

12.3 BIOLOGICAL PATHWAYS OF PARTICULATE MATTER (PM)

The upper respiratory tract actively removes the inhaled PM through the clearance mechanisms of the nasal mucociliary [4]. However, the clearance mechanisms are ineffective barriers to ultrafine particles (UFP) [5]. Previous studies have highlighted two main biological pathways of exposure to ambient PM, particularly UFP: olfactory and systemic. The olfactory pathway is a direct translocation to the brain via the olfactory nerve. This is due to the UFP deposition in the olfactory mucosa of the nasopharyngeal region [6]. A study by Oberdörster et al. (2004) investigated the exposure of UFP through inhalation, using rats, and showed PM deposits in multiple brain areas, largely deposited in the olfactory bulb [6]. The study also identified that small particles translocate through the olfactory pathway much faster than bigger particles. The results revealed (approximately) 100% nasal deposition efficiency for particles below 5 nm in diameter and 10% nasal deposition efficiency for particles larger than 30 nm in diameter. This is supported by a pioneering study which investigated the transport mechanism by intranasally infusing colloidal gold nanoparticles in squirrel monkeys. It reported that the gold nanoparticles were transported through the olfactory nerve axons to the olfactory bulbs [8]. This mechanism showed that the initial deposition in the olfactory mucosa had induced an endocytic uptake by the olfactory rods (retrograde transport within olfactory dendrites and anterograde axonal translocation) to reach the cells in the olfactory bulb. As a result, the olfactory pathway bypasses the blood-brain barrier (BBB). The same conclusion of particles breaking the BBB and potentially reaching the brain was found in three different studies done with young adults at the mean age of 20–25 years, with two of the studies using autopsied humans' frontal cortex [9] and brain tissue [10], and one study testing the olfactory function of young adults with a mean age of 21, living in extreme air pollution [11].

Studies have found that not all UFPs are deposited in the olfactory mucosa. Particles that travel to the lower respiratory tract may reach the brain through the systemic pathway. Several animal studies have found deposits of UFPs in extrapulmonary pathways [7, 12–14]. UFPs have unique characteristics such as stronger inflammatory effects and a large surface area to mass ratio, compared to larger particles, that allow UFPs to penetrate the lung tissues. The UFPs then travel to epithelial barriers, to peripheral organs, and then reach the central nerve system (CNS) parenchyma [5, 15, 16]. The inflammatory effect leads to oxidative stress, inducing and eliciting pulmonary inflammation [17]. In addition, systemic inflammation from UFPs increases the permeability of the BBB, which potentially allows effective entry to the brain tissue [18]. However, the transport mechanism through systemic pathways has not been fully understood with human exposure to UFPs. The author has summarised the known key biological pathways from PM inhalation in Table 12.1.

Table 12.1 Summary of key biological pathways from PM inhalation.

Organ	Biological pathways	References
Lungs	Inflammation	[6, 17]
	Oxidative stress	
	Increased respiratory symptoms	
	Reduced lung function	
Blood	Altered rheology	[19, 20]
	Increased coagulability	
	Translocated particles	
	Reduced oxygen saturation	
Heart	Altered cardiac autonomic function	[21]
	Increased dysrhythmic susceptibility	
	Altered cardiac repolarisation	
	Increased myocardial ischemia	
Brain	Increased cerebrovascular ischemia	[8, 10]

12.4 BIOLOGICAL PATHWAYS OF VOCS

Excessive exposure to VOCs is associated with short-term as well as long-term adverse health effects such as cancer [22] and CNS disorders [23]. There are three possible routes of VOC exposure in humans: inhalation (respiratory system), dermal absorption (dependent on the transdermal permeability), and consumption (drinking water and diet) [24–26]. Studies have stated that among these three, the main route of exposure to VOCs is the respiratory system [27]. The barriers present in the upper respiratory tract to remove pollutants become ineffective to VOCs due to their chemical properties of low polarity and poor solubility in water [28]. Thus, VOCs can easily access the lower respiratory tract, particularly the alveoli [29,30]. As a result, several epidemiological studies have highlighted that the adverse health risks associated with the lungs are due to exposure to VOCs [31–33].

The common methods used to detect exposure to VOCs are via exhaled breath, blood, and urine samples. Filipiak et al. (2013) speculated a plausible transport mechanism of VOCs from the circulating blood to the peripheral vasculature to the alveolar space, while another hypothesis claims that epithelial cells line the lungs' surface or form tissues which introduce VOCs to the exhaled breath [35]. The lack of knowledge of multiple compounds and their toxicity has made it difficult to explain the cause of VOC exposure [36]. However, the exchange of gases in the lungs occurs through the passive diffusion of gases between the blood and the alveolus in the lung capillaries found in the pulmonary alveoli [34]. In addition, Yoon et al. (2010) found that VOCs impair pulmonary function and enhance oxidative stress [37], and the finding was supported by other studies [38,39]. Therefore, VOCs can potentially reach the brain, compromising the integrity of the blood-brain barrier, and affecting the CNS through the enhanced oxidative stress and neuro-inflammation [40–42]. Most of the toxicology studies are based on individual VOCs instead of TVOCs. The author has summarised the known key biological pathways from VOC inhalation in Table 12.2.

Table 12.2 Summary of key biological pathways from VOC inhalation.

Organ	Biological pathways	VOC	References
Lungs	Irritation of the respiratory tract Alveoli affected Impaired pulmonary function Oxidative stress	Nonane	[43] [24–26] [31–33]
Blood		—	—
Heart	Irregular heartbeat	Benzene	[43]
Brain	Central nerve system disorders Neuroinflammation	Styrene Toluene	[43], [23] [40–42]

12.5 BIOLOGICAL PATHWAYS OF CO_2

There are two possible CO_2 exposure pathways: produced by a human's intracellular metabolism, and inhaled from the environment. The CO_2 produced within the cells is transported to the blood, and the blood carries it through the venous system to reach the lungs, where it is transferred to the alveoli and exhaled [44]. The accumulation of CO_2 due to the metabolism rate of a human or inhalation decreases the blood pH, which in turn increases the partial pressure of CO_2. The human body discharges CO_2 to maintain its acid-alkaline equilibrium. Therefore, an increase in partial pressure of CO_2 in the alveoli results in CO_2 freely diffusing through the alveolar membrane into the bloodstream and disrupting the acid-alkaline equilibrium; this is known as respiratory acidosis [45]. Previous studies have found that low-level CO_2 exposure (between 400 and 2000 ppm) induces systemic inflammation and worsens intracellular oxidative stress in human neutrophils [46–48]. It has also been reported that respiratory acidosis causes CO_2 to rapidly diffuse into the brain across the BBB, where a high concentration of CO_2 is found in the CNS [49]. The author has summarised the known key biological pathways from CO_2 inhalation in Table 12.3.

Table 12.3 Summary of key biological pathways from CO_2 inhalation.

Organ	Biological Pathways	References
Blood	Decreased blood pH	[44]
	Reduced oxygen saturation	[45]
	Respiratory acidosis	[46–48]
	Systemic inflammation in neutrophils	
	Intracellular oxidative stress in neutrophils	
Brain	Deposits in the central nerve system from respiratory acidosis	[49]

CONCLUSION

It is evident from the literature discussed in previous sections thatexposure to high PM levels affects the lungs, heart, and brain using blood as the media for PM transportation through the biological pathways. Despite the unclear modes of biological pathways deduced from works of literature for CO_2 and VOCs, there is evidence indicating the effects to have been observed in the lungs and the brain collectively. There is a lack of information regarding the transportation of CO_2 and VOCs' key biological pathways to the heart remain unclear. Hence, more studies are required to well define the biological pathways for VOCs and CO_2 pollutants in order to have a better understanding of how these pollutants are transported from one organ to another.

REFERENCES

[1] WBDG. 2021. Sustainable | WBDG—Whole Building Design Guide. Https://Www. Wbdg.Org/Design-Objectives/Sustainable. Accessed: Sep. 25, 2023. [Online]. Available: https://www.wbdg.org/design-objectives/sustainable

[2] WHO. 2014. Household air pollution. *In*: Textbook of Children's Environmental Health. 210–221. doi: 10.1093/med/9780199929573.003.0023.

[3] WHO. World Health Statistics 2022. World Health. 1–177. 131. 2022. Accessed: Sep. 25, 2023. [Online]. Available: https://www.who.int/news/item/20-05-2022-world-health-statistics-2022

[4] Wanner, A. Salathe, M. and O'Riordan, T.G. 1996. Mucociliary clearance in the airways. Am. J. Respir. and Crit. Care Med. 154(6). American Public Health Association. 1868–1902. doi: 10.1164/ajrccm.154.6.8970383.

[5] Geiser, M., Rothen-Rutishauser, B., Nadine Kapp, Schürch, S., Kreyling, W., Schulz, H., et al. 2005. Ultrafine particles cross cellular membranes by nonphagocytic mechanisms in lungs and in cultured cells. Environ. Health Perspect. 113(11): 1555–1560. doi: 10.1289/ehp.8006.

[6] Oberdörster, G., Sharp, Z., Atudorei, V., Elder, A., Gelein, R., Kreyling, W., et al. 2004. Translocation of inhaled ultrafine particles to the brain. Inhal Toxicol. 16(6–7): 437–445. doi: 10.1080/08958370490439597.

[7] Kawanaka, Y., Tsuchiya, Y., Yun, S.J. and Sakamoto, K. 2009. Size distributions of polycyclic aromatic hydrocarbons in the atmosphere and estimation of the contribution of ultrafine particles to their lung deposition. Environ. Sci. Technol. 43(17): 6851–6856. doi: 10.1021/es900033u.

[8] De Lorenzo, A.J.D. 2008. The olfactory neuron and the blood-brain barrier. 151–176. doi: 10.1002/9780470715369.ch9.

[9] Calderón-Garcidueñas, L., Solt, A.C., Henríquez-Roldán, C., Torres-Jardón, R., Nuse, B., Herritt, L., et al. 2008. Long-term air pollution exposure is associated with neuroinflammation, an altered innate immune response, disruption of the blood-brain barrier, ultrafine particulate deposition, and accumulation of amyloid β-42 and α-synuclein in children and young adult. Toxicol. Pathol. 36(2): 289–310. doi: 10.1177/0192623307313011.

[10] Maher, B.A., Ahmed, I.A.M., Karloukovski, V., MacLaren, D.A., Foulds, P.G., Allsop, D., et al. 2016. Magnetite pollution nanoparticles in the human brain. Proc. Natl. Acad. Sci. U.S.A. 113(39): 10797–10801. doi: 10.1073/pnas.1605941113.

[11] Calderón-Garcidueñas, L., Franco-Lira, M., Henríquez-Roldán, C., Osnaya, N., González-Maciel, A., Reynoso-Robles, R., et al. 2010. Urban air pollution: Influences on olfactory function and pathology in exposed children and young adults. Exp. Toxicol. Pathol. 62(1): 91–102. doi: 10.1016/j.etp.2009.02.117.

[12] Oberdörster, G., Sharp, Z., Atudorei, V., Elder, A., Gelein, R., Lunts, A., et al. 2002. Extrapulmonary translocation of ultrafine carbon particles following whole-body inhalation exposure of rats. J. Toxicol. Environ. Heal.—Part A 65(20): 1531–1543. doi: 10.1080/00984100290071658.

[13] Nemmar, A., Hoet, P.H., Vanquickenborne, B., Dinsdale, D., Thomeer, M., Hoylaerts, M.F., et al. 2002. Passage of inhaled particles into the blood circulation in humans. Circulation. 105(4): 411–414. doi: 10.1161/hc0402.104118.

[14] Chen, J., Tan, M., Nemmar, A., Song, W., Dong, M., Zhang, G., et al. 2006. Quantification of extrapulmonary translocation of intratracheal-instilled particles in vivo in rats: Effect of lipopolysaccharide. Toxicology. 222(3): 195–201. doi: 10.1016/j.tox.2006.02.016.

[15] Möller, W., Felten, K., Sommerer, K., Scheuch, G., Meyer, G., Meyer, P., et al. 2008. Deposition, retention, and translocation of ultrafine particles from the central airways and lung periphery. Am. J. Respir. Crit. Care Med. 177(4): 426–432. doi: 10.1164/rccm.200602-301OC.

[16] Oberdorster, G., Ferin, J., Gelein, R., Soderholm, S.C. and Finkelstein, J. 1992. Role of the alveolar macrophage in lung injury: Studies with ultrafine particles. Environ. Health Perspect. 97: 193–199. doi: 10.1289/ehp.97-1519541.

[17] Hassanvand, M.S., Naddafi, K., Kashani, H., Faridi, S., Kunzli, N., Nabizadeh, R., et al. 2017. Short-term effects of particle size fractions on circulating biomarkers of inflammation in a panel of elderly subjects and healthy young adults. Environ. Pollut. 223: 695–704. doi: 10.1016/J.ENVPOL.2017.02.005.

[18] Elwood, E., Lim, Z., Naveed, H. and Galea, I. 2017. The effect of systemic inflammation on human brain barrier function. Brain. Behav. Immun. 62: 35–40. doi: 10.1016/j.bbi.2016.10.020.

[19] Ferin, J., Oberdörster, G. and Penney, D.P. 1992. Pulmonary retention of ultrafine and fine particles in rats. Am. J. Respir. Cell Mol. Biol. 6(5): 535–542. doi: 10.1165/ajrcmb/6.5.535.

[20] Geiser, M., Rothen-Rutishauser, B., Kapp, N., Schürch, S., Kreyling, W., Schulz, H., et al. 2005. Ultrafine particles cross cellular membranes by nonphagocytic mechanisms in lungs and in cultured cells. Environ. Health Perspect. 113(11): 1555–1560. doi: 10.1289/ehp.8006.

[21] Craig, L., Brook, J.R., Chiotti, Q., Croes, B., Gower, S., Hedley, A., et al. 2008. Air pollution and public health: A guidance document for risk managers. J. Toxicol. Environ. Health—Part A: Current Issues. 71(9–10): 588–698. doi: 10.1080/15287390801997732.

[22] Mølhave, L., Clausen, G., Berglund, B., Ceaurriz, J.De, Kettrup, A., Lindvall, T., et al. 1997. Total volatile organic compounds (TVOC) in indoor air quality investigations. Indoor Air. 7(4): 225–240. doi: 10.1111/j.1600-0668.1997.00002.x.

[23] Jumpponen, M., Rönkkömäki, H., Pasanen, P. and Laitinen, J. 2013. Occupational exposure to gases, polycyclic aromatic hydrocarbons and volatile organic compounds in biomass-fired power plants. Chemosphere. 90(3): 1289–1293. doi: 10.1016/j.chemosphere.2012.10.001.

[24] Manisalidis, I., Stavropoulou, E., A. Stavropoulos, and Bezirtzoglou, E. 2020. Environmental and health impacts of air pollution: a review. Front. Public Health. 8: 14. doi: 10.3389/fpubh.2020.00014.

[25] He, J., Sun, X. and Yang, X. 2019. Human respiratory system as sink for volatile organic compounds: Evidence from field measurements. Indoor Air. 29(6): 968–978. doi: 10.1111/ina.12602.

[26] Raffy, G., Mercier, F., Glorennec, P., Mandin, C. and Le Bot, B. 2018. Oral bioaccessibility of semi-volatile organic compounds (SVOCs) in settled dust: A review of measurement methods, data and influencing factors. J. Hazard. Mater. 352: 215–227. doi: 10.1016/j.jhazmat.2018.03.035.

[27] Gong, Y., Wei, Y., Cheng, J., Jiang, T., Chen, L. and Xu, B. 2017. Health risk assessment and personal exposure to Volatile Organic Compounds (VOCs) in metro carriages—A case study in Shanghai, China. Sci. Total Environ. 574: 1432–1438. doi: 10.1016/j.scitotenv.2016.08.072.

[28] Zhao, Q., Li, Y., Chai, X., Xu, L., Zhang, L., Ning, P., et al. 2019. Interaction of inhalable volatile organic compounds and pulmonary surfactant: Potential hazards of VOCs exposure to lung. J. Hazard. Mater. 369: 512–520. doi: 10.1016/j.jhazmat.2019.01.104.

[29] Haick, H., Broza, Y.Y., Mochalski, P., Ruzsanyi, V. and Amann, A. 2014. Assessment, origin, and implementation of breath volatile cancer markers. Chem. Soc. Rev. 43(5): 1423–1449. doi: 10.1039/c3cs60329f.

[30] Gasparri, R., Santonico, M., Valentini, C., Sedda, G., Borri, A., Petrella, F., et al. 2016. Volatile signature for the early diagnosis of lung cancer. J. Breath Res. 10(1): 016007. doi: 10.1088/1752-7155/10/1/016007.

[31] Pappas, G.P., Herbert, R.J., Henderson, W., Koenig, J., Stover, B. and Barnhart, S. 2000. The respiratory effects of volatile organic compounds. Int. J. Occup. Environ. Health. 6(1): 1–8: doi: 10.1179/oeh.2000.6.1.1.

[32] Arif, A.A. and Shah, S.M. 2007. Association between personal exposure to volatile organic compounds and asthma among US adult population. Int. Arch. Occup. Environ. Health 80(8): 711–719. doi: 10.1007/s00420-007-0183-2.

[33] Cakmak, S., Dales, R.E., Liu, L., Kauri, L.M., Lemieux, C.L., Hebbern, C., et al. 2014. Residential exposure to volatile organic compounds and lung function: results from a

population-based cross-sectional survey. Environ. Pollut. 194: 145–151. doi: 10.1016/j.envpol.2014.07.020.

[34] Filipiak, W., Sponring, A., Filipiak, A., Baur, M., Troppmair, J. and Amann, A. 2013. Volatile Biomarkers: Non-Invasive Diagnosis in Physiology and Medicine. Accessed: Aug. 04, 2021. [Online]. Available: https://books.google.com.sg/books?hl=en&lr=&id=gmls9dj1rEoC&oi=fnd&pg=PA129&dq=thin+capillary+and+alveolar+type+1+cell+voc&ots=KolRlHZMer&sig=4vxho2j2o9qOOMvFP3nLGGeBoWY&redir_esc=y#v=onepage&q&f=false

[35] Filipiak, W., Sponring, A., Filipiak, A., Baur, M., Troppmair, J. and Amann, A. 2023. Volatile biomarkers: Non-invasive diagnosis in physiology and medicine. 2013. Accessed: Sep. 25, 2023. [Online]. Available: https://books.google.com.sg/books?hl=en&lr=&id=gmls9dj1rEoC&oi=fnd&pg=PA463&dq=Filipiak+et+al.+(2013+voc&ots=KosOrIWFcv&sig=T5rp8ZYPaqxJT9yW697DMPmaTA8&redir_esc=y#v=onepage&q=Filipiak et al. (2013 voc&f=false

[36] Aufderheide, M. 2008. An efficient approach to study the toxicological effects of complex mixtures. Exp. Toxicol. Pathol. 60(2–3): 163–180. doi: 10.1016/j.etp.2008.01.015.

[37] Yoon, H.I., Hong, Y.C., Cho, S.H., Kim, H., Kim, Y.H., Sohn, J.R., et al. 2010. Exposure to volatile organic compounds and loss of pulmonary function in the elderly. Eur. Respir. J. 36(6): 1270–1276. doi: 10.1183/ 09031936.00153509.

[38] Bentayeb, M., Billionnet, C., Baiz, N., Derbez, M., Kirchner, S. and Annesi-Maesano, I. 2013. Higher prevalence of breathlessness in elderly exposed to indoor aldehydes and VOCs in a representative sample of French dwellings. Respir. Med. 107(10): 1598–1607. doi: 10.1016/j.rmed.2013.07.015.

[39] Bönisch, U., Böhme, A., Kohajda, T., Mögel, I., Schütze, N., von Bergen, M., et al. 2012. Volatile organic compounds enhance allergic airway inflammation in an experimental mouse model. PLoS One. 7(7): e39817. doi: 10.1371/journal.pone.0039817.

[40] Thawani, V.R., Chakraborty, M. and Firodiya, A.D. 2016. Adverse effects of indoor air pollution on nervous system. Indian J. Environ. Prot. 36(11): 931–939. Accessed: Aug. 05, 2021. [Online]. Available: https://www.hindawi.com/journals/JT/2012/782462/

[41] Butterfield D.A. and Kanski, J. 2001. Brain protein oxidation in age-related neurodegenerative disorders that are associated with aggregated proteins. Mech. Ageing Dev. 122(9): 945–962. doi: 10.1016/S0047-6374(01)00249-4.

[42] Gella A. and Durany, N. 2009. Oxidative stress in alzheimer disease. Cell Adhes. Migr. 3(1): 88–93. doi: 10.4161/cam.3.1.7402.

[43] Rajabi, H., Mosleh, M.H., Mandal, P., Lea-Langton, A. and Sedighi, M. 2020. Emissions of volatile organic compounds from crude oil processing – Global emission inventory and environmental release. Sci. Total Environ. 727(1): 138654. doi: 10.1016/J.SCITOTENV.2020.138654.

[44] Lifson, N. and Gordon, G.B. 1949. The fate of utilized molecular oxygen and the source of the oxygen of. J. Biol. Chem. 180(2): 803–811. doi: 10.1016/S0021-9258(18)56700-4.

[45] Guais A., Brand, G., Jacquot, L., Karrer, M., Dukan, S., Grévillot, G., et al. 2011. Toxicity of carbon dioxide: A review. Chem. Res. Toxicol. 24(12): 2061–2070. doi: 10.1021/tx200220r.

[46] Coakley, R.J., Taggart, C., Greene, C., McElvaney, N.G. and O'Neill, S.J. 2002. Ambient pCO_2 modulates intracellular pH, intracellular oxidant generation, and

interleukin-8 secretion in human neutrophils. J. Leukoc. Biol. 71(4): 603–610. doi: 10.1189/jlb.71.4.603.

[47] Jacobson, T.A., Kler, J.S., Hernke, M.T., Braun, R.K., Meyer, K.C. and Funk, W.E. 2019. Direct human health risks of increased atmospheric carbon dioxide. Nat. Sustain. 2(8): 691–701. doi: 10.1038/s41893-019-0323-1.

[48] Zappulla, D. 2008. Environmental stress, erythrocyte dysfunctions, inflammation, and the metabolic syndrome: adaptations to CO_2 increases? J. Cardiometab. Syndr. 3(1): 30–34. doi: 10.1111/j.1559-4572.2008.07263.x.

[49] Goldberg, M.A., Barlow, C.F. and Roth, L.J. 1961. The effects of carbon dioxide on the entry and accumulation of drugs in the central nervous system. J. Pharmacol. Exp. Ther. 131: 308–318. Accessed: Aug. 06, 2021. [Online]. Available: http://www.pennem.com/wp-content/uploads/Goldberg-1961.pdf

Chapter 13

Additive Manufacturing of Shape Memory Alloys

Anil Chouhan[1], Elango Natarajan*[1,2], Santhosh Mozhuguan Sekar[1], Ang Chun Kit[1] and Kanesan Muthusamy[1]

[1]Faculty of Engineering, Technology and Built Environment, UCSI University, Kuala Lumpur, Malaysia

[2]Department of Mechanical Engineering, PSG Institute of Technology and Applied Research, Coimbatore, Tamilnadu, India

13.1 INTRODUCTION

Shape memory and super elasticity are two of the distinctive qualities that shape-memory alloys have. When heated, alloys with shape memory can return to their previous state, whereas those with super elasticity may withstand enormous deformations with little to no residual strain. When alloys often undergo phase change, they have a higher susceptibility towards energy dissipation than normal metallic substances. Alloys are now a practical choice for many circumstances in structures and infrastructure [1]. Over the past few decades, materials science has undergone significant improvement. The creation of novel alloys and composites has been made possible by the capacity to tailor various material properties (mechanical, thermal, electrical, etc.) for a range of purposes [2]. The most recent advancements in shape memory alloys have been discussed in this chapter, with

*For Correspondence: elango@ucsiuniversity.edu.my

a focus on Ti-Ni, Cu-based and ferrous alloys that are thought to be among the more useful materials for semiconductor applications [3].

For many decades, traditional materials like metals and alloys have been essential structural elements. By using the traditional engineering approach of analysing the material's macroscopic properties and choosing the right one to meet the intended functionality based on the application, engineers have created components and chosen alloys [4]. The scientific community widely acknowledges that martensitic modifications (MTs) in shape memory alloys (SMAs) are primarily distinguished by the shear distortion of the crystal lattice that occurs during the process of transformation. Although a minor volume change may also occur during MT, it is usually considered as an additional effect that can be overlooked when examining the fundamental characteristics of MTs and useful features of SMAs [5].

Despite the fact that a substantial number of metals show the shape memory effect, only those that recover quickly or have a pronounced restoration pressure are significant from an economic perspective. Particularly important amongst them are alloys made of Ni-Ti and Cu, including Cu-Zn-Al and Cu-Al-Ni. The most common alloys used in industrial uses are SMAs based on Ni-Ti since they provide good mechanical characteristics and form memory. [6]. Hartl et al. reported that shape memory alloys (SMAs) are a class of active materials, particularly as employed in aerospace applications. They gave a broad description of SMAs, their beneficial qualities, engineering consequences, and possible applications [7].

A shape memory alloy (SMA) is a unique functional material that is finding more and more uses across a wide range of industries. Recently, research has expanded to include the control of civil constructions utilising SMAs [8]. An overview of the principal uses for Ni-Ti alloys in the domains of dentistry, orthopaedics, vascular medicine, neurology, and surgery is given in this chapter. The primary qualities and benefits of employing SMAs are explained specifically for each device as well [9]. Although shape memory alloys (SMA) are now primarily used in medical applications, there are several industrial applications for SMA that have reached large volume manufacturing and far outpace the material's use in medical professions.

Fasteners and couplings, mostly in the military industry were the most significant applications of shape memory alloy technology in its early stages. Industrial uses are now becoming more widespread as technology develops and alloys become more widely available. medicine

13.2 SMA Fabrication Methods

The most common method for fabricating SMAs is casting. This involves melting the alloy and then pouring it into a mold. The mold can be made of a variety of materials, including metals, plastics or ceramics. Once the alloy has solidified, it is then heat-treated to induce phase transformation. Another method for fabricating SMAs is by powder metallurgy. This involves mixing alloy powders together and then compacting them into a shape. The compacted powder is then sintered, which is a process that uses heat and pressure to fuse the powders together. The sintered material is then heat-treated to induce phase transformation.

Other methods for fabricating SMAs include wire drawing, rolling and extrusion. These methods are used to produce SMAs in specific shapes and sizes. The choice of fabrication method depends on the specific application of the SMA. For example, casting is often used to produce large SMA parts, while powder metallurgy is often used to produce small SMA parts with complex shapes. The fabrication of SMAs is a complex process, but it is essential for producing SMAs with desired properties. The optimum procedure for an instance relies on the unique needs. The numerous ways of producing SMAs each have their own benefits and drawbacks. The following are a few factors to be taken into account while selecting a manufacturing technique for SMAs.

- size and shape of the SMA part
- desired properties of the SMA part
- cost of the fabrication method
- availability of the fabrication equipment

The fabrication of SMAs is a rapidly developing field, and new methods are being developed all the time. This is leading to the development of new and improved SMAs with a wider range of applications. Figure 13.1 depicts a schematic chart of the basic SMA fabrication process. Shape memory alloys are used in various applications. There are three types of SMAs existing in the industry.Cu-based, Ni-Ti based, and Fe-based. Also, with various fabrication methods available, shape memory alloy fabrication process vary according to the application, like CVD (Chemical Vapor Deposition), PVD (Physical Vapor Deposition), heating in electric furnaces, and 3D printing. The production method follows three steps as shown in Figure 13.1.

Figure 13.1 Schematic representation of the SMA fabrication process.

13.2.1 Electric Furnace Heating

Heating using an electric furnace is a common method for fabricating shape memory alloys (SMAs). This method involves melting the alloy in a furnace and then pouring it into a mold. The mold can be made of a variety of materials, including metal, plastic or ceramic. Once the alloy has solidified, it is then heat-treated to induce phase transformation. The furnace used for heating the SMA alloy must be able to reach a high enough temperature to melt the alloy. The furnace must also be able to maintain a constant temperature during the melting process. The type of furnace used depends on the size and shape of the SMA part being fabricated.

The mold used for casting the SMA alloy must be made of a material that is compatible with the alloy. The mold must also be able to withstand the high temperatures that are used during the casting process. The shape of the mold depends on the desired shape of the SMA part. The heat treatment process is used to induce phase transformation in the SMA alloy. It involves heating the alloy to a specific temperature and then cooling it at a specific rate. The heat treatment process depends on the specific SMA alloy being used.

The fabrication of SMAs using an electric furnace is a relatively simple process. However, it is important to follow the correct procedures to ensure that the SMA part is fabricated correctly. Figure 13.2 depicts the flowchart of the Ni-Ti fabrication process in an electric furnace. Some of the steps involved in fabricating SMAs using an electric furnace are given below:

- Cleaning the mold and the furnace
- Weighing the SMA alloy
- Adding the SMA alloy to the furnace
- Heating the furnace to the melting temperature of the SMA alloy
- Pouring the melted alloy into the mold
- Allowing the alloy to solidify
- Removing the SMA part from the mold
- Heat treating the SMA part

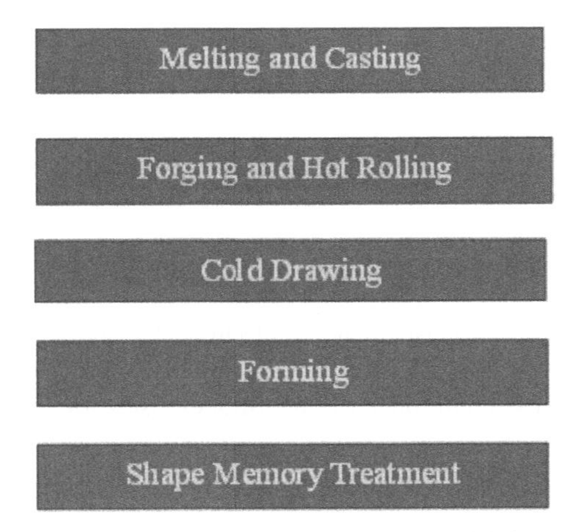

Figure 13.2 Ni-Ti shape memory alloy fabrication.

Fabrication of SMAs using an electric furnace is a versatile process that can be used to fabricate SMA parts of a variety of sizes and shapes. This process is relatively simple and can be performed with relatively inexpensive equipment. However, it is important to follow the correct procedures to ensure that the SMA part is fabricated correctly. The Ni-Ti alloy is an equiatomic intermetallic compound, and a deviation from stoichiometry in the composition has a significant impact on the alloy's properties. Temperature during transformation is very responsive to composition.

13.2.2 Physical Vapor Deposition (PVD)

PVD is a method for depositing thin films of material onto a substrate. This method is often used to fabricate shape memory alloys (SMAs) because it can be used to create films with precise thickness and composition. The PVD process for fabricating SMAs typically involves the following steps:

Step 1: The substrate is cleaned and heated to a high temperature.

Step 2: The SMA material is vaporized using a variety of methods, such as thermal evaporation, sputtering, or laser ablation.

Step 3: The vaporized material is then deposited onto the substrate.

Step 4: The deposited film is annealed to induce the phase transformation.

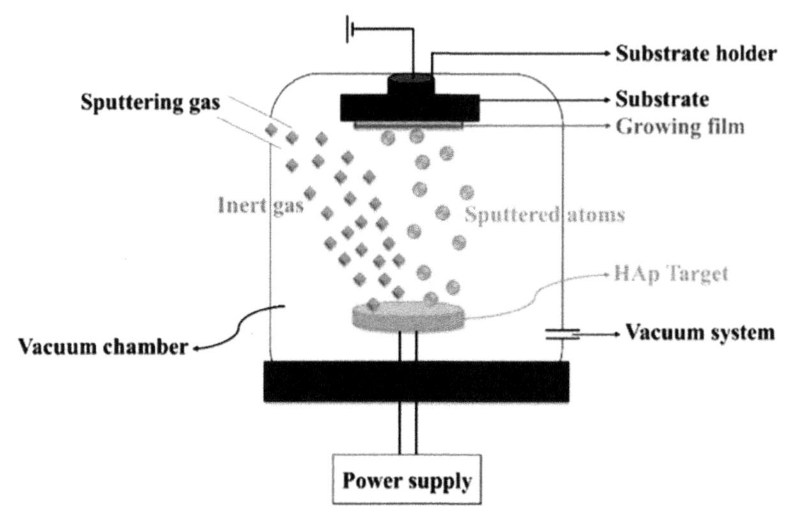

Figure 13.3 Schematic diagram of physical vapor deposition [29].

Some of the most common PVD techniques used for fabricating SMAs are listed below:

- **Thermal evaporation:** This technique involves heating the SMA material to a high temperature until it vaporizes. The vaporized material is then deposited onto the substrate.
- **Sputtering:** This technique involves bombarding the SMA material with a stream of ions. The ions knock atoms off the SMA material, which are then deposited onto the substrate.
- **Laser ablation:** This technique involves using a laser to ablate, or vaporize, the SMA material. The vaporized material is then deposited onto the substrate.

The choice of PVD technique depends on the specific application of the SMA. For example, thermal evaporation is often used to fabricate SMA films for applications in the medical industry, while sputtering is often used to fabricate SMA films for applications in the aerospace industry. The fabrication of SMAs using PVD is a versatile and effective process. This process can be used to fabricate SMA films of a variety of thicknesses, compositions, and shapes. The PVD process is often used to fabricate SMAs for applications in a wide range of industries.

13.2.3 Thermal Evaporation

In evaporation processes, vapors are created from a substance that is present in a source that is heated using a variety of techniques. Figure 13.6 depicts a schematic of an evaporation system. It consists of substrates that are positioned appropriately away from the evaporation source, and an evaporation source to vaporize the desired material. Possible heat sources for evaporation include resistance, induction, electric arcs, electron beams, and lasers. Using a DC/RF power supply, the substrate can be heated and/or biased to the desired potential.

Vacuum conditions with pressures between 5×10^{-5} and 5×10^{-7} torr are used for evaporation. When compared to the distance from the source to the substrate, the mean free path (MFP) in this pressure range is very large: 5×10^{-2} cm to 5×10^{-7} cm. Therefore, before condensing on the substrate, the evaporated atoms essentially go through a collision-less line-of-sight transport. This causes thickness to build up directly above the source and decrease steeply away from it. Therefore, planetary substrate holders are sometimes used to balance the vapor flux across multiple substrates. Also, to lower the MFP, a suitable gas, such as argon at pressures of 5–200 mtor, is introduced into the chamber sometimes. This causes multiple collisions of the vapor species as they travel from the source to the substrate, resulting in coatings of reasonably uniform thickness. This process is also known as pressure plating or gas scattering evaporation [10]. A vacuum is created to avoid contamination and reduce scattering to get a mean free path, and so that deposition can be done at a lower melting point. The main advantage of the thermal evaporation process is that we can get a high deposition rate and line of sight. However, thermal evaporation is an indirect heating process and its composition control is poor.

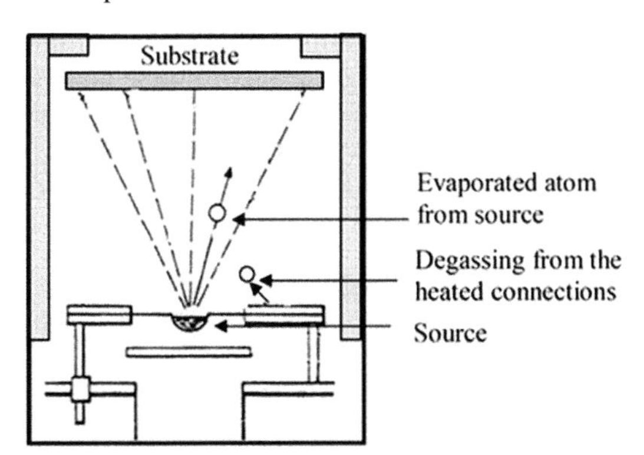

Figure 13.4 Thermal Evaporation Method [30].

13.2.4 Chemical Vapor Deposition (CVD)

CVD is a method for depositing thin films of material onto a substrate. This method is often used to fabricate shape memory alloys (SMAs) because it can be used to create films with precise thickness and composition. The CVD process for fabricating SMAs typically involves the following steps:

Step 1: The substrate is cleaned and heated to a high temperature.

Step 2: The SMA precursors are introduced into the reaction chamber.

Step 3: The precursors react to form the SMA material.

Step 4: The SMA material is deposited onto the substrate.

Step 5: The deposited film is annealed to induce phase transformation.

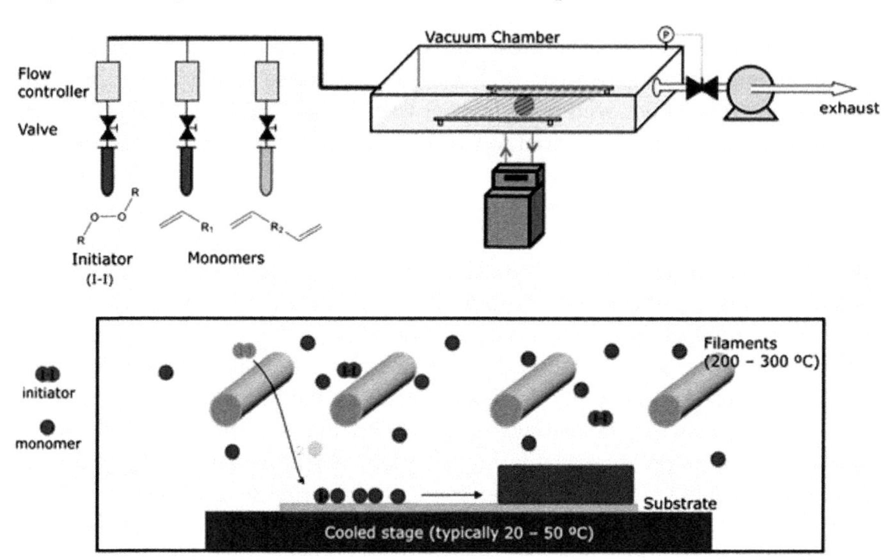

Figure 13.5 Chemical vapor deposition [31].

Figure 13.5 depicts a schematic illustration of the CVD process. Therein, deposition of a thin film takes place through a chemical reaction. Mainly five types of techniques are used to get high-quality films so that the thin films formed will be held in a better quality than PVD [11]. Various types of chemical vapor deposition: Plasma Enhanced (PECVD), Atmospheric Pressure (APCVD), Low Pressure (LPCVD), Very Low Pressure (VLCVD), and Metal-Organic (MOCVD).

13.2.5 3D Printing Technology

3D printing is a rapidly developing technology that is being used to fabricate a wide variety of materials, including shape memory alloys (SMAs). Over its forty-year existence, the industrial technology of three-dimensional (3D) printing has significantly advanced. Modern manufacturing techniques like 3D and 4D printing allow for quick and accurate production of items made of anything – from plastics to metals. Khorsandi evaluated various printing materials along with 3D printing technologies for the use of dentistry applications [12]. Meisel's software that slices a 3D CAD model of the component to be built is used to compute the pattern that the laser beam will follow [13]. Mahmood performed state of the art analysis on various uses and limitations of 3D and 4D printing technologies [14]. The findings demonstrate the effectiveness and adaptability of multi-material 3D printing in customizing the distorted shape of SMA-based soft actuators, which cannot be done with traditional production techniques like molding [15]. There are a number of different 3D printing techniques that can be used to fabricate SMAs at present.

13.2.5.1 Fused Deposition Modelling (FDM)

This technique uses a heated nozzle to extrude a filament of the SMA material. The filament is deposited layer by layer to create the desired shape. FDM is a versatile process that can be used to fabricate parts of a variety of sizes and shapes. This process is relatively simple and can be performed with a relatively inexpensive equipment. However, it is important to follow the correct procedures to ensure that the part is fabricated correctly.

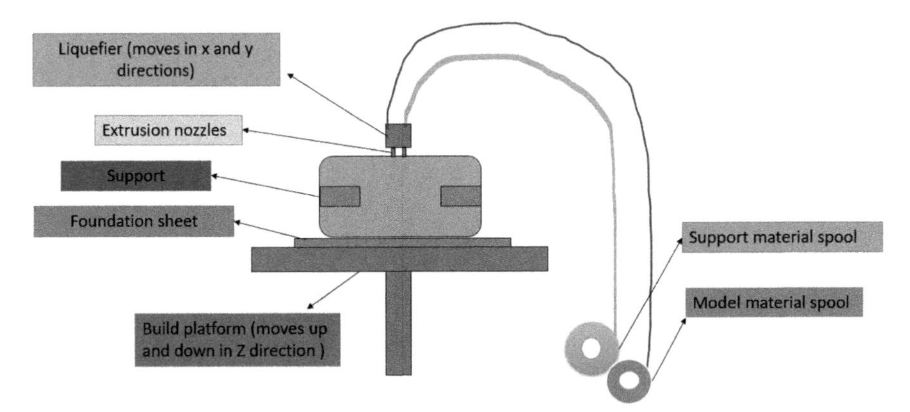

Figure 13.6 Schematic diagram of fused deposition modelling.

13.2.5.2 Selective Laser Sintering (SLS)

In this method, SMA particulates are sintered, or fused, using laser. To get the required form, the fine powder is applied in layers. Using a laser beam to sinter a powdered substance (usually nylon or polyamide), SLS is an additive manufacturing (AM) process. The light from the laser is autonomously directed at spots in the space specified by a 3D model, bonding the substance collectively to form a solid structure. Similar to selective laser melting, it implements the same idea but differs in several technological aspects. SLS is an emerging technology that is currently primarily deployed for rapid prototypes and low-volume manufacture of constituent components, along with the other AM processes listed.

SLS is an additive Layer production technique that involves fusing minute powdered pieces of plastic, metal, ceramic or glass into a mass with a specified three-dimensional form, using a high-power laser. By screening the cross-sections created from a 3D digital description of the component on the surface of a powder bed, the laser precisely fuses granular material. The powdered bed decreases by one layer width after each cross-section scan, a fresh layer of material is added on top, and the procedure is repeated until the component is finished. SLS is a flexible method that may be used to create components of different forms and sizes. This procedure is rather straightforward and it may be carried out with reasonably affordable tools. To guarantee that the item is manufactured appropriately, it is crucial to adhere to the right methods.

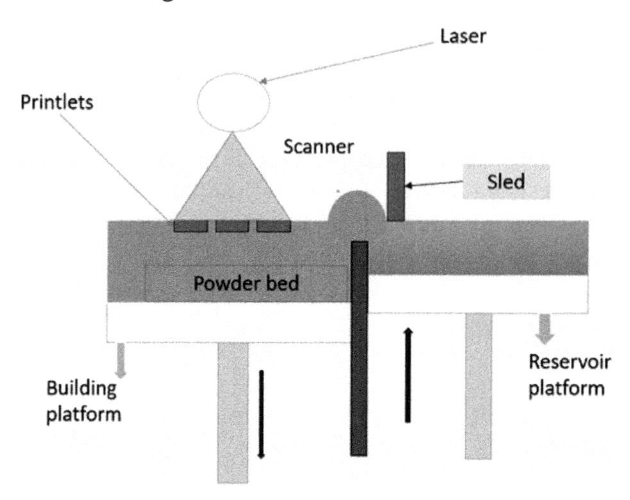

Figure 13.7 Illustration of Selective Laser Sintering.

13.2.5.3 Direct Metal Laser Sintering (DMLS)

In this process, SMA metallic powder particles are sintered, or fused, using laser. To get the required form, the powder is applied in layers. Direct metal laser sintering (DMLS) is an AM process that melts metal powder while automatically targeting the laser at locations in the space specified by a three-dimensional (3D) model, fusing the material to form a solid structure. It is comparable to selective laser melting (SLM), albeit the two are technological variations of the same principle. DMLS is a relatively new technology that has so far been mostly used for quick prototyping and low-volume manufacture of components, along with the other AM processes listed. The technique of fabricating components in a range of sizes and forms using DMLS is flexible. However, this procedure requires specialized tools and is rather difficult. Nonetheless, it is able to create components with complicated geometries and great levels of precision.

Subash reported that the anisotropic behavior of the technology and the static and inanimate nature of 3D printed parts serve as obstacles, but 4D printing eliminates them by giving 3D designs a temporal dimension and providing them vitality by using a stimulus to cause transformation in smart materials [16]. Muthe et al. reported that due to its bio-derived nature, desirable mechanical qualities, and great processability, poly-lactic acid (PLA) has become a widely used polymer for additive manufacturing. Various 3D printers and materials can be used for the fabrication of pressurized air bending-type SPAs [17]. Kafle reported that to make polymeric parts, three distinct early adopted but frequently used 3D printing techniques are used: fused deposition modelling (FDM), selective laser sintering (SLS), and stereolithography (SLA) [18].

13.3 COMPARISON OF DIFFERENT 3D PRINTING TECHNOLOGIES

Throughout the building process, the user has access to the complete volume of the part because of additive manufacturing's layer-by-layer construction method. Meisel et al. described the method for inserting shape memory alloy actuation wire within direct Poly Jet 3D printed objects. For generating successful and consistent embedding results, several "Design for Embedding" considerations have been offered [13]. Some of the advanced 3D printing technologies and their comparison are discussed in Table 13.1.

Table 13.1 Comparison of 3D printing techniques for SMA fabrication.

Technique	Description	Dominance	Limitations	References
Selective Laser Sintering (SLS)	A laser-assisted fusion of powder beds method for fusing powdered metals layer-by-layer	High accuracy and surface finish Wide range of materials available Good for complex geometries	Slow printing speed Requires post processing	[25, 27, 28]
Direct Metal Laser Sintering (DMLS)	A powder-bed fusion procedure comparable to SLS that melts the metal powder using a stronger laser	Faster printing speed than SLS Higher strength parts Better for large parts	More expensive than SLS Requires post processing	[10, 12]
Electron Beam Melting (EBM)	Melting layer by layer using electron beam.	Very high strength parts Good for complex geometries No post-processing required	Very slow printing speed Limited range of materials available	[14, 21]
Wire Arc Additive Manufacturing (WAAM)	A wire-based additive manufacturing process that uses an electric arc to melt a metal wire and deposit it layer by layer	Very high deposition rates Good for large parts No post-processing required	Lower accuracy and surface finish than other methods Limited range of materials available	[17, 15]
Binder Jetting	A powder-based additive manufacturing process that uses a binder to bind metal powder together layer by layer	Low cost Good for complex geometries No post-processing required	Lower accuracy and surface finish than other methods Limited range of materials available	[13]

Caputo et al. reported that in order to create Ni-Mn-Ga parts that undergo reversible martensitic transformation during heating and cooling, employed thermo-magneto-mechanically trained 3D printing. This transformation is a need for the shape-memory behavior. The authors discovered that the design problems associated with the functional parts of Ni-Mn-Ga magnetic SMAs could perhaps be solved through additive manufacturing [20]. Lu et al. used selective laser melting (SLM) to create an ultrahigh-performance Ti50.6Ni49.4 shape memory alloy (SMA) with sufficient energy inputs. The results demonstrated that while B2 austenite and B19' martensite are present in the SMA samples as-printed, the fraction of B19' martensite and the peak temperatures for the austenite and martensitic transformation drop as laser energy is reduced [21].

Lee et al. reported that four-dimensional (4D) printing, which is a more advanced form of 3D printing, has been made possible by the discovery of 3D-printable "smart" materials. A combination of 3D printing and the fourth dimension is known as 4D printing [19]. The different metal 3D printing techniques offer a range of advantages and disadvantages, and the best choice for a particular application depends on the specific requirements. SLS and DMLS are the most common powder bed fusion techniques and offer high accuracy and surface finish [22, 23]. EBM is a more expensive option but produces very strong parts. WAAM is a fast and cost-effective technique for large parts but has lower accuracy and surface finish. Binder jetting is a low-cost option for complex geometries but has lower accuracy and surface finish [24, 26]. In addition to the factors listed above, there are several other considerations that may be important when choosing a metal 3D printing technique, such as:

- size and complexity of the part to be printed
- required strength and surface finish of the part
- availability of materials
- cost of the equipment and materials
- post-processing requirements

It is thus important to consult with a metal 3D printing expert to determine the best technique for a particular application.

Even though there are numerous researchers working on 3D printing technologies for the manufacturing of shape memory alloys, there are a number of difficulties also involved, such as the sensitivity of SMAs to processing conditions. SMAs are very sensitive to the processing conditions used in 3D printing, such as temperature, laser power, and scanning speed. This can make it difficult to achieve the desired properties in the printed parts. Similarly, there is a limited availability of SMA powders. Not all SMAs are available as powders—the raw materials used in 3D printing. This can limit the range of SMAs that can be used in 3D printing. The important aspect that should be taken into consideration is that the printed parts often need to be post-processed to improve their properties. This can be a time-consuming and expensive process.

Despite these difficulties, there has been significant progress in the development of SMAs using 3D printing. In recent years, researchers have developed new methods for 3D printing SMAs that overcome some of the challenges mentioned above. For example, new methods have been developed for printing SMAs with complex geometries and for printing SMAs with improved mechanical properties.

13.4 CONCLUSION

The best additive manufacturing technology for the production of SMAs depends on the specific application. For example, if accuracy is critical, then SLM or EBM would be the best choice. If cost is a major factor, then FDM would be the best option. Ultimately, the best way to choose an AM technology for the production of SMAs is to consider the specific requirements of the application and to compare the different technologies to find the one that best meets the needs. It has been strongly acknowledged in the literature that the following factors should be considered while choosing the specific 3D printing technologies:

- Size and complexity of the part to be printed: Some technologies are better suited for printing large or complex parts, while others are better for printing small or simple parts.
- Materials available for printing: Different technologies support different materials, so it is important to choose a technology that can print the materials needed.
- Post-processing requirements: Some 3D printing technologies require more post-processing than others. This is something to consider if one is not able to do their own post-processing.

As AM technologies continue to develop, it is likely that the difficulties involved in the development of SMAs using 3D printing will be further reduced. This will make it possible for us to use 3D printing to create SMAs with a wider range of properties and for a wider range of applications.

REFERENCES

[1] Chang, W.-S. and Araki, Y. 2016. Use of shape-memory alloys in construction: a critical review. Proc. Inst. Civ. Eng.—Civ. Eng. 169(2): 87–95. doi:10.1680/jcien.15.00010.

[2] Kumar, P.K. and Lagoudas, D.C. 2008. Introduction to Shape Memory Alloys. pp. 1–51. *In*: Shape Memory Alloys., vol 1. Springer, Boston, MA. https://doi.org/10.1007/978-0-387-47685-8_1.

[3] Stachiv, I., Alarcon, E. and Lamac, M. 2021. Shape memory alloys and polymers for MEMS/NEMS applications: Review on recent findings and challenges in design, preparation, and characterization. Metals (Basel). 11(3): 415. doi: 10.3390/met11030415.

[4] Rao, A., Srinivasa, A.R. and Reddy, J.N. 2015. Introduction to Shape Memory Alloys. pp. 1–31. *In*: Design of Shape Memory Alloy (SMA) Actuators. SpringerBriefs in Applied Sciences and Technology. Springer, Cham. https://doi.org/10.1007/978-3-319-03188-0_1.

[5] Chernenko, V., L'vov, V., Cesari, E., Kosogor, A. and Barandiaran, J. 2013. Transformation volume effects on shape memory alloys. Metals (Basel). 3(3): 237–282. doi: 10.3390/met3030237.

[6] Machado, L.G. and Savi, M.A. 2003. Medical applications of shape memory alloys. Brazilian J. Med. Biol. Res. 36 (6): 683–691. doi: 10.1590/S0100-879X2003000600001.

[7] Hartl, D.J. and Lagoudas, D.C. 2007. Aerospace applications of shape memory alloys. Proc. Inst. Mech. Eng. Part G.J. Aerosp. Eng. 221(4): 535–552. doi: 10.1243/09544100JAERO211.

[8] Song, G., Ma, N. and Li H-N. 2006. Applications of shape memory alloys in civil structures. Eng. Struct. 28(9): 1266–1274. doi: 10.1016/j.engstruct.2005.12.010.

[9] Petrini, L. and Migliavacca, F. 2011. Biomedical Applications of Shape Memory Alloys. J. Metall. 1–15. doi: 10.1155/2011/501483.

[10] Bunshah, Rointan F. 2001. Handbook of Hard Coatings Deposition Technologies. Properties and Applications.

[11] Spear, K.E. and Frenklach, M. 1994. High-temperature chemistry of CVD (chemical vapor deposition) diamond growth. Pure Appl. Chem. 66(9): 1773–1782. doi: 10.1351/pac199466091773.

[12] Khorsandi, D., Fahimipour, A. and Abasian, P. 2021. 3D and 4D printing in dentistry and maxillofacial surgery: Printing techniques, materials, and applications. Acta Biomater. 122: 26–49. doi: 10.1016/j.actbio.2020.12.044.

[13] Meisel, N.A., Elliott, A.M. and Williams, C.B. 2015. A procedure for creating actuated joints via embedding shape memory alloys in PolyJet 3D printing. J. Intell. Mater. Syst. Struct. 26(12): 1498–1512. doi: 10.1177/1045389X14544144.

[14] Mahmood, A., Akram, T., Chen, H. and Chen, S. 2022. On the Evolution of Additive Manufacturing (3D/4D Printing) Technologies: Materials, Applications, and Challenges. Polymers (Basel). 14(21): 4698. doi: 10.3390/polym14214698.

[15] Akbari, S., Sakhaei, A.H., Panjwani, S., Kowsari, K., Serjouei, A. and Ge, Q. 2019. Multimaterial 3D printed soft actuators powered by shape memory alloy wires. Sensors Actuators A Phys. 290: 177–189. doi: 10.1016/j.sna.2019.03.015.

[16] Subash, A. and Kandasubramanian, B. 2020. 4D printing of shape memory polymers. Eur. Polym. J. 134: 109771. doi: 10.1016/j.eurpolymj.2020.109771.

[17] Muthe, L.P., Pickering, K. and Gauss, C. 2022. A review of 3D/4D printing of poly-lactic acid composites with bio-derived reinforcements. Compos. Part C, Open Access. 8: 100271. doi: 10.1016/j.jcomc.2022.100271.

[18] Kafle, A., Luis, E., Silwal, R., Pan, H.M., Shrestha, P.L. and Bastola, A.K. 2021. 3D/4D printing of polymers: fused deposition modelling (FDM), selective laser sintering (SLS), and stereolithography (SLA). Polymers (Basel). 13(18): 3101. doi: 10.3390/polym13183101.

[19] Lee, A.Y., An, J. and Chua, C.K. 2017. Two-way 4D printing: A review on the reversibility of 3D-printed shape memory materials. Engineering. 3(5): 663–674. doi: 10.1016/J.ENG.2017.05.014.

[20] Caputo, M.P., Berkowitz, A.E., Armstrong, A., Müllner, P. and Solomon, C.V. 2018. 4D printing of net shape parts made from Ni-Mn-Ga magnetic shape-memory alloys. Addit. Manuf. 21: 579–588. doi: 10.1016/j.addma.2018.03.028.

[21] Lu, H.Z., Yang, C., Luo, X., Ma, H., Song, B., Li, Y., et al. 2019. Ultrahigh-performance TiNi shape memory alloy by 4D printing. Mater. Sci. Eng. A. 763: 138166. doi: 10.1016/j.msea.2019.138166.

[22] Safavi, M.S., Walsh, F.C., Surmeneva, M.A., Surmenev, R.A. and Khalil-Allafi, J. 2021. Electrodeposited hydroxyapatite-based biocoatings: recent progress and future challenges. Coatings. 11: 110. https: //doi.org/10.3390/coatings11010110.

[23] Almas, B., Tahir, I.A., Aqsa, T., Muhammad, B.T. and Mohsin, I. 2020. Interfaces and surfaces. pp. 51–87. *In*: Chemistry of Nanomaterials, Fundamentals and Applications, Elsevier. https: //doi.org/10.1016/B978-0-12-818908-5.00003-2.

[24] Gebisa, A.W. and Lemu, H.G. 2018. Investigating effects of Fused-Deposition Modeling (FDM) processing parameters on flexural properties of ULTEM 9085 using designed experiment. Mater. (Basel). 11(4): 500. https: //doi.org/10.3390/ma11040500.

[25] Gueche, Y.A., Sanchez-Ballester, N.M., Cailleaux, S., Bataille, B. and Soulairol, I. 2021. Selective laser sintering (SLS), a new chapter in the production of solid oral forms (SOFs) by 3D printing. pharmaceutics. 13: 1212. https: //doi.org/10.3390/pharmaceutics13081212.

[26] Antonio, S.P. 2021. Design and Manufacturing of Plastics Products Integrating Traditional Methods with Additive Manufacturing A volume in Plastics Design Library.

[27] Syed, H.R., Syed, H.M., Rizwan, A.R.R. and Sanjeet, C. 2020. Selective laser sintering in biomedical manufacturing, Woodhead Publishing Series in Biomaterials, 193–233.

[28] Xinpeng, G., Guoxia, F., Jinzhi, W., Zhanhua, W., Marino, L. and Hesheng, X. 2020. Powder quality and electrical conductivity of selective laser sintered polymer composite components. Woodhead Publishing Series in Composites Science and Engineering Part 2. 149–185.

[29] Safavi, M.S., Frank C.W., Maria, A.S., Roman A.S. and Jafar K.-A. 2021. Electrodeposited hydroxyapatite-based biocoatings: recent progress and future challenges. Coatings 11. 1: 110. https: //doi.org/10.3390/coatings11010110

[30] Kumar, V., Sharma, D.K., Sharma, K. and Dwivedi. D.K. 2016. Structural, optical and electrical characterization of vacuum-evaporated nanocrystalline CdSe thin films for photosensor applications. Appl. Phys. A 122: 960.

[31] Saleh, B., Jiang, J., Fathi, R., Al-Hababi, T., Xu, Q., Wang, L., et al., 2020. 30 Years of functionally graded materials: An overview of manufacturing methods,applications and future challenges. Composites Part B: Engineering. 201: 108376. https: //doi.org/10.1016/j.compositesb.2020.108376.

Chapter **14**

Recycled High-Density Polyethylene Plastics (HDPE) Reinforced with Natural Fibers for Floor Tiles

Ammar A. Al-Talib*, Zhou Yi, Santhosh Mozhuguan Sekar, Ang Chun Kit and C.S. Hassan

Faculty of Engineering, Technology and Built Environment, UCSI University, 56000 Kuala Lumpur, Malaysia

14.1 INTRODUCTION

Plastic bags, films, and containers are often made up of polyethylene, a kind of polymer. It was initially created by the German chemist Hans von Pechmann by accident in 1898 while exploring diazomethane. However, experts considered it of little use at the time. At ICI (Imperial Chemical Industries) in the United Kingdom, scientists Eric Fawcett and Reginald Gibson formulated the first industrially useful synthesis method for polyethylene on 1933 [1]. Since then, the material has been produced abundantly as researchers saw many potential uses for it. Although polyethylene makes up about 34% of the entire plastics industry [2], the manufacture of polyethylene has led to ecological issues since it is not biodegradable [3]. Hence, experts are currently looking for ways to use it as a matrix for organic materials.

*For Correspondence: ammart@ucsiuniversity.edu.my

Fibres made in nature may be utilised as parts in composite materials and are produced from the bodies of wildlife and plants. High-tech products like composite vehicle components use natural fibres. Natural fibres provide a number of benefits over industrial fibres, the notable ones being affordability, minimal environmental impact, biodegradability, and the fact that they are renewable [4]. They are particularly good at isolating noise and vibrations [5]. Furthermore, after being dismissed, natural fibres, especially glass fibres, are vulnerable to bacterial deterioration. In contrast to other fibres, natural fibres have a really modest market share. While animal fibres like wool and silk account for barely more than 1% of yearly output, plant-based fibres like jute, flax and hemp account for 5.7% of the world market (6 metric tonnes) [6]. As the future for organic fibres is further explored, their market share is rising.

On the other hand, plastic garbage is a significant worldwide ecological issue right now. It has been determined that the production of plastic waste puts human life at risk in a variety of ways, including pollution of water and land [7], and the spread of disease [8]. This issue may be resolved by using plastic products less, recycling waste plastic for new uses, and creating environmentally friendly plastics that are more resistant to deterioration than ordinary polymers (such as biopolymers). Plastics are organic polymers that may be used to create a variety of objects such as containers and clothing [9]. They were accidentally found in 1856 by Alexander Parkes, but it was not until World War II that they were extensively used. The production of plastic reached 368 million tonnes globally in 2019 [10], and this figure is steadily increasing. The primary driver for such large volume manufacturing is the cheap cost of the raw material [11]. Its longevity is another important factor [12]. Although a variety of materials may be utilised to substitute plastic, they do not hold up as well as plastics under the same conditions. Fortunately, it is now understood that all of these benefits come at a price: contamination.

For decades, plastic pollution has been a big issue, particularly in marine habitats [13]. Plastics are being used at an increasing rate and degrading them might take hundreds of years [14]. Plastic garbage output is expected to be over 6 billion tonnes, with over 300 million tonnes of plastic waste generated globally each year, only 10% of which is recycled [15]. According to UN statistics, plastic fragments have been discovered in oceans all around the world, with a concentration of up to 580,000 pieces per square kilometre of sea surface [16]. Furthermore, many marine organisms misinterpret plastics as food and consume it. The earliest evidence of marine turtles and seabirds ingesting plastic was discovered in the 1960s [17]. Plastic bags are frequently mistaken for food by animals and livestock, and they frequently die as a result of consuming discarded plastic bags [18]. In addition, the substantial consumption of fisheries (fish and shellfish) in certain nations, the prevalence of plastic particles in marine species suitable for consumption by humans, and other factors have raised concerns about the possible health consequences of ingesting microplastic-contaminated fisheries [19].

Recycled plastic waste and turning it into beneficial synthetic goods helps lessen pollution by plastics. There are two categories of plastic polymers: thermosets and thermoplastics. HDPE is a recyclable thermoplastic polymeric material, in contrast to thermoset plastics [20]. Excellent biological compatibility, stiffness, strength, and good creep characteristics are all features of HDPE [21, 22]. HDPE, although lacking flexibility, beats LDPE and LLDPE in terms of durability, endurance and chemical resistivity [23]. Bottles, pipe structures, wrapping, and other things are often made from HDPE [24]. From 11.9 million metric tonnes in 1990 to 43.9 billion tonnes in 2017 [25], the need for HDPE resins increased at a rate of 3.3% annually.

Dehusking, retting, defibring and polishing are the steps used to remove and purify coconut fibre [26]. Coconut fibre, which is mostly composed of cellulose, lignin and hemicellulose, is utilised in a variety of industries, notably cords, floors, and composite products [27]. The world produces about 250,000 tonnes of coconut fibre annually, and its mechanical and physical properties are considered to have a lot of promise for enhancing composite ductility, flexural toughness, and energy absorption. Reinforced composite materials can exhibit enhanced post-cracking actions, thanks to the outstanding durability and versatility of these fibres [28].

Coconut fibre-based materials (CFRP) have a greater noise absorption index than other fibre-reinforced composites. Additionally, CFRP have superior impact resistance than hybrids made of sugarcane fibre as coconut fibre is more resilient than sugarcane fibre. In this proposed study HDPE was combined with different proportions of coconut fibres and tested for its mechanical properties, hardness and water absorption rate.

14.2 PRIOR ART RESEARCH

HDPE and high-density polyethylene nano clay (PMON) have been used by investigators to modify lubricants utilising different plastic wastes [29]. Investigations have been made on the structural and wear characteristics of MWCNT/HDPE nanocomposites and the results revealed that these nanocomposites had higher decomposition temperatures and thermal stability than pure HDPE, and that 2 wt.% MWCNT/HDPE nanocomposites are potential gear materials [30]. Using ABAQUS and ANSYS, an HDPE double-walled curved pipe has been accurately modelled in three dimensions, and coupling prediction of tube-soil-fluid under actual circumstances has also been accomplished by researchers [31].

The properties of birch-fibre reinforced HDPE produced by injection moulding, including drop weight impact, Izod impact strength, hardness, tensile strength, and elastic modulus, were tested by Koffi [32]. Also, Mendes Juliana Farinassi and colleagues aimed to create ecological composites made of high-density polyethylene (HDPE) and wasted coffee grounds (SCG), which were subsequently tested for thermophysical and mechanical properties [33]. Different doses of green coconut fibre in the growth substrate have been evaluated for the early development and physiology of two soybean varieties in another study [34]. Moreover, the effect of liquid smoke treatment on the tensile strength of single fibres and coconut fibre (CF) reinforced composites have also been studied. The results showed that fibre saturated in liquid smoke may work as a good way to enhance the structural characteristics of CF [35]. Alkalinizing coconut fibres with a 5% NaOH solution at 70°C for 30 minutes yields a maximum mechanical strength of 295.13 MPa [36].

The surface characteristics of coconut shell fibres treated with limestone water have been evaluated by researchers. The results demonstrate that after 8 hours of soaking in limestone water, the surface of coconut shell fibres is smooth, rough and grooved, and the fibres are better linked to the matrix; thus, limestone water can be utilised as a natural fibre treatment medium [37]. Researchers have also employed coconut fibre as a green inhibitor to see how different inhibitor concentrations, immersion times, and inhibition efficiency of coconut husk extracts affect the results [38].

Lawrence et al. produced tiles from recycled HDPE plastic and used rubber tyres [39]. Chandra Sekhara et al. manufactured tiles using High Density Polyethylene (HDPE) plastic waste and fine graded Wollastonite powder and tested them for water absorption, hardness, compression and chemical resistance [40]. The purpose of their was to analyse the mechanical features (tensile strength, flexural strength and impact strength), hardness, and water absorption rate of neat, 5%, 10%, 15%, and 20% coconut fibres reinforced recycled HDPE. By contrasting the physical characteristics and water intake rate of various amounts of coconut fibre fillers in the HDPE utilized, the best mixing ratio was determined.

14.3 FABRICATION AND CHARACTERISATION OF HYBRID COMPOSITES

The chemical compounding procedure is used to create samples for testing. First, powdered coconut fibres are created using a grinder. The coconut fibres are soaked in a sodium hydroxide solution for 24 hours, and then dried for 24 hours at 50 degrees in a drier. The mechanical characteristics of the chemically treated coconut fibre reinforced with HDPE is subsequently examined. Second, in a built-in blender (BRABENDER Palatograph EC, Germany), the HDPE and the chemically modified coconut fibres are combined.

Table 14.1 Composition of HDPE Hybrid Composites.

Sample No.	HDPE (Wt.%)	Coconut Fibre (Wt.%)
Neat HDPE	100	0
HCHDPE1	95	5
HCHDPE2	90	10
HCHDPE3	85	15
HCHDPE4	80	20

*HCHDPE – Hybrid Coconut Fibre Reinforced HDPE

The cylinder temperature is raised to 190°C and the screw is rotated at 50 rpm. Following this, a certain quantity of HDPE flakes and an appropriate amount of coconut fibre powder (5%, 10%, 15% and 20%) are added. Third, a compression moulding machine (Techno press – 40HC – B, Malaysia) with a heated mould is used to compress the laminate The substance is heated for ten minutes, compacted for ten minutes, and then moved to a cold plate for thirty minutes of cooling. The samples thus obtained are 150 mm × 150 mm × 4 mm. Finally, a cutting machine is used to form the samples into the desired shape.

14.3.1 Mechanical Characterisation

Following the test standards ASTM D638–04, the generated samples are tension tested on the Gotech Testing Machine (Al – 3000), with the crosshead speed set at 50 mm/min and the temperature set at 23°C. In this test, three samples are examined, and the results are averaged. The flexural modulus and flexural strength of the composite are examined using the three-point bending flexural test. The specimens are evaluated in accordance with ASTM-D790, utilising the Gotech Testing Machine (Al - 3000) with a crosshead speed of 2 mm/min. The Advanced Pendulum Impact System (RR/IMT, Ray-Ran, UK) is used to perform Izod impact tests in accordance with ASTM D256. Figure 14.1(a), (b) and (c) show the measurements of the specimens ready for tensile, flexural and impact testing.

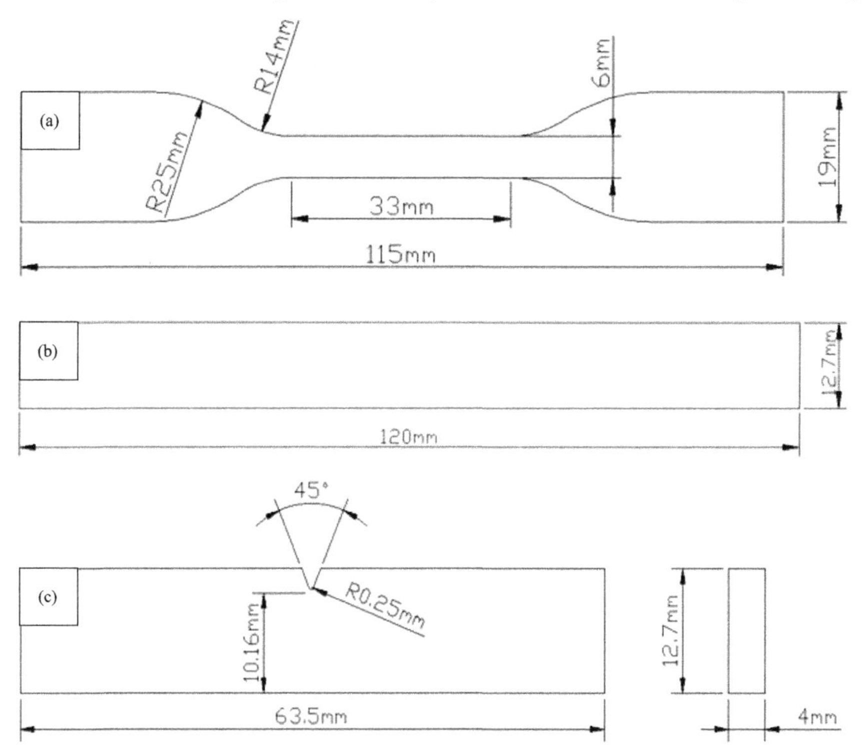

Figure 14.1 Dimensions of (a) tensile sample (b) flexural sample (c) impact test sample.

14.3.2 Hardness Testing

In this investigation, the Shore hardness test method is applied. A Shore hardness durometer is a tool for determining the hardness of materials such as polymers, elastomers and rubber. Hardness tests is carried out using a shore D hardness durometer, based on the standard ASTM D2240.

14.3.3 Water Absorption Test

The water absorption test measures the total weight gain of the specimen and the behaviour change brought on by water uptake at an appropriate temperature and for a particular amount of time. Here is the test methodology. Prior to weighing the samples, they are cut into a certain dimension in accordance with ASTM D570. The samples are then submerged in pure purified water after being dried in a dryer for 24 hours at 55 degrees Celsius. Third, using a computerised scale, the weights of the samples are determined after 24 hours, 2 days, 3 days, 7 days, and 2 weeks. Finally, using formula 1, the water uptake rate was determined for the samples.

$$C = \frac{m_2 - m_1}{m_1} \tag{1}$$

Where,

C – water absorption rate
m_1 – weight of the sample before immersion
m_2 – weight of the sample after immersion

14.4 RESULTS AND DISCUSSION

14.4.1 Tensile Testing

The specimen's modulus of elasticity and tensile strengths are shown in Table 14.2 and the stress-strain graphs below. According to the test findings, HDPE's tensile modulus improves when 5% and 20% coconut fibres are added, with the HDPE reinforced with 20% coconut fibre having the greatest tensile modulus (0.870 GPa). In HDPE, addition of 5% and 20% coconut fibre raises the tensile modulus by 5.1% and 13% respectively. The stress-strain graphs of the hybrid samples from the investigation are shown in Figure 14.2.

Table 14.2 Tensile modulus and strength of neat and hybrid samples.

	Neat HDPE	HCHDPE1	HCHDPE2	HCHDPE3	HCHDPE4
Tensile Modulus (GPa)	0.7698	0.8088	0.5488	0.5238	0.870
Tensile Strength (MPa)	24.1	12.2	16.2	13.3	13.2

Adding coconut fibre drastically decreases the tensile strength, comparable to the research by Hidalgo et al. 2020 [41] wherein pure HDPE had the maximum tensile strength (24.1 MPa). Assuming that HDPE is a hydrophobic substance and coconut fibre is a hydrophilic material [42], the weak interfacial bond between hydrophilic fillers and hydrophobic matrix and the interfacial discontinuity are likely to have affected the mechanical properties. As the percentage of coconut fibre in the polymeric matrix grows, the weak interfacial region between the polymeric matrix and the coconut fibre increases, lowering the tensile strength values [41].

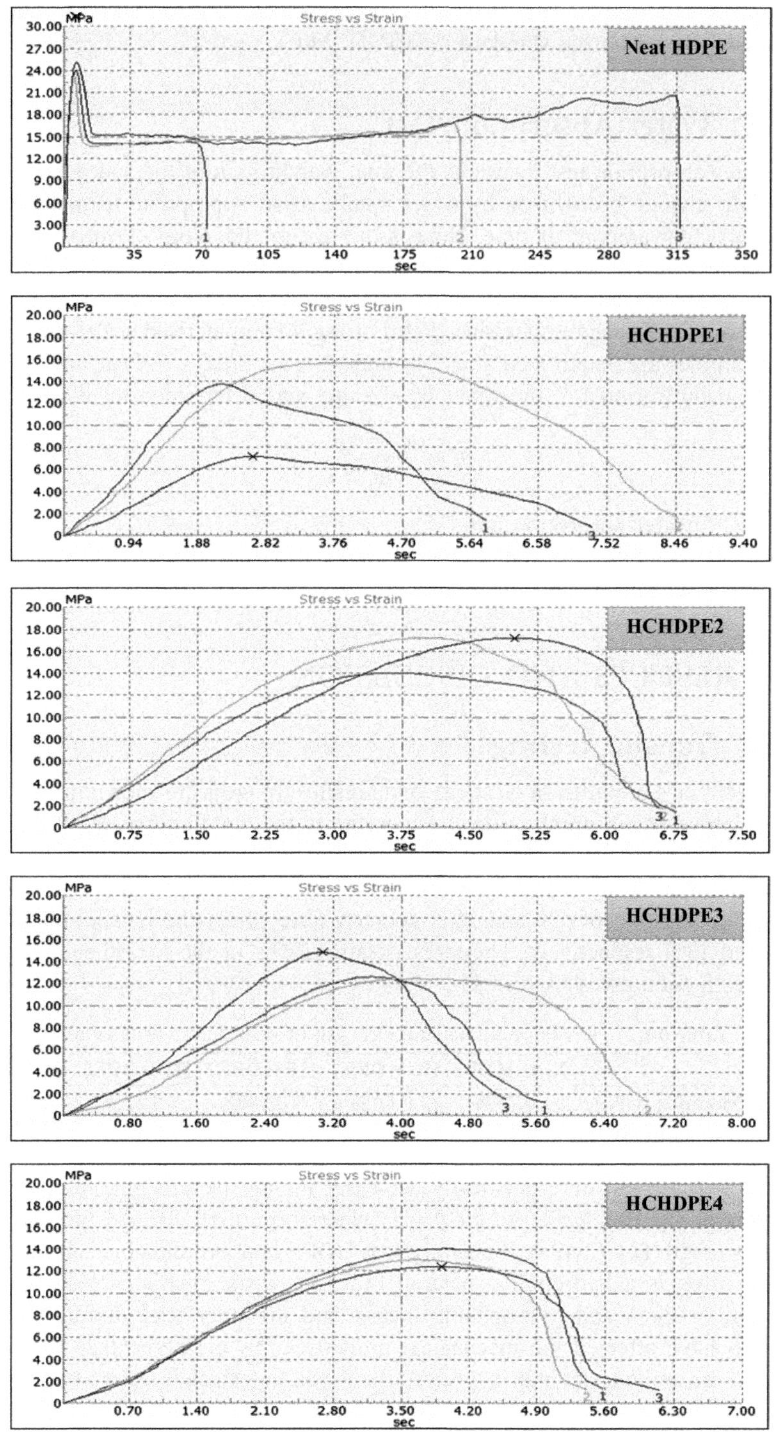

Figure 14.2 Tensile stress-strain relationship of neat and hybrid samples.

14.4.2 Flexural Test Results

Table 14.3 displays the average readings for flexural strength as well as the modulus of elasticity of the HDPE/coconut fibre composites created for the research discussed in this section. It is clear from this table that adding 10% to 20% of coconut fibres increases the bending modulus of HDPE. The 20% coconut fibre combination (HCHDPE4) has the greatest flexural modulus than any material (0.647 GPa), 10.2% greater than that of HDPE. This is because the ability of the polymer chains to move is inhibited by the coconut fibres, thus stiffening the material [41]. The maximum flexural strength, however, is found in pure HDPE (19.88 MPa). Figure 14.3 reports the flexural force-deflection relationship of the hybrid samples developed during the experiment.

The addition of coconut fibres, however, leads to a decrease in the flexural strength of HDPE, which is different from the results of the tests carried out by others [44]. Perhaps, as with tensile strength, it is the poor interfacial bonding that occurs between the hydrophilic material (coconut fibre) and the hydrophobic material (HDPE) that leads to a reduction in flexural strength. Compared to pure HDPE, the flexural strength of the composite samples decreased by 32.64% (13.39 MPa), 16.90% (16.52 MPa), 21.18% (15.67 MPa) and 20.82% (15.74 MPa) respectively.

Table 14.3 Flexural modulus and flexural strength of neat and hybrid samples.

	Neat HDPE	HCHDPE1	HCHDPE2	HCHDPE3	HCHDPE4
Flexural Modulus (GPa)	0.581	0.448	0.584	0.618	0.647
Flexural Strength (MPa)	19.88	13.39	16.52	15.67	15.74

14.4.3 Impact Test Results

The inclusion of coconut fibre weakens the sample's ability to withstand impacts, according to the findings of the IZOD impact test with notches. These findings are strikingly comparable to those of impact testing conducted by other authors on PP-HDPE-CF [41, 43] and Coir-EFB-PP [45]. Figure 14.4 and Table 14.4 depicts the confidence intervals of impact resistance (kJ/m^2) of the various samples developed and the overall impact resistance observation of the test samples respectively.

The findings may be attributed to a variety of factors, including the previously described poor interfacial interaction between coconut fibres and HDPE and the fact that the integration of fibres affects fracture development, crack width, and crack spacing [46]. Pure HDPE exhibits the maximum impact strength in the test (68.55 kJ/m^2), followed by HDPE reinforced with 10 wt.% coconut fibre (17.82 kJ/m^2) and 15 wt.% coconut fibre (10.75 kJ/m^2).

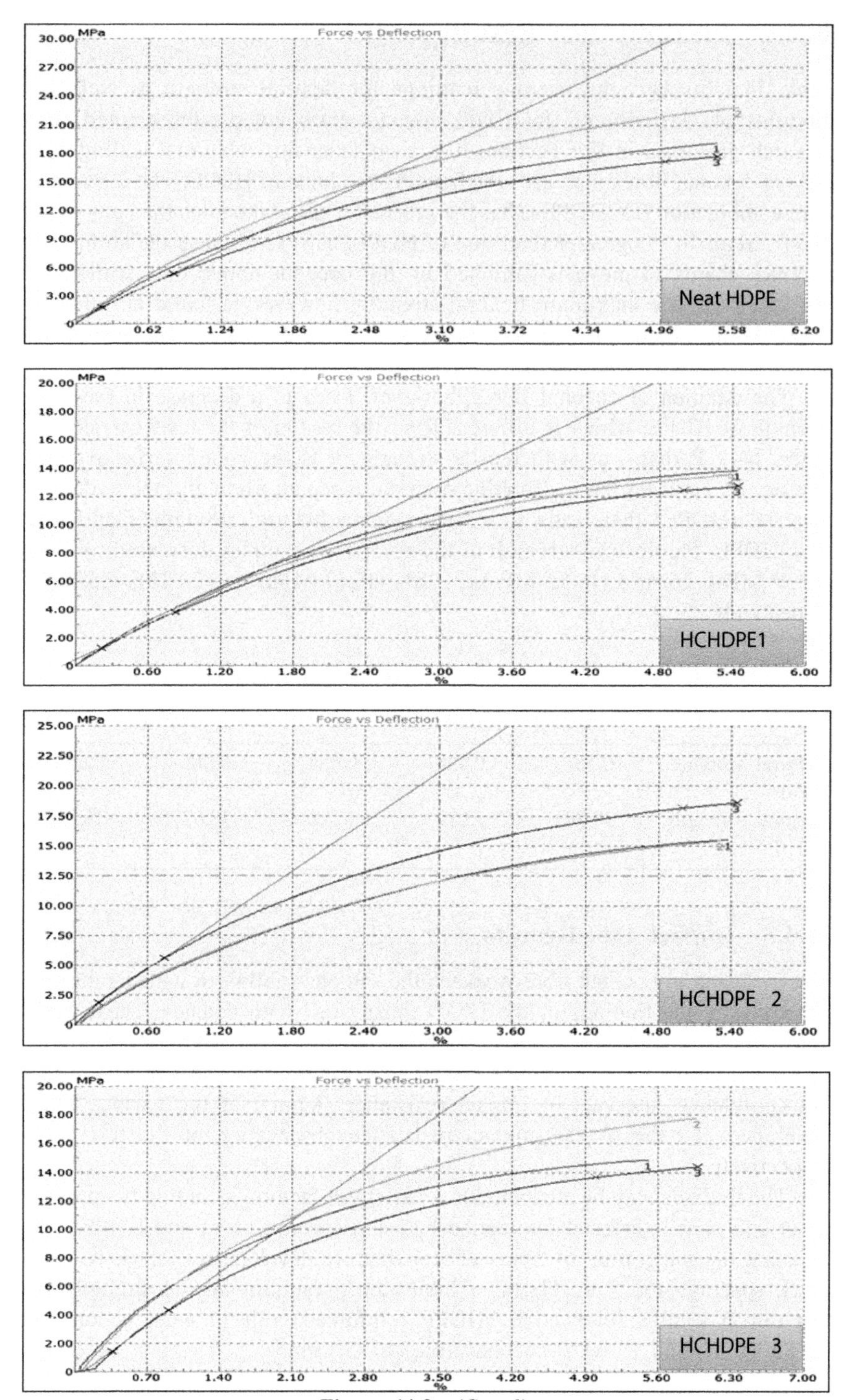

Figure 14.3 (*Contd.*)

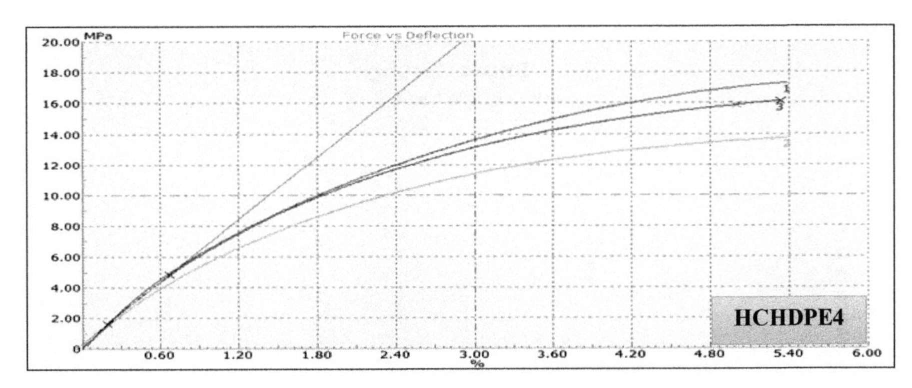

Figure 14.3 Flexural force-deflection relationship of neat and hybrid samples.

Table 14.4 Impact Resistance.

Impact Test		***C = Complete Break, H = Hinge Break, P = Partial Break, NB = Non-break**		
Sample Code	**Test No.**	**Impact Resistance**		**Type of Failure**
		kJ/m	**kJ/m^2**	
Neat HDPE	1	0.71	75.02	P
	2	0.68	66.44	P
	3	0.72	64.18	P
HCHDPE1	1	0.15	14.51	P
	2	0.151	13.52	P
	3	0.143	15.38	P
HCHDPE2	1	0.253	24.78	H
	2	0.167	15.23	H
	3	0.139	13.46	H
HCHDPE3	1	0.082	7.802	H
	2	0.173	13.17	H
	3	0.121	11.28	H
HCHDPE4	1	0.135	12.99	H
	2	0.128	11.85	H
	3	0.094	11.27	H

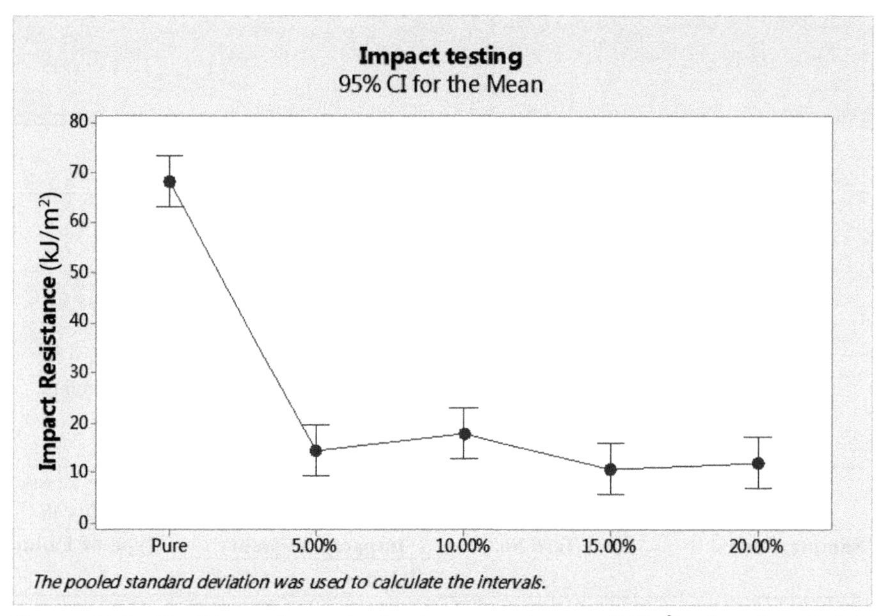

Figure 14.4 Confidence intervals for impact resistance (kJ/m^2) of various hybrid samples.

14.4.4 Hardness Test Results

Figure 14.6 depicts a range chart of hardness. It can be observed that as the fibre content rises, so does the hardness of specimens. Figure 14.5 displays the Shore D hardness of all samples. In this test, pure HDPE had the lowest hardness (62.33 HD), whereas the composite specimen including 20% CF displayed the maximum hardness (64 HD).

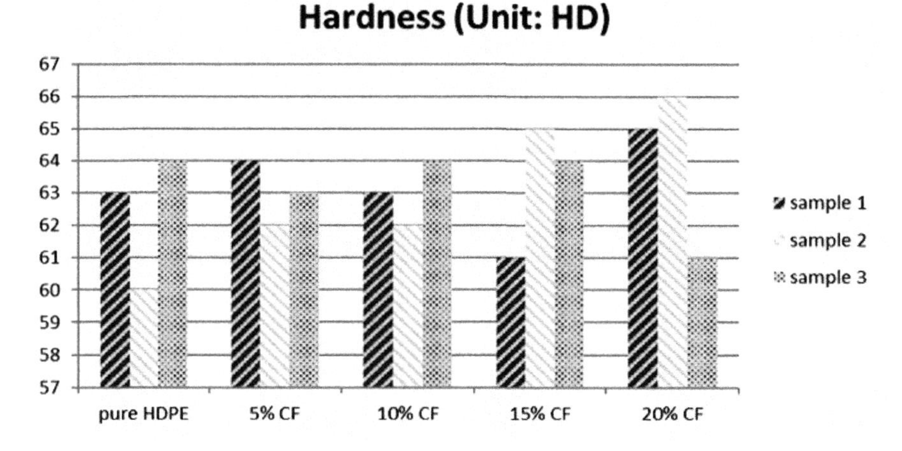

Figure 14.5 Shore D hardness results of the developed samples.

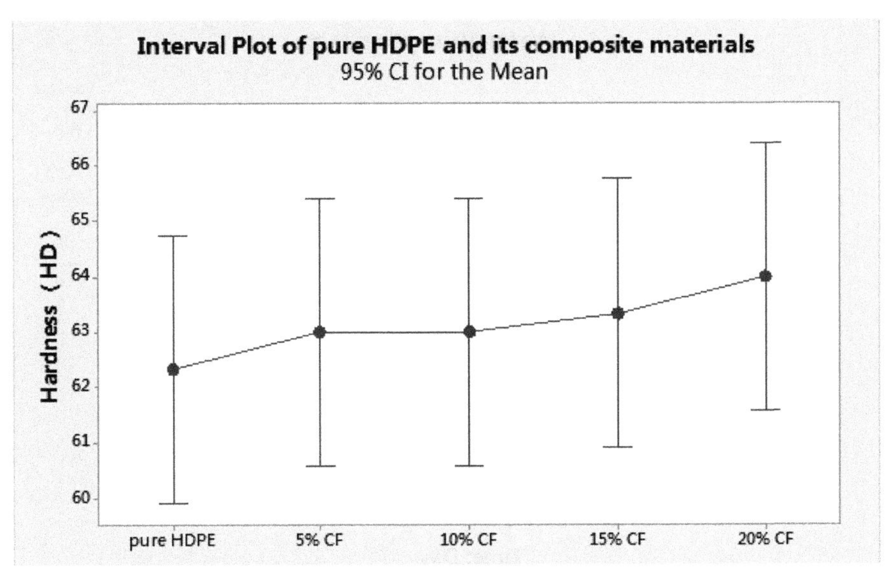

Figure 14.6 Average values of the Shore D hardness of the samples.

In actuality, the hardness of HDPE is not significantly affected by a rise in fibre content. However, Sathish et al. claimed that a natural fibre content of 20% to 40% would raise the sample's hardness, whereas a concentration of 50% would cause a reduction [47]. As a result, future research will likely examine the hardness of composites with increasing coconut fibre content. Despite this, a harder tile will generally be of greater quality; hence, the 20% CF hybrid specimen performs the best in this testing.

14.5 WATER ABSORPTION RATE

As a part of the ASTM-D570 water immersion test, the specimens are put through a constant increase in water absorption at an ambient room temperature of 23°C. The HDPE sample (HCHDPE4) reinforced with 20% coconut fibre has the maximum retention of water (4.78%) in this experiment. Figure 14.7 illustrates how samples containing 15% (HCHDPE3) and 20% (HCHDPE4) of coconut fibre show greater water absorption. However, for floor tiles, the lower the water absorption, the better the quality of the floor tiles. Therefore, in this test, the 10% coconut fibre composite sample had the lowest water absorption (0.39%) and best met the requirements for a good floor tile.

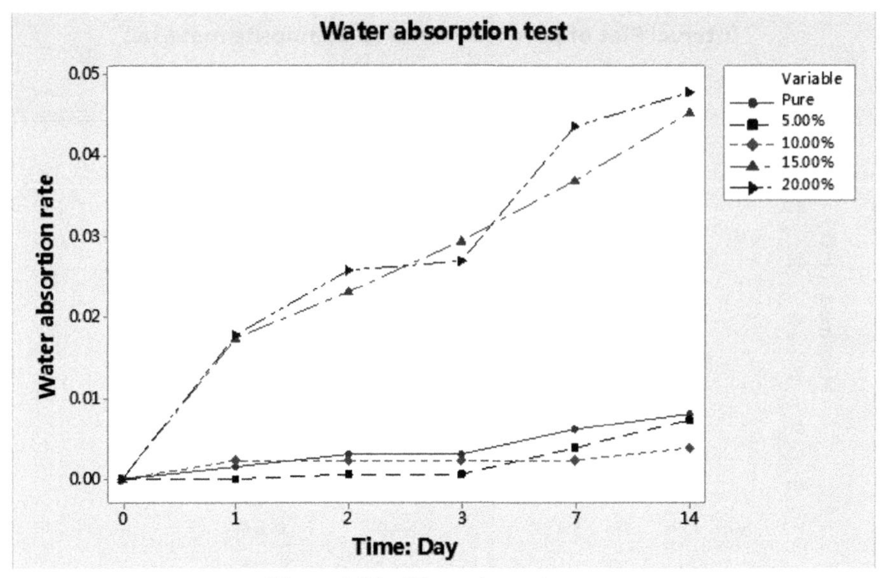

Figure 14.7 Water absorption rate.

14.6 CONCLUSION

One method of decreasing plastic waste is to recycle it and using it to make new things. Plastic garbage, for instance, may be utilised to create building and industrial goods, such floor tiles. The use of floor tiles created from waste plastic will continue in the future due to their cheap cost and high level of durability. In the study discussed in this chapter, the HDPE (HCHDPE4) composite reinforced with 20% coconut fibre showed the highest tensile and flexural modulus and the highest hardness, but the material had a high water absorption rate due to the hydrophilic nature of the coconut fibre. Neat DPE showed good mechanical properties, but the mechanical properties of HDPE deteriorated after multiple recycling.

Therefore, considering various factors, HDPE (HCHDPE2) composite reinforced with 10% coconut fibre is the most suitable material for floor tiles as it has the highest tensile, flexural and impact strengths than HDPE, is harder than HDPE and demonstrates the lowest water absorption. The contamination and adverse effects resulting from HDPE scrap may be mitigated if floor tiles created from discarded HDPE plastic, and HDPE composites can be produced in significant numbers. In an age of long-term viability, products produced from recyclable and reusable materials may have a positive impact on the economy and the environment.

REFERENCES

[1] Seymour, R.B., and Cheng, T. 1986. History of Polyolefins. Springer, eBooks. https://doi.org/10.1007/978-94-009-5472-4.

[2] Natrayan, L. and Santhosh, M.S. 2021. Effect of pineapple/coconut sheath fiber reinforced with polyester resin matrix on mechanical and microstructure properties of hybrid polyester composite. *In*: Rajmohan, T., Palanikumar, K., Davim, J.P. (eds). Advances in Materials and Manufacturing Engineering. Springer Proceedings in Materials. 7. Springer, Singapore. https://doi.org/10.1007/978-981-15-6267-9_37.

[3] Tran, T.N., Mai, B.T., Setti, C. and Athanassiou, A. 2020. Transparent bioplastic derived from CO_2-based polymer functionalized with oregano waste extract toward active food packaging. ACS Appl. Mater. Interfaces. 12(41): 46667–46677.

[4] Parcesepe, E., De Masi, R.F. Lima, C., Mauro, G.M., Pecce, M.R. and Maddaloni, G. 2021. Assessment of mechanical and thermal properties of hemp-lime mortar. Materials (Basel, Switzerland). 14(4): 882. doi:10.3390/ma14040882.

[5] Sarkar, F., Akonda, M. and Shah, D.U. 2020. Mechanical properties of flax tape-reinforced thermoset composites. Materials (Basel, Switzerland). 13(23): 5485. doi:10.3390/ma13235485.

[6] Rathinasabapathy, S., Santhosh, M.S. and Asokan, M. 2020. Significance of boron nitride in composites and its applications. IntechOpen. doi: 10.5772/intechopen.81557.

[7] Wang, J.-M., Wang, H., Chen, E.-C., Chen, Y.-J. and Wu, T.-M. 2020. Enhanced photodegradation stability in poly(butylene adipate-co-terephthalate) composites using organically modified layered zinc phenylphosphonate. Polymers. 12(9): 1968. doi:10.3390/polym12091968

[8] Nyathi, B. and Chamunorwa, A.T. 2020. Overview of legal and policy framework approaches for plastic bag waste management in african countries. J. Environ. Public Health. 8892773. doi:10.1155/2020/8892773

[9] Schrader, P., Gosch, A., Berer, M. and Marzi, S. 2020. Fracture of thin-walled polyoxymethylene bulk specimens in Modes I and III. 2020. Materials (Basel, Switzerland). 13(22): 5096. doi:10.3390/ma13225096

[10] Kovačević, Z., Flinčec Grgac, S. and Bischof, S. 2021. Progress in biodegradable flame retardant nano-biocomposites. Polymers. 13(5): 741. doi:10.3390/polym13050741

[11] Singh, R.K., Ruj, B., Sadhukhan, A.K. and Gupta, P. 2020. Conventional pyrolysis of plastic waste for product recovery and utilization of pyrolytic gases for carbon nanotubes production. Environ. Sci. Pollut. Res. Int. 10.1007/s11356-020-11204-1. doi:10.1007/s11356-020-11204-1

[12] Xanthos, D. and Tony, R.W. 2017. International policies to reduce plastic marine pollution from single-use plastics (plastic bags and microbeads): A review. Mar. Pollut. Bull. 118(1–2): 17–26. doi:10.1016/j.marpolbul.2017.02.048

[13] Elango, N., Ramesh, S., Kalaimani, M., Saravanakumar, N., Anto Dilip, A. and Abdul Rahim Sadiq Batcha. 2023. Enhanced mechanical, tribological, and acoustical behavior of polyphenylene sulfide composites reinforced with zero-dimensional alumina. J. Appl. Polym. Sci., 140(16): e53748.

[14] Sanes, J., Sánchez, C., Pamies, R. and Avilés, M.-D. 2020. Extrusion of polymer nanocomposites with graphene and graphene derivative nanofillers: An overview of recent developments. Materials (Basel, Switzerland). 13(3): 549. doi:10.3390/ma13030549.

[15] Chen, T., Zhang, Y., Yang, J., Cong, G., Jiang, G. and Li, G. 2021. Behavior strategy analysis based on the multi-stakeholder game under the plastic straw ban in China. Int. J. Environ. Res. Public Health. 18(23): 12729. doi:10.3390/ijerph182312729

[16] Napper, I.E. and Richard, C.T. 2020. Plastic debris in the marine environment: History and future challenges. Global challenges (Hoboken, N.J.). 4(6): 1900081. doi:10.1002/gch2.201900081

[17] Nyathi, B. and Chamunorwa, A.T. 2020. Overview of legal and policy framework approaches for plastic bag waste management in African countries. J. Environ. Public Health. 8892773. doi:10.1155/2020/8892773

[18] Barboza, L.G.A., Vethaak, A.D., Lavorante, B.R., Lundebye, A.K. and Guilhermino, L. 2018. Marine microplastic debris: An emerging issue for food security, food safety and human health. Mar. Pollut. Bull. 133: 336–348. doi:10.1016/j.marpolbul.2018.05.047

[19] Rajendran Royan, N.R., Leong, J.S., Chan, W.N., Tan, J.R. and Shamsuddin, Z.S.B. 2021. Current state and challenges of natural fiber-reinforced polymer composites as feeder in FDM-based 3D printing. Polymers. 13(14): 2289. doi:10.3390/polym13142289.

[20] Peng, W., Peng, Z., Tang, P., Sun, H., Lei, H., Li, Z., et al. 2020. Review of plastic surgery biomaterials and current progress in their 3D manufacturing technology. Materials (Basel, Switzerland). 13(18): 4108. doi:10.3390/ma13184108.

[21] Amjadi, M. and Ali, F. 2020. Tensile Behaviour of High-Density Polyethylene Including the Effects of Processing Technique, Thickness, Temperature, and Strain Rate. Polymers. 12(9): 1857. doi:10.3390/polym12091857.

[22] Chao, H.-W., Chen, H.-H. and Chang, T.-H. 2021. Measuring the complex permittivities of plastics in irregular shapes. Polymers. 13(16): 2658. doi:10.3390/polym13162658.

[23] Renner, J.S., Mensah, R.A., Jiang, L., Xu, Q., Das, O. and Berto, F. 2021. Fire behavior of wood-based composite materials. Polymers. 13(24): 4352. doi:10.3390/polym13244352

[24] Amjadi, M. and Ali, F. 2020. Tensile behavior of high-density polyethylene including the effects of processing technique, thickness, temperature, and strain rate. Polymers. 12(9): 1857. doi:10.3390/polym12091857

[25] Palanikumar, K., Elango, N., Kalaimani, M., Chun, K.A. and Gérald, F. 2023. Targeted pre-treatment of hemp fibers and the effect on mechanical properties of polymer composites. Fibers. 11(5): 43. https://doi.org/10.3390/fib11050043.

[26] Elango, N., Santhosh, M.S., Kalaimani, M., Sasikumar, R., Saravana, K.N. and Dilip, A.A. 2022. Mechanical and wear behaviour of PEEK, PTFE and PU: review and experimental study. J. Polym. Eng. 42(5): 407–417. https://doi.org/10.1515/polyeng-2021-0325.

[27] Al-Talib, A.A., Chen, R.S., Natarajan, E. and Chai, A.D. 2023. Improved mechanical properties and use of rice husk-reinforced recycled thermoplastic composite in safety helmets. pp 17–30. *In*: Natarajan, E., Vinodh, S. and Rajkumar, V. (eds). Materials, Design and Manufacturing for Sustainable Environment. Lect. Notes Mech. Eng. Springer, Singapore. https://doi.org/10.1007/978-981-19-3053-9_2.

[28] Hassan, T., Jamshaid, H., Mishra, R., Khan, M.Q., Petru, M., Novak, J., et al., 2020. Acoustic, mechanical and thermal properties of green composites reinforced with natural fibers waste. Polymers. 12(3): 654. doi:10.3390/polym12030654.

[29] Kamal, R.S., Shaban, M.M., Raju, G. and Farag, R.K. 2021. High-density polyethylene waste (HDPE)-waste-modified lube oil nanocomposites as pour point depressants. ACS omega. 6(47): 31926–31934. doi: 10.1021/acsomega.1c04693.

[30] Dabees, S., Tirth, V., Mohamed, A. and Kamel, B.M. 2021. Wear performance and mechanical properties of MWCNT/HDPE nanocomposites for gearing applications. J. Mater. Res. Technol. 12: 2476–2488. https://doi.org/10.1016/j.jmrt.2020.09.129.

[31] Fang, H., Tan, P., Du, X., Li, B., Yang, K. and Zhang, Y. 2021. Mechanical response of buried HDPE double-wall corrugated pipe under traffic-sewage coupling load. Tunnelling Underground Space Technol. 108: 103664.

[32] Koffi, A., Demagna, K. and Lotfi, T. 2021. Mechanical properties and drop-weight impact performance of injection-molded HDPE/birch fiber composites. Polymer Testing. 93: 106956.

[33] Mendes, J.F., Martins, J.T., Manrich, A., Luchesi, B.R., Dantas, A.P.S., Vanderlei, R.M., et al. 2021. Thermo-physical and mechanical characteristics of composites based on high-density polyethylene (HDPE) e spent coffee grounds (SCG). J. Polym. Environ. 29(9): 2888–2900. https://doi.org/10.1007/s10924-021-02090-w.

[34] Silva, J.d.S., Mendes, J.d.O., Costa, R.S.D., Oliveira, A.R.F., Braga, M.D.M. and Mesquita, R.O. 2021. Development of soybean plants using a substrate based on green coconut fiber. Agronomía Colombiana. 39(1): 47–58. https://doi.org/10.15446/agron.colomb.v39n1.91203.

[35] Mukhlis, M., Witono, H. and Rulan, M. 2021. The effect of treatment of coconut fiber with liquid smoke on mechanical properties of composite. E3S Web of Conferences. 328. EDP Sciences.

[36] Ghufron, M. 2021. Effect of sonification in the alkalization process of coconut fiber to improve fiber strength. IOP Conf. Ser.: Earth Environ. Sci. 743(1): 012041. IOP Publishing.

[37] Prabhudass, J.M., Palanikumar, K., Elango, N. and Kalaimani, M. 2022. Enhanced thermal stability, mechanical properties and structural integrity of MWCNT filled bamboo/Kenaf hybrid polymer nanocomposites. 15(2): 506. https://doi.org/10.3390/ma15020506.

[38] Sandy, M.K., Ervan, W. and Desi, H. 2021. Coconut fiber extraction using soda pulping method as green corrosion inhibitor for ASTM A36 steel. 2049. J. Phys. Conf. Ser. 2049(1). IOP Publishing.

[39] Barcala, L., Macanan, A., Macauba, R. and Ricardo, M.B. 2016. Production and marketing of tiles from recycled high—density polyethylene (Hdpe) plastic and scrap rubber tire. Ani: Letran Calamba Res. Rep. 3(1). Web. 29 December 2016.

[40] Rao, P.V., Chandra, S., Kumar, V.M. and Arun. A. 2019. Fabrication and testing of composite tile made from plastic waste and mineral admixture for aggressive environments. Mater. Today: Proceedings. 15: 90-95.

[41] Elango, N., Mohd Faudzi, A.A., Muhammad Razif, M.R. and Mohd Nordin, I.N.A. 2013. Determination of non-linear material constants of RTV silicone applied to a soft actuator for robotic applications. KEM. 594–595: 1099–1104. https://doi.org/10.4028/www.scientific.net/kem.594-595.1099.

[42] Khalid, H.U., Mokhtar, C.I. and Norlin, N. 2020. Permeation damage of polymer liner in oil and gas pipelines: A review. Polymers. 12(10): 2307.

[43] Hopkin, L., Broadbent, H. and Ahlborn, G.J. 2022. Influence of almond and coconut flours on ketogenic, gluten-free cupcakes. Food Chem.: X. 13: 100182.

[44] Sekar, A. and Gunasekaran, K. 2018. Optimization of coconut fiber in coconut shell concrete and its mechanical and bond properties. Materials. 11(9): 1726.

[45] Zainudin, E.S., Yan, L.H., Haniffah, W.H., Jawaid, M. and Alothman, O.Y. 2014. Effect of coir fiber loading on mechanical and morphological properties of oil palm fibers reinforced polypropylene composites. Polym. Compos. 35(7): 1418–1425. https://doi.org/10.1002/pc.22794.

[46] Ghoneim, M., Yehia, A., Yehia, S. and Abuzaid, W. 2020. Shear strength of fiber reinforced recycled aggregate concrete. Materials. 13(18): 4183. doi: 10.3390/ma1318 4183.

[47] Sathish, T., Mohanavel, V., Velmurugan, P., Saravanan, R., Raja, T., Ravichandran, M., et al., 2022. Investigating Influences of Synthesizing Eco-Friendly Waste-Coir-Fiber Nanofiller-Based Ramie and Abaca Natural Fiber Composite Parameters on Mechanical Properties. Bioinorg. Chem. Appl. 2022: 6557817. doi: 10.1155/2022/6557817. Retraction in: Bioinorg Chem Appl. 2024 Jan 24;2024:9872890. doi: 10.1155/2024/9872890.

Chapter **15**

Emerging Contaminants in Water: An Overview of Causes, Metrics, and Treatment Methods

Elango Natarajan*[1], Ganesh Ramasamy[1,2], Haslinda Abdullah[3], Santhosh Mozhuguan Sekar[1] and A.R. Abd Hamid[1]

[1]Faculty of Engineering, Technology and Built Environment, UCSI University, Kuala Lumpur, Malaysia

[2]Faculty of Business and Communications, INTI International University, Malaysia

[3]Faculty of Management and Strategic Studies, National Defence University of Malaysia

15.1 INTRODUCTION

15.1.1 Water Contamination

Water contamination refers to the presence of harmful substances in water that can make water unsafe for use or consumption. Contaminants can include bacteria, viruses, chemicals or heavy metals. The most common source of water contamination is industrial, where factories release wastewater containing harmful chemicals and pollutants into the environment. Moreover, from agricultural sources, fertilisers and pesticides used in farms can be washed into nearby waterways,

*For Correspondence: elango@ucsiuniversity.edu.my

thus contaminating the water supply. Most sewage treatment plants are designed to remove harmful substances from wastewater; however, some contaminants can still make it into the environment. Oil spills can also contaminate water bodies, making the water unsafe for drinking, swimming and fishing. Similarly, hydraulic fracturing, or fracking, a method of extracting natural gas from shale rock formations, can contaminate groundwater with chemicals and pollutants.

15.1.2 Classification of Water Contamination

Water contamination can be split into four subcategories: physical, chemical, biological and radiological. Physical contamination refers to the visible pollutants that directly affect the properties of water, due to their physical appearance. This includes most of the daily artificial waste, excrement, and other organic substances resulting from landslides and soil erosion. Chemical contamination pollutes the water by affecting the water quality, causing it to be unsuitable or harmful as a habitat and for the daily needs of organisms. Chemical contaminants are artificial, the main sources being chemical emissions from factories, pesticides from livestock and farming, and pharmaceutical and medical wastes. They can also be natural, such as organic compounds released through biological degradation. These contaminants can visible or invisible, depending to their chemical properties [1]. On the other hand, biological contamination involves microorganisms such viruses, bacteria, protozoa, parasites and invasive organisms in the environment as pollutants. These microorganisms may cause harm to organisms upon intake, while invasive species may affect the entire ecosystem of an area. Lastly, radiological contamination is a sub-category of chemical contamination, but its harm is significantly more serious. Hence, the dedicated category. Radioactive elements contain an unbalanced number of protons and neutrons, which results in a high probability of radiation emission that may cause acute radiation syndrome or dermal radiation injuries, and impacts on human physiology, such cancer, leukemia, genetic mutations, cataracts, or even death (Environmental Protection Agency, n.d.). Figure 15.1 shows the various ways of groundwater contamination. Various pathways such as pharmaceuticals, personal care products, pesticides, and industrial chemicals are some of the emerging contaminants worldwide. Discharge from wastewater treatment plants, agricultural runoff, and atmospheric deposition are some routes for these substances to enter the environment [2].

15.1.3 Effects of Water Contamination

The presence of and increase in emerging contaminants, and the resultant water pollution can cause significant negative impacts on the environment. In terms of environmental impacts, this leads to the degradation of aquatic ecosystems and the loss of biodiversity and has caused a decline in fish populations in Malaysian rivers. Additionally, eutrophication is brought on by the discharge of industrial effluents and farm runoff into water bodies. Eutrophication can result in harmful algal blooms and oxygen depletion, both of which can be alarming for the aquatic population or even kill aquatic creatures [3]. Water pollution also impacts agriculture. Polluted water for irrigation may interfere with the ability of plants to absorb nutrients, or have toxic side effects that result in stunted development, low yields, or poor produce. Additionally, livestock consuming polluted crops may get sick or even die.

Water pollution has also had negative impacts on human health, as these emerging pollutants may get into the drinking water supply and endanger our health. Studies have shown that exposure to contaminated water can lead to various diseases such a sinfections, shigellosis and diarrhoea. Moreover, studies have revealed a link between water pollution and a higher risk of illness, especially in regions where people live close to industrial areas and rivers [3]. Endocrine disruptors that exist in some of these pollutants can impact the human reproductive system and result in developmental defects as well. Pollutants also affects the quantity and quality of clean water that is accessible; as the amount of pure water available for human use and consumption decreases, it can cause serious negative effects on both human health and the ecosystem. The fishing and tourism businesses of Malaysia, which rely heavily on clean water and robust ecosystems, may suffer because of the contamination of the country's seas as well.

Similarly, microplastics can cause a detrimental impact on waters through the pollution of aquatic ecosystems. Microplastics can be ingested by small aquatic animals—microplastics have been discovered in the intestines and stomachs of fish, which may result from accidental or intentional ingestion of marine fauna and other smaller fish [4]. The ingestion of microplastics by marine animals is harmful as the contaminant may leach into the organs of marine fauna, into the fish and subsequently into humans, upon consumption [5, 6]. This issue is worrying as microplastics have the affinity to adsorb contaminants from the aquatic environment, such as dioxin, polychlorinated biphenyls (PCBs), polybrominated diphenyl ethers (PBDEs), etc. [7]. Ergo, solution through treatment technology is necessary tackle this escalating issue. To summarise, water contamination can have numerous negative effects on human health, the environment and the economy. Some of the most common effects of water contamination include:

- **Diseases:** Water-borne diseases can be caused by bacteria, viruses and parasites that are found in contaminated water.
- **Ecosystem damage:** Water contamination can harm aquatic ecosystems, including fish populations, wetlands and coral reefs.
- **Economic loss:** Water contamination can lead to lost productivity, increased healthcare costs, and damage to infrastructure.

15.1.3.1 Global Research on Water Contamination

Lin et al. [8] investigated the effects of water contamination on the human diseases and health hazards, and reported that the diseases caused by contaminants in water mostly vary based on age, climate and temperature changes. Also, diarrhoea is the most common disease caused by enteroviruses developing in the aquatic environment. Levallois and Villanueva [9] investigated the impact of drinking water quality on human health and interestingly reported that there were numerous water contaminants which cause good as well adverse impacts on human health. Notably, minerals like calcium, zinc and copper in the drinking water improve bone strength, boost immunity and enhance red cell production.

Warren [10] reported the importance of establishing new methodologies for treating inorganic and organic pollutants that are predominantly important, present high toxicity and persistence, and are difficult to treat using current methodologies. Dwivedi [11] reported in his research on fresh water around India that over 70% of the fresh water in liquid form was becoming unfit for consumption due to the addition of contaminants. Veronica [12] investigated and reported that 50% of the drinking water comes from wells in the United States. Hence, groundwater contamination due to industrial wastes directly affects the quality of drinking water.

Water pollution research by the World Health Organization (WHO) reports that contaminated water and poor sanitation are linked to the transmission of diseases such as cholera, diarrhoea, dysentery, hepatitis A, typhoid and polio. Absent, inadequate or inappropriately managed water and sanitation services expose individuals to preventable health risks. This is particularly the case in healthcare facilities where both patients and staff are placed at additional risk of infection and disease when water, sanitation and hygiene services are lacking. Globally, 15% of patients develop an infection during a hospital stay, with the proportion much greater in low-income countries.

15.1.3.2 Sustainable Development Goals (SDGs) and the United Nations' Views on Water Contamination

The Sustainable Development Goals (SDGs) adopted by the United Nations General Assembly (UNGA) act as a shared blueprint to guide countries and organisations to form a global partnership towards achieving peace and prosperity. The SDGs address global concerns such as education, poverty, and clean water and sanitation. Furthermore, SDGs are universal and are applied to the global community regardless of race, religion and background [13]. The SDGs also have lists of targets and indicators whereby each goal has 8 to 12 targets and each target has 1 to 4 indicators to evaluate the progress of achieving the said targets and goals [14]. Currently, the SDGs have a total of 169 targets and 231 indicators to monitor the progress towards the SDGs and the expected outcomes for a more sustainable world and community. However, the indicator framework is extensively reviewed from time to time, whereby some indicators are replaced, revised or deleted to reduce the difficulties of measuring the progress towards the SDGs. In relation to water contamination and management, some of the SDGs are as follows:

Goal 3: Good health and well-being

Goal 6: Clean water and sanitation

Goal 12: Responsible consumption and production

Goal 14: Life below water

According to the World Health Organization (WHO), water pollution is responsible for 1.2 million deaths each year. This includes deaths from diarrhoea, cholera, dysentery, typhoid fever, and other water-borne diseases. The United Nations Children's Fund (UNICEF) estimates that 2.2 billion people do not have access to safely managed drinking water. In 2017, an estimated 842,000 deaths were attributable to unsafe water, sanitation, and hygiene. In the United States, the Center for Disease Control and Prevention (CDC) estimates that 4.5 million people get sick from contaminated water each year. Of these, 180,000 are hospitalised and 3,000 die.

15.2 SOURCES OF WATER CONTAMINATION

Water contamination can be divided into three major categories. The first category of emerging contaminants are the compounds that were just recently introduced into the environment. The second category includes substances that may have been present in the environment for a long time in the past but have been discovered only recently, and their significance only recently began to draw attention. The third category includes substances that have been around for a while but whose potential harm to people and the environment has only lately come to light [15].

Microplastics are tiny plastic fragments, typically less than 5 millimeters in size, that are produced when larger plastic objects break down or when microbeads from personal care products are released into the environment. Due to their pervasive presence in the ecosystem and possible harm to aquatic creatures, they have grown to become a significant environmental problem.

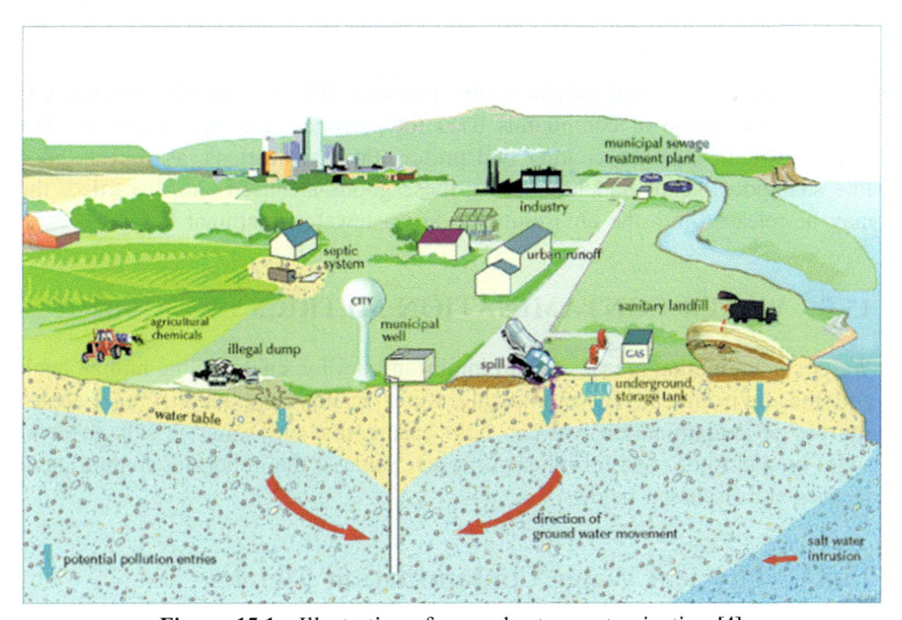

Figure 15.1 Illustration of groundwater contamination [4].

Microplastics can infiltrate the environment from several sources, including plastic waste, wastewater discharge, and industrial effluents. On the other hand, chemicals found in pharmaceutical and personal care products infiltrate the environment through a few different channels, including wastewater treatment facilities and agricultural runoff. The following are some of the causes of pharmaceutical and microplastic contaminants in waters.

- **Domestic and industrial wastewater:** Domestic and industrial wastewater is a significant source of emerging contaminants. Wastewater treatment plants (WWTPs) are not designed to remove all emerging contaminants, which means that these chemicals and particles can enter the water cycle.
- **Aquaculture:** Techniques used in aquaculture, such as using fish feed laced with microplastics and prescription drugs, can add to the discharge of these pollutants into waterways.
- **Plastic waste:** When plastic waste is disposed of improperly, microplastics are released into the environment and may ultimately find their way into waterways.
- **Industrial pollution:** Due to Malaysia's quick industrialisation, there are now more industrial pollutants than before. These pollutants include heavy metals, hazardous chemicals and organic compounds that are released into the water by industrial activities like mining, manufacturing and processing.
- **Runoff from agriculture:** Agriculture is a significant business in Malaysia, and the use of fertilisers and pesticides in farming operations can result in runoff that contains pollutants including nitrates, phosphates and insecticides. Antibiotics, hormones and other veterinary medications can also find their way into water supplies through animal waste from livestock and poultry farms.

Pharmaceuticals and personal care products (PPCPs) are an umbrella term for a diverse group of compounds used for personal care and cosmetics. They consist of both over the counter and prescription medications, as well as goods like shampoos, soaps and lotions. As these goods are used and discarded, they may be released into the environment via wastewater treatment facilities.

15.3 WATER CONTAMINATION METRICS

Water contamination metrics are an important tool for safeguarding human health and the environment. By monitoring water quality and identifying potential sources of contamination, we can help ensure that our water is safe to drink, swim in, and to be used for other purposes. Some of the important contamination metrics are discussed below.

15.3.1 Physical Water Contamination Metrics

These metrics measure the physical properties of water, such as turbidity, temperature and pH. Turbidity is a measure of the cloudiness of water, and it can be caused by suspended particles or dissolved organic matter. Temperature can affect the solubility of chemicals in water, and pH can affect the toxicity of some chemicals. The colour of water can be affected by the presence of dissolved organic matter, minerals, or algae. Some colours, such as brown or red, can indicate the presence of pollutants. The taste of water can be affected by the presence of dissolved minerals, algae, or bacteria. Certain tastes, such as metallic or salty, can indicate the presence of pollutants.

15.3.2 Chemical Water Contamination Metrics

These metrics measure the presence of chemicals in water, such as heavy metals, pesticides, and fertilisers. Heavy metals are a group of metals that are toxic to humans and animals. They can be found in water as a result of industrial waste, agricultural runoff, and from natural sources. Some of the most common heavy metals found in water include lead, mercury, cadmium and arsenic. Pesticides are chemicals that are used to kill pests such as insects, rodents and weeds. They can be found in water as a result of agricultural runoff and improper disposal of pesticides. Some of the most common pesticides found in water include DDT, chlorpyrifos and atrazine.

Fertilisers are chemicals that are used to promote plant growth. They can be found in water as a result of agricultural runoff and improper disposal of fertilisers. Some of the most common fertilisers found in water include nitrogen, phosphorus and potassium. Volatile Organic Compounds (VOCs) are a group of chemicals that evaporate easily. They can be found in water due to industrial emissions, gasoline leaks, and other sources. Some of the most common VOCs found in water include benzene, toluene and ethylbenzene.

15.3.3 Biological Water Contamination Metrics

These metrics measure the presence of bacteria, viruses and other microorganisms in water. Total Coliform Bacteria is a measure of the total number of coliform bacteria in water. Coliform bacteria are a group of bacteria that are found in the intestines of humans and animals. Their presence in water can indicate that the water is contaminated with faecal matter. Faecal Coliform Bacteria is a measure of the number of faecal coliform bacteria in water. Faecal coliform bacteria are a subset of coliform bacteria that are more likely to come from human faecal matter. Their presence in water can indicate that the water is contaminated with sewage. *E. coli* is a type of bacteria that is found in the intestines of humans and animals. It can cause a variety of health problems, including diarrhoea, vomiting, and kidney failure. The presence of *E. coli* in water is a serious health hazard.

Viruses are a group of pathogens that can cause several diseases. They can be found in water, as a result of contamination with sewage and from other sources. Some of the most common viruses found in water include rotavirus, norovirus, and hepatitis A. Algae are a type of plant that can grow in water. Some algae can produce toxins that can be harmful to humans and animals. Algal blooms are a major problem in many water bodies around the world. These are just a few of the biological water contamination metrics that are used to assess the quality of water. By monitoring these metrics, we can help ensure that our water is safe to drink, swim in, and to be used for other purposes.

15.3.4 Radiological Water Contamination Metrics

These metrics measure the presence of radioactive materials in water. Radioactive materials can be harmful to humans and animals, and they can also pollute water bodies. Total alpha activity measures the total amount of alpha radiation emitted by all the radioactive contaminants in a sample of water. Total beta activity measures the total amount of beta radiation emitted by all the radioactive contaminants in a sample of water. Gross alpha count measures the number of alpha particles emitted per minute by a sample of water. Gross beta count is the number of beta particles emitted per minute by a sample of water. Specific activity represents the amount of radioactive material in a sample of water, expressed as the activity per unit volume of water.

The EPA has set maximum contaminant levels (MCLs) for a number of radiological contaminants. These MCLs are expressed in units of picocuries per litre (pCi/L). For example, the MCL for radium-226 in drinking water is 5 pCi/L. If the levels of radiological contaminants in water exceed the MCLs, then the water is considered to be contaminated. The health risks associated with exposure to radiological water contaminants depend on the type of contaminant, the amount of exposure, and the individual's health status. In general, exposure to high levels of radiological water contaminants can increase the risk of cancer, birth defects, and other health problems. Some of the other most common metrics include:

- **Total Suspended Solids (TSS):** This is a measure of the number of suspended particles in water. TSS is a result of a variety of factors, including soil erosion, industrial waste, and agricultural runoff.
- **Total Dissolved Solids (TDS):** This is a measure of the number of dissolved solids in water. TDS comprises a variety of factors, including salts, minerals and organic matter.
- **Biochemical Oxygen Demand (BOD):** This is a measure of the amount of oxygen that is required by bacteria to break down organic matter in water. BOD can be used to assess the level of organic pollution in water.
- **Chemical Oxygen Demand (COD):** This is a measure of the amount of oxygen that is required to oxidise organic matter in water. COD is a more comprehensive measure of organic pollution than BOD.
- **Nitrates:** Nitrates are a form of nitrogen that can be found in water. Nitrates can be harmful to humans and animals, and they can also contribute to the formation of algal blooms.
- **Phosphates:** Phosphates are a form of phosphorus that can be found in water. Phosphates can promote the growth of algae, which can lead to algal blooms.

Water contamination metrics are used to assess the quality of water and to identify potential sources of contamination. They are also used to monitor the effectiveness of water treatment processes. The specific metrics that are used to assess water quality can vary depending on the source of water, the intended use, and regulatory requirements.

15.4 WATER CONTAMINATION METRICS

The type of water treatment method used depends on the specific contaminants present in water. For example, if the water is contaminated with suspended particles, then coagulation and flocculation would be used. If the water is contaminated with bacteria, then it is disinfected. In addition to these common methods, there are a number of other water treatment methods that are used in specific applications. For example, distillation is used to remove dissolved salts from water, and reverse osmosis is used to remove dissolved ions from water.

The choice of water treatment method also depends on the intended use of water. For example, water to be used for drinking must be treated to remove all harmful contaminants. Water used for irrigation, however, may not need to be treated as thoroughly. Water treatment is an important process that helps ensure that water is safe for human consumption and use. By using a variety of water treatment methods, it is possible to remove a wide range of contaminants from water, making it safe and healthy for people to use. Some of the most common water contaminants (American continent) are depicted in Figure 15.2.

15.4.1 Coagulation and Flocculation

Coagulation and flocculation are two important water treatment processes that are used to remove suspended particles from water. Coagulation is the first step in the process, and it involves adding a chemical, called a coagulant, to water. The coagulant causes the particles to clump together, forming larger particles called flocs. Flocculation is the second step in the process; it involves agitating the water to help the flocs form and settle down.

The coagulants that are used in coagulation and flocculation are typically inorganic chemicals, such as alum, ferrous sulfate and polyaluminum chloride. These chemicals work by neutralizing the charges on the particles, which causes them to clump together. The flocs that are formed are large enough to settle down in the water, allowing the clear water to be separated from the flocs.

Coagulation and flocculation are used in a variety of water treatment applications, including drinking water treatment, wastewater treatment, and industrial water treatment. These processes are an important part of ensuring that water is safe for human consumption and use. Here are some of the benefits of using coagulation and flocculation in water treatment:

- It is a relatively inexpensive and effective way to remove suspended particles from water.

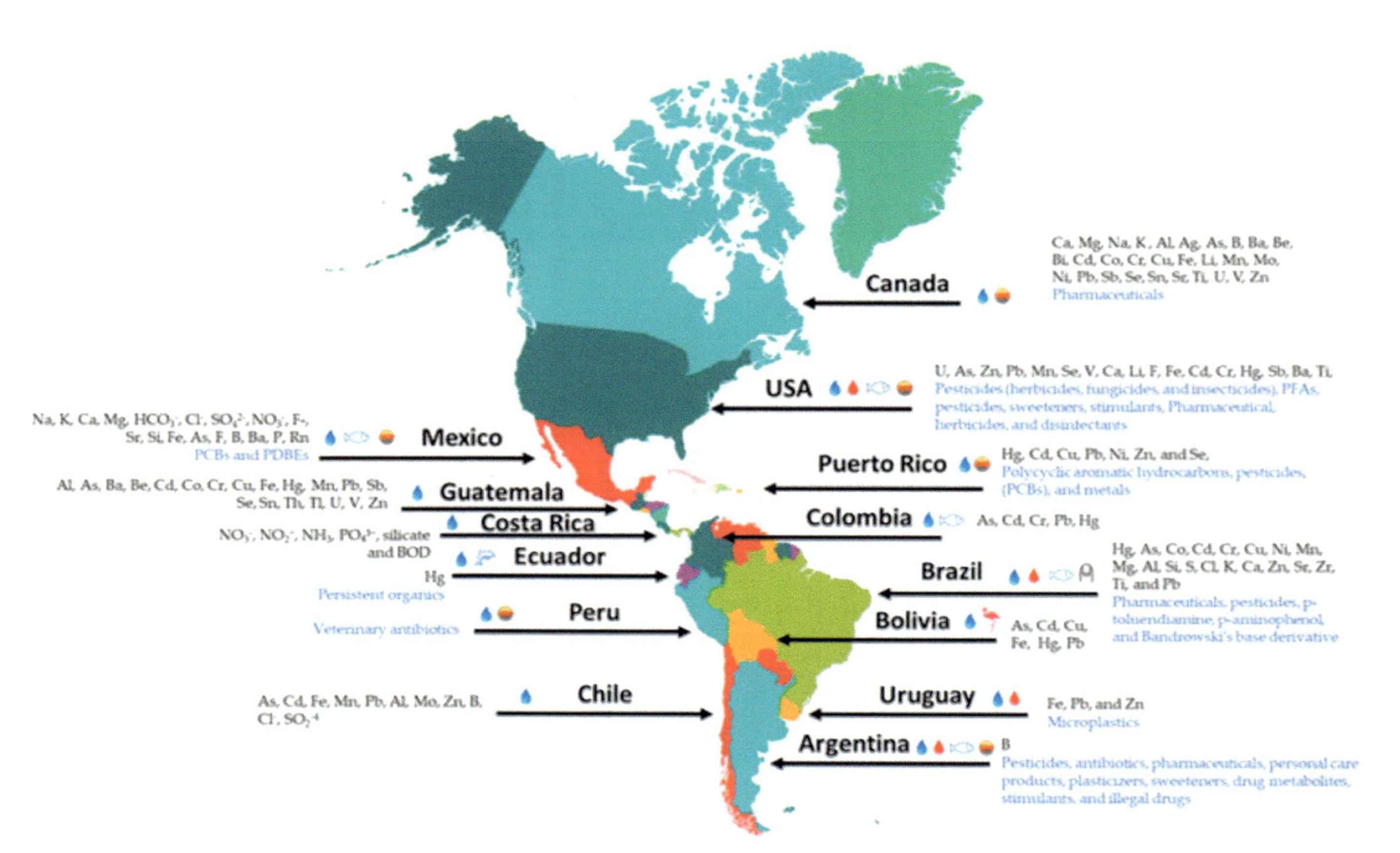

Figure 15.2 Common pollutants detected in the American continents [10].

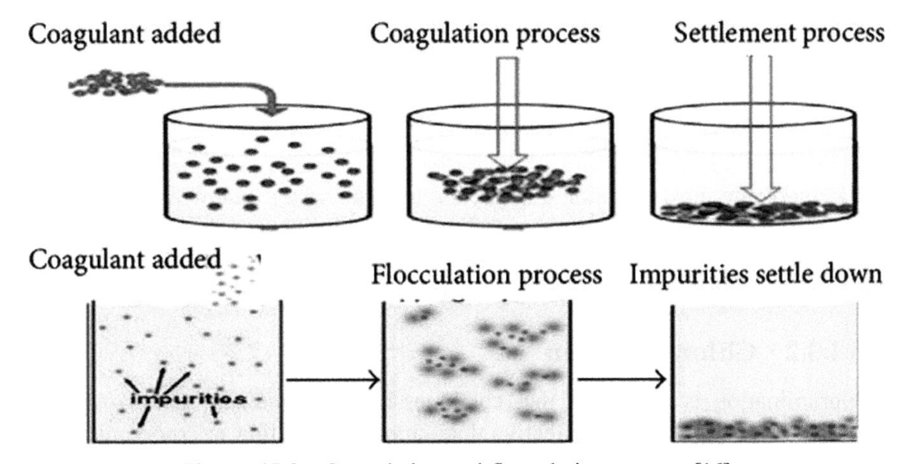

Figure 15.3 Coagulation and flocculation process [16].

- It can be used to remove a wide range of suspended particles, including organic matter, inorganic matter, and bacteria.
- It is a relatively simple process to operate and maintain.

15.4.2 Filtration Process

Filtration is a water treatment process that removes suspended particles from water by passing it through a filter medium. The filter medium can be made up of a variety of materials, including sand, gravel, activated carbon, and membrane filters. There are two main types of filtration:

- **Granular filtration:** This type of filtration uses a bed of granular material, such as sand or gravel, to remove suspended particles from water. The particles are trapped in the spaces between granules.
- **Membrane filtration:** This type of filtration uses a membrane with very small pores to remove suspended particles from water. The particles of interest are too large to pass through the pores of the membrane.

Filtration is an effective way to remove a wide range of suspended particles from water, including sediments, bacteria, and viruses. It is also a relatively inexpensive and easy-to-operate water treatment process.

15.4.3 Disinfection

Disinfection is a water treatment process that kills pathogenic microorganisms in water. Pathogenic microorganisms are those that can cause diseases. The most common disinfectants used in water treatment are chlorine, chloramine and ozone. Chlorine is the most widely used disinfectant in the world. Chloramine is a combination of chlorine and ammonia. Ozone is a gas that is more effective than chlorine in terms of killing microorganisms, but it is also more expensive.

Disinfection is an important step in water treatment as it helps ensure that water is safe for human consumption and use. The process is effective, inexpensive, and easy-to-operate, but it has certain limitations.

15.4.3.1 Chlorination

Chlorine is the most common disinfectant used for water treatment. It is added to water in the form of chlorine gas, sodium hypochlorite, or calcium hypochlorite. Chlorine kills microorganisms by reacting with their cell walls. Figure 15.4 depicts the chlorination process.

15.4.3.2 Chloramination

Chloramination is a process that combines chlorine and ammonia to disinfect water. Chloramine is more effective than chlorine at killing microorganisms, but is also more expensive. Chloramination is often used in water treatment plants that serve large populations.

15.4.3.3 Ozonation

Ozone is a gas that is more effective than chlorine at killing microorganisms. Ozone is also a strong oxidizing agent, which means that it can break down the organic matter in water. Ozonation is often used in water treatment plants that serve small populations or have special requirements, such as treating water for use in hospitals. The detailed ozonation treatment of contaminated water is depicted in Figure 15.5.

15.4.4 Aeration

Aeration is a water treatment process that adds oxygen to water. This is done by bubbling air through the water, which helps in removing dissolved gases such as carbon dioxide, and in improving the taste of water. Aeration is also used to oxidise dissolved metals such as iron and manganese, and to precipitate them out of the solution. Aeration is an important water treatment process that is used in a variety of applications, including drinking water treatment, wastewater treatment, and industrial water treatment. Some of the benefits of using aeration for water treatment are listed below.

- It helps remove dissolved gases from water, which can improve the taste of water.
- It helps in oxidising dissolved metals, which can improve the clarity of water.
- It helps precipitate dissolved solids, improving the clarity of water.
- It helps in improving the biological activity in water, which helps remove contaminants.

15.5 EVALUATION OF WATER TREATMENT TECHNOLOGIES

It is very important to evaluate water treatment technologies and their effectiveness. Water treatment methods are evaluated based on a number of factors, including:

- **Effectiveness:** The ability of the method to remove contaminants from water
- **Cost:** The cost of the method, including the cost of equipment, chemicals and labour
- **Ease of Operation:** The ease with which the method can be operated and maintained
- **Environmental Impact:** The impact of the method on the environment
- **Public Acceptability:** The acceptability of the method by the public

The relative importance of these factors varies depending on the specific application. For example, the effectiveness of the method is more important for drinking water treatment than for industrial water treatment. The cost of the method is more important for small water treatment plants than for large water treatment plants.

15.6 CONCLUSION

Emerging water contaminants are man-made or naturally occurring substances that are introduced into water bodies. They predominantly exist in low concentrations and are proving to be a wellspring of worry targeted not just at the well-being of citizens but towards the environment as well. Through wastewater treatment plants, some emerging contaminants are reduced in concentration, but a complete elimination of these substances is yet to be achieved. Some emerging contaminants even after undergoing wastewater treatment remain present in substantial amounts and ultimately make their way back into the environment, possibly in even more toxic forms due to their several chemical interactions. Although treatment technologies have seen a great deal of advancement since they were introduced, they still lack in some respects in terms of effectiveness.

Continents all over the world are affected by the prevalence of these new contaminants, some of which include pharmaceuticals and microplastics. The existence of these emerging contaminants is explained by the respective discharges from hospitals and industrial and agricultural establishments, as well as the ineffectiveness of treatments used in wastewater treatment plants. The earth's need for water keeps on increasing with population growth, urbanisation and resource exploitation. The consequences of such pollution can be seen on both the environment and the human health. Analysis through social sciences is useful in determining the growth of these contaminants and establishing possible interventions to control their distribution in water bodies. One way of achieving this is by decreasing the sources of these contaminants by limiting their consumption.

ACKNOWLEDGEMENT

The authors would like to express their gratitude to the Ministry of Higher Education, Malaysia for funding this research project through Fundamental Research Grant Scheme (FRGS), with the project code FRGS/1/2020/SS0/UCSI/02/2.

REFERENCES

[1] Stefanakis, A.I. and Becker, J.A. 2016. A review of emerging contaminants in water: classification, sources, and potential risks. pp. 55–80. *In*: A. McKeown and G. Bugyi (eds). Impact of Water Pollution on Human Health and Environmental Sustainability. IGI Global. https://doi.org/10.4018/978-1-4666-9559-7.ch003.

[2] Arman, N.Z., Salmiati, S., Aris, A., Salim, M.R., Nazifa, T.H., Muhamad, M.S., et al. 2021. A review on emerging pollutants in the water environment: Existences, health effects and treatment processes. Water. 13(22): 3258. https://doi.org/10.3390/w13223258.

[3] Zhang, Y., Shao, T., Yang, X., He, Y., Wu, Z. and Zhang, X. 2019. Removal of organic contaminants from drinking water by activated carbon: A review. J. Water Supply Res. Technol. AQUA. 68(1): 1–20. doi: 10.2166/aqua.2018.020.

[4] EPA. 2015. Getting up to speed ground water contamination. US EPA Seminar Publication. https://www.epa.gov/sites/production/files/2015-08/documents/mgwc-gwc1.pdf.

[5] Nduko, J.M. and Taguchi, S. 2021. Microbial production of biodegradable lactate based polymers and oligomeric building blocks from renewable and waste resources. Front. Bioeng. Biotechnol. 8: 618077. https://doi.org/10.3389/fbioe.2020.618077.

[6] Qiao, R., Deng Y., Zhang S., Wolosker M., Zhu Q., Ren H., et al. 2019. Accumulation of different shapes of microplastics initiates intestinal injury and gut microbiota dysbiosis in the gut of zebrafish. Chemosphere. 236: 124334. https://doi.org/10.1016/J.CHEMOSPHERE.2019.07.065.

[7] Sarkar, D.J., Sarkar, S.D., Manna, R.K., Samanta, S. and Das, B.K. 2020. Microplastics pollution: An emerging threat to freshwater aquatic ecosystem of India. J. Inland. Fish. Soc. India. 52(1): 5–15. https://doi.org/10.47780/jifsi.52.1.2020.106513.

[8] Li Lin, H.Y. and Xiaocang, X. 2022. Effects of water pollution on human health and disease heterogeneity: A review. Front. Environ. Sci. 10: 880246. doi: 10.3389/fenvs.2022.880246.

[9] Levallois, P. and Villanueva, C.M. 2019. Drinking water quality and human health: An editorial. Int. J. Environ. Res. Public Health. 16(4): 631. doi: 10.3390/ijerph16040631.

[10] Warren-Vega, Walter M., Armando Campos-Rodríguez, Ana I. Zárate-Guzmán, and Luis A. Romero-Cano. 2023. A current review of water pollutants in american continent: Trends and perspectives in detection, health risks, and treatment technologies. Int. J. Environ. Res. Public Health. 20(5): 4499. https://doi.org/10.3390/ijerph20054499.

[11] Dwived, A.K. 2017. Researches in water pollution: A review. Int. Res. J. Nat. Appl. Sci. 4(1): 118–142.

[12] Veronica I.P. and Patrick, R. 1983. Ground water contamination in the United States. Science. 221(4612): 713–718. doi: 10.1126/science.6879171.

[13] United Nations. The 17 goals | sustainable development. United Nations. Retrieved March 24, 2023, from https://sdgs.un.org/goals.

[14] Jeong, Y., Gong, G., Lee, H., Seong, J., Hong, S.W. and Lee, C. 2022. Transformation of microplastics by oxidative water and wastewater treatment processes: A critical review. J. Hazard. Mater. 443: 130313. https://doi.org/10.1016/j.jhazmat.2022.130313.

[15] Li, Y., Xu, Q., Yao, J., Han, G. and Zhuang, Y. 2019. Evaluating the effects of urbanization on water quality in the Shaying River Basin, China. Water. 11(4): 845. https://doi.org/10.3390/w11040845.

[16] Fabregat, A., Beyene, H.D., Hailegebrial, T.D. and Dirersa, W.B. 2016. Investigation of coagulation activity of cactus powder in water treatment. 7815903: 2356–7171, https://doi.org/10.1155/2016/7815903.

Chapter 16

Fundamentals and Applications of Geopolymers

Mian Umer Shafiq*[1,2], Maryam Jamil[3], Momna Khan[4], Lei Wang[5], Hisham Ben Mahmud[6] and Talha Mujahid[7]

[1]Chemical and Petroleum Engineering Department, UCSI University, Malaysia

[2]UCSI-Cheras Low Carbon Innovation Hub Research Consortium, Kuala Lumpur, Malaysia

[3]College of Physics and Optoelectronic Engineering, Shenzhen University, China

[4]NFC Institute of Engineering and Technology Pakistan

[5]College of Energy, State Key Laboratory of Oil and Gas Reservoir Geology and Exploitation, Chengdu University of Technology, China

[6]Petroleum Engineering Department, Universiti Teknologi Petronas Malaysia

[7]Mari Gas Company Limited, Mari Field Dharki, Pakistan

16.1 GEOPOLYMERS

Geopolymers belong to a category of inorganic, amorphous aluminosilicate materials that are formed through a process called 'polymerization'. These materials are considered as an alternative material to conventional cement and are important because of their exceptional mechanical characteristics, fire resistance, and environmental advantages over Portland cement. The process of

*For Correspondence: shafiq@ucsiuniversity.edu.my

polymerization needs an activation source consisting of aluminosilicate minerals such as silica fume, fly ash, slug, and metakaolin, and uses some alkaline solutions like potassium or sodium hydroxide or solutions of silicate. Through this activation process, the raw materials undergo a chemical transformation that causes the formation of a three-dimensional polymeric network [1].

16.1.1 Manufacturing of Geopolymers

The process of forming geopolymers closely resembles cementation. Geopolymers are composed of aluminosilicates made up of inorganic ceramic polymers that are cross-linked with alkali metal ions. In the fabrication process, a low water content is employed ($H_2O/M_2O \sim$ 10–25 wt%) to promote the formation of an amorphous geopolymer structure, rather than crystalline zeolites. This is important as the formation of zeolites would result in the production of hydroceramic waste forms. By controlling the water content, the desired amorphous geopolymer structure can be achieved, offering advantages in terms of material properties and waste management [2]. To represent the geopolymer matrix, a nominal composition of $4SiO_2 \bullet Al_2O_3 \bullet M_2O$ is commonly used, although the Si:Al ratio may vary depending on the intended application, ranging from 1 to 3. For applications such as cement and concrete, a ratio of 2:1 is typically used [3]. The alkali component in geopolymers can be sodium (Na), potassium (K), or cesium (Cs). Geopolymers demonstrate excellent binding properties at low temperatures, making them more environmentally friendly compared to cement waste forms. Unlike the clinkering process required for cement production at temperatures of 1,400–1,500°C, the starting materials for geopolymers need to be heated to only approximately 700°C. This lower temperature requirement contributes to the environmental acceptability and energy efficiency of geopolymers as binder materials. Geopolymers and geopolymeric cement, which include fly ash-based geopolymeric concrete, offer exceptional suitability for various environmental applications, such as permanent storage of radioactive species and some hazardous wastes [4]. Geopolymers can be deployed as capping materials, sealants, barriers, and other essential structures required at containment sites. Their excellent durability, chemical resistance, and ability to immobilize harmful substances make them ideal for effectively managing and protect from hazardous materials/environments like electrical hazards, fire hazards, and hazardous chemicals. Geopolymers present a promising solution for environmental challenges, providing long-term stability and protection in critical applications involving the encapsulation of radioactive or hazardous waste.

In Europe, pilot-scale demonstrations have been conducted on mining wastes and uranium mill tailings to explore the potential of geopolymers. Geopolymers were initially studied in the mid to late 1990s for the disposal of radioactive wastes [2]. Since then, further research has been conducted to investigate additional applications of geopolymers in this field. These ongoing investigations aim to expand our understanding of how geopolymers can be effectively utilized for various radioactive waste management applications. By exploring these applications, we can continue to develop safe and efficient methods for the disposal of radioactive waste using geopolymers. Geopolymers with Si:Al ratios of 1:1 and 2:1 have been explored for the stabilization of hazardous Resource Conservation and Recovery Act (RCRA) metals, including Se, Ni, Hg, Ba, Cr, Cd, and Pb. To simulate the treatment process, the RCRA spike was created by mixing the RCRA components at a concentration 60 times higher than the Universal Treatment Standards (UTS) limits [2]. It is important to note that the mixture used for this purpose was highly acidic, with a pH below 1. The research discussed in this chapter aims to assess the effectiveness of geopolymers in stabilizing and immobilizing hazardous RCRA metals under extreme conditions, providing valuable insights for hazardous waste management and remediation efforts.

In the study, the RCRA simulant was used to replace half of the 10 wt% water in the geopolymer formulation. Surprisingly, the resulting geopolymers were found to meet the liquid percentage limits established by the Environment Protection Agency Toxicity Characteristic Leaching Procedure (EPA TCLP) test, even though they contained concentrations of RCRA components 60 times higher than the UTS limits. However, it remains unclear whether the RCRA components interacted with the geopolymer matrix, i.e., whether the process involved encapsulation or embedding of the hazardous constituents. Further investigation is necessary to fully understand the mechanisms by which geopolymers effectively immobilize hazardous elements and whether long-term stability can be achieved. Nonetheless, these findings present promising possibilities for the use of geopolymers in the stabilization and safe disposal of hazardous RCRA metals. ANSTO conducted a study on geopolymers derived from metakaolin and alkaline silicate solutions, with nominal Na/Al and Si/Al molar ratios of 1 and 2. The aim of the study was to investigate the potential of these geopolymers for the stabilization of 37Cs and 90Sr. Through transmission electron microscopy analysis, it was observed that these geopolymers exhibited an amorphous structure on a scale of approximately 1 nm, after curing at 40 °C. Within the amorphous phase, Cs was found to be uniformly distributed. However, Sr was only partially incorporated into the amorphous phase and showed a preference for partitioning into crystalline $SrCO_3$. These findings shed light on the behavior of 37Cs and 90Sr in geopolymers, highlighting the potential for their effective stabilization and immobilization. Further research in this area can contribute to the development of advanced waste management strategies for radioactive isotopes [5]. Geopolymer formulations, commercially known as DuraLith, have been patented for the stabilization of 129I and 99Tc. Testing has shown promising results, particularly in the retention of technetium when rhenium is used as a surrogate for Tc. However, the same level of success has not been achieved for iodine [6].

To address concerns such as radiolytic H_2 production and freeze-thaw problems, geopolymers can be heated at about 300°C. This effectively removes the said issues without significantly impacting the strength or leachability of the geopolymers. This thermal treatment provides a practical solution to enhance the performance and stability of geopolymers in various applications. Geopolymers have also demonstrated excellent fire resistance, making them a reliable choice for fire protection and prevention [7]. Their ability to withstand high temperatures and resist fire further contributes to their suitability for a wide range of applications, including those in which fire safety is crucial.

16.1.2 Geopolymer Synthesis

Geopolymers are formed by the polymerization of aluminosilicate substances mixed in an alkali activator solution at high temperatures or room temperature, producing a three-dimensional silicaaluminate network structure and amorphous phase [8]. However, research show that different reaction procedures are involved in polymerization. It can be broken down into three phases.

- The disintegration of aluminosilicate substances in the alkali solution produces the alumina tetrahedron unit and free silica.
- The movement, solidification of substances, the condensation procedure of silica hydroxyl, and alumina yield inorganic geopolymer in the gel phase. Due to the hydrolysis process, water is removed from this structure.
- As the gel phase solidifies, it condenses to produce a geopolymer.

16.2 THE SOURCE OF MATERIAL

16.2.1 Kaolin-based Geopolymers

Clay is a natural material source with very small particles (<2 μm). It is a silicate that is composed of layers of octahedral alumina and tetrahedral oxygen. Kaolin is the most common clay mineral, followed by zeolite, and so on. Kaolin, also known as dolomite, is a white and soft fine clay with fire resistance and good plasticity. Through dehydration, Kaolin is converted into Metakaolin (anhydrous aluminum silicate) at high temperatures. Geopolymers based on metakaolin exhibit robust bonding, high compressive strengths, and remarkable thermal insulating properties [9]. Researchers have attempted to preserve these exceptional mechanical properties while reducing costs and ensuring stability for recycling [10]. Studies indicate that kaolinite-based geopolymers possess dual functionalities of noise reduction and heat insulation [11]. Zeolites, composed of alkali earth metals or alkali metals, exhibit ion exchange capabilities, acid resistance, heat resistance, catalytic properties, and adsorption capacities. The geopolymers produced by the alkali excitation of natural zeolite retain the properties of porous arranged zeolite, while having the mechanical properties of geopolymer gel [12].

16.2.2 Laterite Soil Geopolymer

Laterite soil is rich in aluminum minerals, iron, and aluminosilicates. It is used in traditional brick, buildings and roads due to its corrosion resistance. It possesses high mechanical strength properties [13, 14]. The microstructure and mechanical properties of laterite soil are affected by the molar oxide ratio of silica and aluminum [15]. Polymers with high strength are formed by combining solid wastes with laterite. These mixed geopolymers have good usage in non-loading bearing building applications [16]. The other natural minerals are mullite, diatomite, bauxite, halloysite, bentonite, etc.

16.3 GEOPOLYMER FROM WASTE PRODUCTS

16.3.1 Blast Furnace Slag

In iron making, blast furnace slag (BFS) is a by-product that can be obtained at 1500°C [17]. BFS and ground granulated blast furnace slag (GGBS) are both chilled in water at cooling conditions. After grinding, BFS is used as a partial alternative for OPC because of its high hardness, pozzolanic activity, and amorphous nature [18, 19]. It can be used to increase concrete's porosity, permeability, and hydration heat, as well as to strengthen concrete over time, consume less water, and be less reactive to sulfate and alkali silicates [17, 20].

16.3.2 Red Mud

It is a by-product of the Bayer Process, a method for industrially purifying aluminum. The Bayer process dissolves bauxite in sodium hydroxide at high pressure and temperature. A small amount of sodium hydroxide increases the pH value [21]. The raw material is used in mud form; it lowers the time and required energy for drying purposes. Using raw materials with high alkalinity also reduces the overall amount of alkali activator, which lowers the cost of producing geopolymers [22].

16.3.3 Fly Ash

Fly Ash is obtained by burning coal, usually categorized in class C or class F. Class F fly ash (FFA), having a very low CaO content, is prepared by the burning of bituminous coal. To produce class C fly ash (CFA) with high calcium content, lignite and sub-bituminous coal are also used [23]. Class C Fly Ash composition is the same as natural volcanic ash. In the early 20th century, FA was used as the main component of concrete and cement [24]. Fly ash is a better replacement for cement as it decreases construction costs and greenhouse gas emissions. It is easily available and is low-cost. High strength geopolymers from fly ash can be formed using an alkali activator solution [25, 26].

16.3.4 WASTE INCINERATION BOTTOM ASH AND BIOMASS ASH

Burning of rice husks produces a by-product known as husk ash (RHA). Agricultural waste such as RHA is rich in silica and is considered as an alternative to enhance the properties of geopolymers [27]. RHA in geopolymer concrete can reduce the need for nano-SiO_2 usage. Additionally, it lessens the environmental problems brought on by the dumping of RHA on land, particularly in nations that produce rice [28]. Due to its high reactivity and extremely high specific surface area, it is frequently used in geopolymer concrete [29]. The primary waste product of municipal solid waste incineration is bottom ash. According to research [28, 30], it is recycled into binders and concrete for construction. Compared to concrete without bottom ash, concrete with bottom ash is stronger [31]. To activate these geopolymers, alkaline and acidic activators are used.

16.4 IMPORTANT CHARACTERISTICS AND ADVANTAGES OF GEOPOLYMERS

16.4.1 Strength and Durability

Geopolymers have the ability to display significant compressive strength, making them suitable for a varied range of applications. These applications include construction materials, infrastructure repair, as well as the fabrication of advanced composites.

16.4.2 Environmental Friendliness

Geopolymers have a huge advantage over traditional Portland cement, with significantly lower carbon dioxide emissions. This is primarily due to their lower energy consumption during production and their ability to incorporate waste materials such as fly ash and other agricultural waste that contain alumino-silicate minerals, thereby reducing the reliance on landfill disposal.

16.4.3 Fire Resistance

Geopolymers exhibit exceptional fire resistance abilities, making them invaluable for applications that require robust fire protection. They can be effectively utilized in the construction of fire-resistant walls and barriers, providing additional safety and security.

16.4.4 Chemical Resistance

Geopolymers possess remarkable resistance when exposed to diverse chemicals and harsh environments. This outstanding capability portrays them as highly suitable for applications in chemical processing plants and waste containment systems. Their durability ensures long-lasting performance and reliability in challenging conditions.

16.4.5 Rapid Setting and Curing

Geopolymers have the impressive ability to achieve rapid setting and curing times. This characteristic significantly reduces construction time and enhances overall efficiency. With faster setting and curing, projects can be completed more swiftly, saving valuable time and resources.

16.4.6 Recycling and Reusability

Geopolymers offer recyclability and reusability features that make them highly desirable for sustainable development. Their ability to be recycled and reused contributes towards waste reduction and promotion of circular economy. This aspect aligns with the principles of environmental sustainability and resource conservation.

16.5 APPLICATIONS OF GEOPOLYMERS

There is a wide range of existing and potential applications of geopolymers. Some applications are in the development stage, while some have already been commercialized and industrialized. The major categories of these applications are as follows:

Geopolymer Concretes and Cement

- Concretes and cements with less-CO_2 emission
- Building materials (bricks)

Archaeology and Arts

- Science history, archaeology, and traditions
- Decoration and arts, artificially decorated stones

Geopolymer Binders and Resins

- High-technology resin systems like binders, grouts and paints
- Medical applications
- Heat insulators, foams, and fire-resistant materials
- Used for the production of organic fiber compounds
- Fewer energy tiles, thermal shock refractories, and refractory products
- Hazardous and radioactive waste components

Geopolymers, being aluminosilicate structures, find a wide range of applications, particularly in construction and cementitious materials industry. However, their potential extends beyond traditional uses. Geopolymers have also been used as support for optical applications, color holders, pH indicators, and even fluorescent materials. Additionally, they have shown promise in the field of photocatalysis, enabling the degradation of volatile organic compounds.

This versatility highlights the exciting possibilities for geopolymers in various innovative applications beyond their conventional use. Geopolymers have experienced growing usage across diverse industries, including construction, aerospace, marine, automotive, and many more. However, it is important to acknowledge that there still remain challenges that need to be considered. These challenges include the initial high cost of production and the limited commercial availability of suitable source materials. Research and development in the field of geopolymers are actively ongoing with the objective of optimizing their properties, reducing production costs, and expanding their range of applications. As is the case with any innovative material, the wider adoption of geopolymers will depend on several factors, such as economic feasibility, scalability, and regulatory acceptance. These considerations are crucial in ensuring that geopolymers can fulfill their potential as a sustainable and versatile material in various industries. Ongoing efforts are focused on addressing these factors to promote the wider use of geopolymers in the future.

16.6 IMPORTANCE OF GEOPOLYMERS

Geopolymers, as alternative binder systems, are gaining significant attention in the field of research and development. Their ability to exhibit exceptional technical properties, including high strength, excellent acid resistance, and/or resistance to high temperatures, makes them particularly attractive. Geopolymers offer a promising alternative to traditional binder systems, showing great potential for various applications that require robust and durable materials. The increasing interest in geopolymers stems from their ability to provide sustainable and innovative solutions in diverse industries.

16.7 UTILIZING INDUSTRIAL WASTE

Several studies have documented that geopolymers can exhibit excellent performance when utilizing secondary raw materials such as industrial wastes like fly ash or slag. This discovery has generated significant interest in this technology, particularly in countries experiencing rapid industrialization. These countries often accumulate substantial amounts of industrial waste and may not have well-established recycling methods in place. Therefore, the utilization of geopolymers presents an appealing solution for these nations, as it allows them to effectively utilize their industrial waste and contribute to sustainable development.

16.7.1 Geopolymer Concrete

The concrete industry is facing difficulties in meeting the increased demand for Portland cement because of the scarcity of limestone resources, the slow growth of manufacturing and the rising cost of carbon. According to studies, India's demand for cement has been continuously increasing as a result of the country's expanding infrastructure development. By 2020, the country's cement needs were predicted to reach 550 million tonnes, with a shortfall of 230 million tonnes (58%) in supply [32]. The creation of substitutes for Portland cement as binders aims to address the shortage by lowering the environmental effect of building, using more waste pozzolan and enhancing concrete performance. As the family of alkali-activated cement grows, alkaline cement is divided into two groups based on the phase composition of the hydration products: R–A–S–H in aluminosilicate-based systems (where R = Na^+ or K^+) and RC–A–S–H in alkali-activated slag or alkaline Portland cement. Geopolymers have recently attracted a lot of attention due to their remarkable compressive strength, low permeability, exceptional chemical resistance, and outstanding fire-resistant behavior. Geopolymers are a promising candidate for replacing regular Portland cement for the creation of a variety of sustainable products for the construction industry, including fire-resistant coatings, concrete, waste immobilization solutions, and fiber-reinforced composites [33]. Look at a number of substitutes, such as magnesium oxycarbonate cement (carbon-negative cement), calcium sulphoaluminate cement, and alkali-activated cement. The right choice of raw ingredients, the right mixture, and the design of processes to fit a particular application can optimize the properties of geopolymers. Due to the significance of the issue, France, Spain and Italy collaborated on a project called 'Cost-effective geopolymeric cement for benign stabilization of toxic materials (GEOCISTEM)', funded by the European Commission. The project's goal was to produce geopolymeric cement using less expensive alkaline volcanic tuffs instead of potassium silicate [32].

Daidovits used the term 'geopolymer' [34] to describe materials with networks or chains of inorganic molecules. Geopolymer cement concrete is created using waste products like ground-granulated blast furnace slag (GGBS) and fly ash. Fly ash is a waste product from thermal power plants, and ground-granule blast furnace slag is a waste product from steel mills. When used for geopolymer concrete building projects, fly ash and GGBS are both properly treated. The use of this concrete reduces waste stock and carbon emissions by reducing the requirement of Portland cement. The primary components of geopolymers are silicon and aluminum, which are obtained from naturally occurring (such as kaolinite) or man-made (such as fly ash or slab) materials that have been thermally activated [35]. Subsequently, these components are polymerized into molecular chains and networks to produce a rigid binder using an alkaline activating solution. Other names for geopolymers include inorganic polymer cement and alkali-activated cement. After lime and normal Portland cement, geopolymer is recognized as the third generation of cement. Amorphous alkali aluminosilicate is referred to as a 'geopolymer' in general, but it is also known as 'inorganic polymer', 'alkali-activated cement', 'geochemist', 'alkali-bonded ceramic', 'hydroceramic', etc. All of these names refer to materials that are produced utilizing the same chemical, notwithstanding the changes in nomenclature. It Amorphous Alkali Aluminosilicate essentially comprises a silicate monomer (–Si–O–Al–O) repeating unit. The geo-polymerization process has used a variety of aluminosilicate minerals as solid raw materials, including feldspar, kaolinite, and industrial solid leftovers such as metallurgical slag, fly ash, mining wastes, etc. The properties of these aluminosilicate sources, such as mineralogical composition, chemical makeup, fineness, morphology and glassy phase content, affect their reactivity. The discussions lead to the conclusion that geopolymer concrete has a lot of promise as a building material in a variety of applications. The design criteria outlined in ACI standards and other National codes for OPC concrete are reportedly equally applicable to geopolymer concrete. Ready-mixed geopolymer concrete can be created, and it signifies success [35].

Geopolymers are created by reacting an alkali hydroxide/alkali silicate with a solid aluminosilicate powder. Figure 16.1 provides a schematic illustration of how fly ash-based geopolymers and concrete are formed. Under extremely alkaline circumstances, polymerization takes place when reactive aluminosilicates are swiftly dissolved and free $[SiO_4]$– and $[AlO_4]$–tetrahedral units are released in the solution. By sharing an oxygen atom with the polymeric precursor, the tetrahedral units are alternately connected, creating polymeric Si–O–Al–O linkages.

Water that would typically be lost during the breakdown is now released. When handling the mixture, the water released from the geopolymer during the reaction gives it workability. In contrast, water in a Portland cement mixture experiences a chemical reaction during the hydration process. According to reports, the sodium aluminosilicate hydrate gels with varied Si/Al ratios are the hydration products of metakaolin/fly ash activation, while calcium silicate hydrate with a low Ca/Si ratio is the main phase formed in slag activation. Geopolymers made from distinct aluminosilicate sources may have many physical characteristics that appear to be similar, yet their microstructures and chemical characteristics differ greatly. The ability to make metakaolin-based geopolymers consistently and predictably is a benefit [36].

Figure 16.1 Fly ash geopolymer [32].

By maximizing the numerous variables that affect their performance, geopolymer can be effectively employed as a building material, according to the discussion above. Less calcium-rich aluminum silicate source materials cause a delay in the geopolymer concrete and slow the rate at which strength increases. Fly ash with a high calcium content can be used to speed up the polymerization reaction and shorten the settling period. Additionally, the use of high-calcium fly ash increases compressive strength. Fly ash and GGBS can be combined for improved strength under ambient curing. Additionally, this combination has been seen to provide compressive strength that is higher than fly ash-based geopolymer concrete. Additionally, the use of GGBS improves the unit's durability properties and increases its size. Also, by replacing a small percentage of fly ash with rice husk ash—roughly 10%—it is possible to achieve acceptable strength [32]. The strength of geopolymers generally increases with binder material fineness. The combination of sodium hydroxide solution and sodium silicate produces the best activation of the polymerization process, out of all the available alkali activators. Additionally, when the concentration of sodium hydroxide rises, the compressive strength also increases. This rise in molarity also causes the brittleness of the concrete to increase. M-sand can be used to create sustainable geopolymer concrete, and if the M-sand is finer, it also boosts the compressive strength. With the use of GGBS, ambient curing is achievable in geopolymer concrete.

16.7.1.1 Significance of Geopolymer Cement Concrete

A tonne of cement requires roughly 2 tonnes of raw materials (shale and limestone), which results in the release of 0.87 tonnes of CO_2, about 3 kg of nitrogen oxide (NOx, a pollutant that contributes to ground-level smog), and 0.4 kg of PM10 (particulate matter of size 10 m, which is harmful to the respiratory system, when inhaled). Around 7% of the world's total CO_2 emissions, or 23 billion tonnes yearly, are attributed to the production of Portland cement. Through advancements in process technology and increases in process efficiency, the cement industry has been making significant progress in lowering CO_2 emissions; however, further advancements are limited because CO_2 generation is inherent to the fundamental calcination of the limestone process [37]. One of the main causes of the industry's high environmental impact is the mining of limestone, which affects local water regimes, air quality, and land-use patterns. It has long been acknowledged that one of the biggest problems businesses face is the emission of dust during the production of cement. Many millions of tons of dry material are handled by the cement sector. Even if merely 0.1% of this is lost to the atmosphere, the environment can suffer greatly. The fact that there is neither a strong economic incentive nor an increased regulatory pressure to prevent emissions makes fugitive emissions a big problem. Because it mines for its raw materials, produces a product that cannot be recycled, and uses non-renewable energy sources, the cement business does not fit the current definition of a sustainable industry. The energy needed in the production process can be significantly decreased through waste management and the use of waste byproducts from thermal power plants, fertilizer companies, and steel mills. This lowers the price of raw materials, energy costs, and greenhouse gas emissions. It can do this by converting readily available wastes like ash and slag into useful goods like geopolymer concretes [37].

16.7.1.2 Recent Developments

Along with developments in these materials, the potential for utilizing geopolymers produced by the alkaline activation of aluminosilicates in building construction has been highlighted in this section. The properties of mortars and concrete made from geopolymer binders have been evaluated with respect to their fresh and hardened states, the interfacial transition zone between the aggregate and the geopolymer, their binding with steel reinforcing bars, and their resistance to high temperatures. The resilience of geopolymer pastes and concrete is demonstrated by how little damage they sustain in a variety of harsh situations. The product advancements have also been briefly described together with the R&D work done on heat and ambient cured geopolymers at CSIR-CBRI. The results of the study show that geopolymer concrete has characteristics similar to those of OPC concrete and has the potential to be utilized in civil engineering applications [38].

It has been demonstrated that geopolymer concrete has high strength, little shrinkage, resistance to reinforcement corrosion, resistance to acid and sulfate attack, resistance to freeze-thaw, resistance to fire, and resistance to alkali-aggregate interaction. The strength properties of geopolymer concrete are affected by a number of factors. These include the types and fineness of the source aluminosilicate material used, the type and concentration of the alkaline activators used, the curing temperature and curing duration, the usage of M-sand, etc. The proper selection of these parameters can lead to high-performance characteristics. The study discussed here focuses on the development of geopolymer concrete as well as the impact of several variables on the characteristics of the material [38].

- **Fresh and Hardened Properties:** With a goal strength of up to 80 MPa, various geopolymer concrete (GPC) mix proportions have been observed. Regarding the water-geopolymer solid ratio, activator potency, water-to-Na_2O ratio, curing duration, temperature, and age hardening, the attributes of the mixtures have also been investigated. The slump of the mixtures change based on the molarity of the activator, workability aids, and additional ingredients.
- **Durability Research on GPCC's:** The long-term durability of OPC concrete, which has always been problematic in harsh conditions, is one of the main issues with this material. Concrete degradation is often evaluated for sulfate attack and chloride-induced corrosion, According to ASTM 1202C, GPCC specimens exhibit low to very low chloride permeability ratings. In general, GPCCs provide embedded steel with superior corrosion protection than CC. When exposed to 2% and 10% sulfuric acid, GPCC has been shown to have very strong acid resistance. The long-term durability of OPC concrete, which has always been problematic in harsh conditions, is one of the main issues with this material. Concrete deterioration is often evaluated one the basis of freeze-thaw damage, alkali-silica interaction, air carbonation, chloride-induced corrosion, and sulfate assault. In light of this, numerous investigations are being conducted to comprehend how geopolymers will behave under these circumstances [39].
- **Effect of Alkali Activators:** The strength, setting time, and bonding of high-calcium fly ash geopolymer concrete have been studied [38]. In the said investigation, the alkaline liquid to fly ash ratio was held constant at 0.5. The ratios of sodium silicate to sodium hydroxide were 1.0 and 2.0. Furthermore, sodium hydroxide solution was employed in three distinct concentrations: 10 M, 15 M and 20 M. Both the 60°C heat

curing for 24 hours and the 23°C room temperature curing regimes were employed. The results showed that the elastic modulus and the strength of geopolymer concrete increased with increasing molarity. A report on the compressive strength and microstructural characteristics of fly ash based geopolymer concrete was proposed by Gaurav Nagalia and Yeonho Park. This study looked into how alkali hydroxide and its concentration affected the growth of compressive strength. In this work, the effects of several alkaline solutions, including NaOH, KOH, $Ba(OH)_2$, and LiOH, on geopolymers formed by combining Class C (9.42% CaO) and Class F-fly ash (1.29% CaO) were investigated. Scanner electron microscopy (SEM) and X-ray diffraction (XRD) were also used to conduct microstructural examinations. The results showed that the only solution to create a high level of activation was NaOH (sodium hydroxide). The molarity of sodium hydroxide was held at 8 M, 12 M and 14 M. The results showed that as molarity increased, compressive strength also increased [39].

- **Effect of Curing Period:** Amran [39] discussed the experimental efforts to assess the impact of different parameters affecting compressive strength. Alkaline liquid to fly ash ratios of 0.35 and 0.4 were used in two concrete mixtures cast, Mix-1 and Mix-2. Additionally, different concentrations of sodium hydroxide and sodium silicate were utilized to make the activator solution. The molarity of the sodium hydroxide solution was changed to 8 M, 10 M, 12 M and 14 M. The test specimens were baked for 24 and 48 hours at 750°C for curing. The results demonstrated that the compressive strength of the geopolymer concrete increased with the lengthening of the curing process. Also, the compressive strength of the geopolymer concrete increased noticeably when it was cured for up to 24 hours at a temperature between 60°C and 90°C [39].

16.7.2 Geopolymer Cement

The term 'geopolymer' was introduced by Joseph Davidovits [34]. When a chemical reaction takes place between materials comprising aluminosilicates (i.e., source material) and alkali hydroxides or soluble silicates, then an inorganic polymeric cementitious binder is formed and is known as geopolymer [40]. The aluminum and silicon present in the source material gets dissolved when it comes in contact with an alkali solution, which results in the formation of oligomers. These oligomers then undergo polycondensation to develop a three-dimensional framework [41] including:

1. Polysialate;
2. Polysialte-siloxo
3. Polysialate-disiloxo

16.7.2.1 Objectives of Cementing in Oil and Gas Wells

The primary objectives of cementing an oil and gas well include:

- Providing integrity to the casing

- Zonal isolation
- Controlling the lost circulation
- Plugging and abandoning the well (dry/depleted well) [42].

16.7.2.2 Significance of Geopolymer Cement in the Oil Industry

Portland cement production causes significant carbon dioxide (CO_2) emissions due to the calcination of limestone with the combustion of hydrocarbons required to generate the energy needed to heat the kiln; about one ton of CO_2 is generated in the production of one ton of Portland cement [43, 44]. The manufacturing process for geopolymer cement is much cleaner [45]. Portland cement is prone to cracking when exposed to pressure and thermal-induced loads. Geopolymers provide an increased compressive strength as the curing temperature increases, whereas increasing the temperature decreases the compressive strength of ordinary Portland cement. The reduction in the strength of Portland cement at high temperatures is usually accompanied with an increase in porosity and permeability. Geopolymer technology has wide applications in the construction industry. However, for oil and gas well cementing, researchers are examining the characteristics of various geopolymer systems under subsurface conditions.

16.7.2.3 Properties of Geopolymer Cement

Cement slurries properties are predictable according to the subsurface situations such as pressure, temperature, and the specific job. Analysts are working to find a substitute for ordinary Portland cement due to the concerns associated with its production and mainly to determine how geopolymers can overcome the shortcomings of Portland cement-based systems.

Low permeability: Acidic and high saline conditions are characteristics of aggressive environments. The loss of a well's integrity occurs due to cement degradation. Capture, injection and depletion of reservoirs (i.e., CO_2 storage wells) are a valuable technique that leads to a decrease in CO_2 ejection [46]. The capability of cement cover to provide a better blockade in these depletion wells is a critical concern. Nasvi et al. [47] determined that the conventional system shows a higher CO_2 permeability range (10^{-20} to 10^{-11} m^2), while fly ash geopolymers have a lower permeability range (2×10^{-21} to 6×10^{-20} m^2). A geopolymer with low permeability would be an excellent substitute for materials used in cementing processes, particularly for blocking carbon dioxide (CO_2). Nasvi et al. [48] also found that with increasing injection pressure, the permeability of the geopolymer decreases, implying that the geopolymer would become an adequate seal in the high depths of CO_2 storage wells. Nasvi et al. [49] conducted another study to determine the consequences of various geopolymers on CO_2 permeability. The geopolymers were found to be two to three times less permeable than the Portland cement system (Class G). Geopolymers consist of 15% residues and yield a permeability product that is 1000 times less than that of the Portland cement system (Class G). Therefore, less permeability is a primary need for adequate CO_2 storage wells.

Durability: Barlet-Gouedard et al. [50] conducted research to determine the durability of metakaolin-based geopolymers in CO_2 surroundings. The geopolymer showed brilliant mechanical characteristics, with rheology investigation indicating no proper degradation. Nasvi et al. [51] studied the mechanical properties of geopolymers under CO_2 blockage. In the experiment, the samples were preserved for up to 6 months at 435 psi. For the test, the geopolymer was not placed in a CO_2 chamber. The results did not show any strength changes in aggressive surroundings. The saturated sample also did not show any microstructure changes. The research indicated that fly ash geopolymer is a good alternative as it provides long-lasting strength.

Well plug as well as abandonment: As the economic limit of the well is attained, it is plugged and abandoned (P&A). This can be temporary or constant. Portland cement is used for this purpose, but it faces several issues such as carbonation. Portland Cement It also includes geopolymers like bentonite, metals and gels [52]. Different experiments were done by Khalifeh et al. [53] to determine the chances of utilizing geopolymers in well plug and abandonment jobs. The geopolymers were produced using fly ash (Class C) as the main component and a mixture of NaOH and $(Na_2O)_x{\cdot}SiO_2$ solutions. Curing was performed at 188°F or 257°F and 5000 psi. By increasing the NaOH concentration, the durability of the material increased. Khalifeh et al. [54] also studied the aplite rock as an origin rock for geopolymers for the well plug procedure. It was observed that aplite rock imparted suitable mechanical characteristics and low permeability (7–30 × 10−1 md), which is required for well plugging and adequate zonal isolation. The research showed that fly ash powder was charged with NaOH or $(Na_2O)_x{\cdot}SiO_2$ alkaline solution, giving a low shrinkage effect in gas migration when compared to OPC. Also, the geopolymer system provided higher compressive strength and shear bond strength.

16.8 CONCLUSION

The chapter discusses the fundamentals and applications of geopolymers, emphasizing their significance as an alternative material to conventional cement, particularly in the context of waste management, strength, and environmental advantages. Geopolymers are inorganic, amorphous aluminosilicate materials formed through a process called 'polymerization', presenting exceptional mechanical characteristics, fire resistance, and environmental benefits over Portland cement. The chapter also gives insight into geopolymer synthesis, the source of materials used in their production, and their important characteristics and advantages, such as strength, environmental friendliness, fire resistance, chemical resistance, rapid setting and curing, and recyclability. Furthermore, the chapter provide details of the significance of geopolymer cement in the oil industry, emphasizing its potential applications in providing integrity to the casing, zonal isolation, controlling lost circulation, and plugging and abandoning oil and gas wells. The chapter also provides insights into recent developments and studies related to geopolymer cement, demonstrating its potential to provide long-lasting strength, effectively seal CO_2 storage wells.

REFERENCES

[1] Ren, D.M., Yan, C.J. and Duan, P. 2017. Durability performances of wollastonite, tremolite and basalt fiber-reinforced metakaolin geopolymer composites under sulfate and chloride attack. Constr. Build. Mater. 134. 56–66.

[2] William E. Lee, Michael I. Ojovan and Carol M. Jantzen. 2013. Radioactive waste management and contaminated site clean-up. Processes, Technologies, and International Experience. Energy. Woodhead Publishing.

[3] Provis, J.L. and Deventer, J.S.J.V. 2007. Geopolymerisation kinetics. Reaction kinetic modeling. Che. Eng. Sci. 62(9): 2318–2329.

[4] Temuujin, J., Surenjav, E. and Ruescher, C.H. 2019. Processing and uses of fly ash addressing radioactivity (critical review). Chemosphere. 216: 866-882.

[5] Duan, P., Yan, C.J. and Luo, W.J. 2016. A novel waterproof, fast setting and high early strength repair material derived from metakaolin geopolymer. Constr. Build. Mater. 124: 69–73.

[6] Silva, P.D., Sagoe-Crenstil, K. and Sirivivatnanon, V. 2007. Kinetics of polymerization: role of Al_2O_3 and SiO_2. Cem. Concr. Res. 37(4): 512–518.

[7] Duxson, P., Fernandez-Jimenez, A. and Provis, J.L. 2007. Geopolymer technology: the current state of the art. J. Mater. Sci. Eng. 42(9): 2917–2933.

[8] Prud'homme, E., Michaud, P. and Joussein, E. 2011. In situ, inorganic foams are prepared from various clays at low temperatures. Appl. Clay Sci. 51(1–2): 15–22.

[9] Sellami, M., Barre, M. and Toumi, M., 2019. Synthesis, thermal properties and electrical conductivity of phosphoric acid-based geopolymer with metakaolin. Appl. Clay Sci. 180: 105192.

[10] Huseien, G.F., Mirza, J. and Ismail, M. 2018. Effect of metakaolin replaced granulated blast furnace slag on fresh and early strength properties of geopolymer mortar. Ain Shams Eng. J. 9(4): 1557–1566.

[11] Gao, H., Liu, H. and Liao, L. 2020. A bifunctional hierarchical porous kaolinite geopolymer with good performance in thermal and sound insulation. Constr. Build. Mater. 251: 118888.

[12] Nikolov, A., Nugteren, H. and Rostovsky, I. 2020. Optimization of geopolymers based on natural zeolite clinoptilolite by calcination and use of aluminate activators. Constr. Build. Mater. 243: 118257.

[13] Subaer, Haris, A. and Irhamsyah, A. 2019. Physico-mechanical properties of geopolymer based on laterite deposit sidrap, south sulawesi. *In*: 3rd International Conference on Mathematics, Sciences, Technology, Education and Their Application. Sulawesi Selatan.

[14] Mathew, G. and Issac, B.M. 2020. Effect of molarity of sodium hydroxide on the aluminosilicate content in laterite aggregate of lateralized geopolymer concrete. J. Buil. Eng. 32: 101486.

[15] Subaer, Haris, A. and Nurhayati, A. 2016. The influence of Si:Al and Na:Al on the physical and microstructure characters of geopolymers based on metakaolin. Mater. Sci. Forum. 841(1): 170–177.

[16] Lemougna, P.N., Wang, K.-T., Tang, Q., Kamseu, K., Billong, N., Melo, U.C., et al. 2017. Effect of slag and calcium carbonate addition on the development of geopolymer from indurated laterite. Appl. Clay Sci. 148: 109–117.

[17] Amran, Y.H.M., Alyousef, R. and Alabduljabbar, H. 2020. Clean production and properties of geopolymer concrete: a review. J. Cleaner Prod. 251: 119679.

[18] Jiang, W., Li, X. and Lyu, Y. 2020. Mechanical and hydration properties of low clinker cement containing high volume superfine blast furnace slag and nano silica. Constr. Build. Mater. 238: 117683.

[19] Silva, G., Kim, S. and Aguilar, R. 2020. Natural fibers as reinforcement additives for geopolymers—A review of potential eco-friendly applications to the construction industry. Sustainable Mater. Technol. 23: 00132.

[20] Li, W. and Yi, Y. 2020. Use of carbide slag from the acetylene industry for activation of ground granulated blast-furnace slag. Constr. Build. Mater. 238: 117713.

[21] Nie, Q., Hu, W. and Huang, B. 2019. Synergistic utilization of red mud for flue-gas desulfurization and fly ash-based geopolymer preparation. J. Hazard. Mater. 369: 503–511.

[22] Yang, Z., Mocadlo, R. and Zhao, M. 2019. Preparation of a geopolymer from red mud slurry and class F fly ash and its behavior at elevated temperatures. Constr. Build. Mater. 221: 308–317.

[23] Guo, X., Shi, H. and Wei, X. 2017. Pore properties, inner chemical environment, and microstructure of nano-modified CFAWBP (class C fly ash-waste brick powder) based geopolymers. Cem. Concr. Compos. 79: 53–61.

[24] Scrivener, K.L., John, V.M. and Gartner, E.M. 2018. Eco-efficient cements: potential economically viable solutions for a low-CO_2 cement-based materials industry. Cem. Concr. Res. 114: 2–26.

[25] Schmücker, M. and Mackenzie, K.J.D. 2005. Microstructure of sodium polysialate siloxo geopolymer. Ceram. Int. 31(3): 433–437.

[26] Gupta, R., Bhardwaj, P. and Mishra, D. 2017. Formulation of mechanochemically evolved fly ash-based hybrid inorganic geopolymers with multilevel characterization. J. Inorg. Organomet. Polym. Mater. 27(2): 385–398.

[27] Tosti, L., van Zomeren, A. and Pels, J.R. 2018. Technical and environmental performance of lower carbon footprint cement mortars containing biomass fly ash as a secondary cementitious material. Resour. Conserv. Recycl. 134: 25–33.

[28] Nuaklong, P., Jongvivatsakul, P. and Pothisiri, T. 2020. Influence of rice husk ash on mechanical properties and fire resistance of recycled aggregate high-calcium fly ash geopolymer concrete. J. Cleaner Prod. 252: 119797.

[29] Raisi, E.M., Amiri, J.V. and Davoodi, M.R. 2018. Mechanical performance of self-compacting concrete incorporating rice husk ash. Constr. Build. Mater. 177: 148–157.

[30] Nagrockiene, D. and Daugela, A. 2018. Investigation into the properties of concrete modified with biomass combustion fly ash. Constr. Build. Mater. 174: 369–375.

[31] Rutkowska, G., Wichowski, P. and Fronczyk, J. 2018. Use of fly ashes from municipal sewage sludge combustion in production of ash concretes. Constr. Build. Mater. 188: 874–883.

[32] Parveen, S.D., Junaid, T., Jindal, B.B. and Mehta, A. 2018. Mechanical and micro-structural properties of fly ash based geopolymer concrete incorporating alccofine at ambient curing. Constr. Build. Mater. 180: 298–307.

[33] Mehta, P.K. and Monteiro, P.J.M. 2006. Concrete: Microstructure, Properties and Materials. The McGraw Hill companies, Inc. USA.

[34] Davidovits, J. 1982. Mineral polymers and methods of making them. United States Patent. p. US4349386A.

[35] Hardjito, D., Wallah, S.E., Sumajouw, D.M.J. and Rangan, B.V. 2004. On the development of fly ash-based geopolymer concrete. ACI Mater J. 101: 467–472.

[36] Cai, J., Li, X. and Tan, J. 2020. Thermal and compressive behaviors of fly ash and metakaolin-based geopolymer. J. Buil. Eng. 30: 101307.

[37] Almutairi, A.L., Tayeh, B.A., Adesina, A., Isleem, H.F. and Zeyad, A.M., 2021. Potential applications of geopolymer concrete in construction: A review. Case Stud. Constr. Mater. 15.

[38] Singh, B., Ishwarya G., Gupta, M. and Bhattacharyya, S.K. 2015. Geopolymer concrete: A review of some recent developments. Constr. Build. Mater. 85: 78–90.

[39] Amran, M., Al-Fakih, A., Chu, S.H., Fediuk, R., Haruna, S., Azevedo, A., et al. 2021. Long-term durability properties of geopolymer concrete: An in-depth review. Case Stud. Constr. Mater. 15: 2021.

[40] Davidovits, J. 1991. Geopolymers—inorganic polymeric new materials. J. Therm. Anal. 37: 1633–1656.

[41] Sitaram, A.S., Bhardwaj, P. and Gupta, R. 2019. Advanced geopolymerization technology. *In*: Geopolymers and Other Geosynthetics. Intech Open.

[42] Xu, B., Yuan, B., Guo, J., Xie, Y. and Lei, B. 2019. Novel technology to reduce the risk lost circulation and improve cementing quality using managed pressure cementing for narrow safety pressure window wells in the sichuan basin. J. Petrol. Sci. Eng. 180: 707–715.

[43] Davidovits, J. 1994. Properties of geopolymer cements. First International Conference on Alkaline Cements and Concretes, SRIBM, Kiev State Technical University. Kiev Ukraine.

[44] McCaffrey, R. 2002. Climate change, and the cement industry. Global Cement and Lime Magazine. Environ. Sp. Issue. 15–19.

[45] Singh, N.B., Saxena, S.K. and Kumar, M. 2018. Effect of nanomaterials on the properties of geopolymer mortars and concrete. Mater. Today Proc. 5(3): Part 1. 9035–9040.

[46] Jenkins, C.R., Cook, P.J., Ennis-King, J., Undershultz, J., Boreham, C., Dance, T., et al. 2012. Safe storage and effective monitoring of CO_2 in depleted gas fields. Proc. Natl. Acad. Sci. U.S.A. 109.

[47] Nasvi, M.C.M., Ranjith, P.G. and Sanjayan, J. 2013a. The permeability of geopolymer at down-hole stress conditions: application for carbon dioxide sequestration wells. Appl. Energy. 102: 1391–1398.

[48] Nasvi, M.C.M., Ranjith, P.G., Sanjayan, J. and Haque, A. 2013b. Sub- and super-critical carbon dioxide permeability of wellbore materials under geological sequestration conditions: an experimental study. Energy. 54: 231–239.

[49] Nasvi, M.C.M., Ranjith, P.G. and Sanjayan, J. 2014. Effect of different mix compositions on apparent carbon dioxide (CO_2) permeability of geopolymer: suitability as well cement for CO_2 sequestration wells. Appl. Energy. 114: 939–948.

[50] Barlet-Gouedard, V., Zusatz-Ayache, B. and Porcherie, O. 2010. Geopolymer composition and application for carbon dioxide storage. US 7846250B2.

[51] Nasvi, M.C.M., Rathnaweera, T.D. and Padmanabhan, E. 2016. Geopolymer as well cement and its mechanical integrity under deep down-hole stress conditions: application for carbon capture and storage wells. Geomech. Geophys. Geo-Energy Geo-Resourc. 2: 245–256.

[52] Vrålstad, T., Saasen, A., Fjær, E., Øia, T., Ytrehus, J.D. and Khalifeh, M. 2019. Plug and abandonment of offshore wells: ensuring long-term well integrity and cost efficiency. J. Petrol. Sci. Eng. 173: 478–491.

[53] Khalifeh, M., Saasen, A., Vrålstad, T. and Hodne, H. 2014. Potential utilization of geopolymers in plug and abandonment operations. *In*: Pap. Present. SPE Bergen One Day Semin. Bergen, Norway. 389–402.

[54] Khalifeh, M., Hodne, H., Korsnes, R.I. and Saasen, A. 2015. Caprock restoration in plug and abandonment operations; possible utilization of rock-based geopolymers for permanent zonal isolation and well plugging. *In*: Pap. Present. Int. Pet. Technol. Conf. Doha, Qatar. December 6–9. IPTC-18454-MS.

Chapter 17

Enhanced Mechanical Properties of Kenaf Fibres with Fly Ash, and Al_2O_3 Nanofillers Epoxy Hybrid Composites

Natrayan, L.
Department of Mechanical Engineering, Saveetha School of Engineering,
SIMATS, Chennai, 602105, India
Email: natrayanphd@gmail.com

17.1 INTRODUCTION

Due to the difficulty of predicting crude oil reserves in the coming years and its ecological impact, the invention of polymeric materials utilising reused or biodegradable polymers is being studied aggressively [1]. Natural fibres are being recommended as an alternative to synthetic composite materials, owing to their advantages like renewability, environmental friendliness, reusability, abundance, permeation, resistance to corrosion, elevated level of flexibility, hydrophilicity, low toxicity, ability to sustain moisture, no discharge of dangerous substances, no annoyance to the epidermis, no allergen impact, competitive mechanical characteristics, reduced emissions, and less obnoxiousness to dispensation apparatus [2–4]. The main disadvantage of organic fibre-based composites is that hydrophilic natural materials are incompatible with hydrophobic polymeric matrix [5]. Only a tiny percentage of the wide variety of crop wastes is used as domestic fuel or fertiliser, while most is burnt in fields. As a consequence of environmental pollution, it harms the environment [6]. Using such agricultural leftovers as reinforcing materials to prepare polymeric composites is a critical solution to this challenge. Given the present concern about climate change and greenhouse gases, natural materials are progressively becoming regarded as an ecologically sound option for synthetic materials in reinforcing thermoplastic composites [7]. Natural fibres are the most often utilised natural materials in biodegradable polymers today, and strengthening such fibres using polymers improves the biomechanical characteristics [8].

Since kenaf bushes grow quickly in various climates and are consequently inexpensive, they constitute a particularly enticing replacement for additional natural assets that have recently seen extensive mining [9]. Recently, Kenaf has become more significant. In order to satisfy the organic fabric demand for composite materials, plant-based fibres are being used increasingly often. This has led to the global growth of kenaf as a textile crop. Kenaf materials are biodegradable because they contain more cellulose [10, 11]. The absence of amorphous silicate, essential in reducing brashness in manufacturing amenities, is an additional benefit of kenaf strands. In addition, it has previously been established that including cellulosic fillers decreases the durability of the substance [12, 13]. For polymers and other sectors, kenaf has become recognised as a beneficial digestion aid. A maximum of forty percent of the kenaf stalk can be used as fibre, roughly twice as much as that of hemp, jute or flax seeds, making the material quite economical [14]. A wide range of unique products, notably newsprint, building supplies, water softeners and animal food, include kenaf. Additionally, there is a growing market for using kenaf filaments, especially for reinforced polymers [15, 16].

Several variables can impact the efficiency of biocomposites. Aside from the hydrophilic character, the amount of fibre or filler can impact the quality of biocomposites. In general, a substantial amount of fibre is necessary to obtain good performance. As the result, the influence of fibre volume fraction on the characteristics of biocomposites is very important [17, 18]. The process settings utilised are also key components that determine the characteristics of the composite. Hence, optimal processing methods and variables should be deliberately crafted to produce good hybrid materials [19, 20].

Today, practically everything has become ‘nano’, including substances such as carbon black which was widely employed as a reinforcing material or filler in elastic bands for over a century [18, 21]. The main idea behind nanomaterials is to create a very broad contact between nanoparticle construction elements and the polymer matrices. The homogenous dispersion of nanoparticles is often challenging. When combined with the appropriate fillers, polymeric matrix elements may be used to create composite materials with a wide range of potential uses in architecture, automobile, aviation and manufacturing sectors [22, 23]. Nano fillers are both natural and synthetic in origin. Synthetic fillers include Al_2O_3, TiO_2, calcium carbonate and graphene. Furthermore, biological nanoparticles like coco fibre nanoparticles, carbon black, and cellulose nanofiller, among many others, are generated biologically and constitute biological nanoparticles. Nanocomposite has several uses in various sectors for its bigger surface region and higher anamorphic widescreen [13, 17]. Combining natural fibers with nano-reinforced bio-based polymers can enhance their characteristics. Nanomaterials are now considered promising fillers for improving the mechanical and tribological characteristics of polymeric materials. Because nano ingredients are typically defect-free, the latest innovations in the domain of polymeric area establishments promise to overcome the limitations of traditional millimetre size particles [24, 25].

The mixture of nanofillers and organic fibres in matrices results in decreased water uptake qualities and enhanced mechanical characteristics. Nanoparticles possess the potential to greatly enhance a material's performance across various aspects, such as heat conductivity, optical or electronic properties, mechanical strength, and more. This enhancement can occur independently or in combination with traditional filler particles in the material design [26, 27]. Nanoparticles are introduced into polymeric matrixes at a scale ranging between 2% and 8% of total mass. Adding a modest quantity of nanoparticles improves the characteristics of rice husk flour/HDPE combinations. Adding nonclay to the high-density polyethylene hay grain polymeric matrix improves its mechanical properties, thermal stability, fuel penetration, durability, and other physical properties [28, 29]. As a result, most research in this field focuses on enhancing the physical features of composites. Al_2O_3, SiO_2 and TiO_2 nanoparticles change the polymer matrices in glass composites to enhance the mechanical behaviour. Al_2O_3-altered epoxy has higher stiffness and impact strength than SiO_2 and TiO_2 modifications. The effects of various amounts of silicon, aluminium oxide and potassium permanganate on E-glass fibre reinforcing filler composite materials has revealed that carbon nanotube-filled composites display good toughness, impact resistance, and flexural strengths because magnesium hypochlorite fluff addition results in the highest tensile strength [30, 31].

In this chapter explores the combination of two reinforcements such as Al_2O_3 and fly ash fillers, integrated with kenaf fibre within an epoxy matrix. The effects of Al_2O_3 and fly ash additives in Kenaf/epoxy composite material on the physical and mechanical properties were subsequently investigated. The composites were fabricated through the normal hand lay-up method to achieve the abovementioned objectives. After fabrication, the mechanical characteristics of the composites were tested.

17.2 EXPERIMENTAL PROCEDURE

17.2.1 Materials

The polymer solution in this investigation comprised synthetic resin AW-106 and curing agent HV-953, both obtained from Ganga Pharmaceutical Companies in India. Natural kenaf fibres acquired from Wood Plastic Company in Salam, India were used as reinforcing material to prepare synthetic epoxy composite. This study used different kinds of inorganic nanoparticles: aluminium oxide from Siva Chemicals in India and fly ash from commercial chimneys of a power plant in Naively, India. Table 17.1 shows the mechanical properties of reinforcement materials and the matrix.

Table 17.1 Mechanical Properties of Kenaf and Epoxy Matrix.

Sl. No.	Particulars	Kenaf	Epoxy Matrix
1.	Density (g/cm^3)	1.39	1.48
2.	Tensile Strength (MPa)	934	67–69
3.	Young's Modulus (GPa)	19.84	8.6
4.	Cellulose Content (%)	1.46	–
5.	Hemicellulose (%)	14–16	–
6.	Lignin (%)	4–9	–

17.2.2 Hybrid Composite Fabrication

Before getting dried in direct sunlight for 70 hours to eliminate any remaining moisture, the kenaf threads are cleaned under flowing hot spring waters to eliminate any clinging accumulation of contaminants. The kenaf filaments are separated into small fragments to create miniature organic fibres and put into typical slicing and milling machinery. These fibres are filtered on a 212-micrometre grid. The binding polymer compounds, drying substances, and strengthening constituents are created by weighing them in various volumetric percentages using computerised measurement equipment. Table 17.2 displays the weight proportions used to create epoxy nanocomposites utilising kenaf microcapsules and charcoal ashes with Al_2O_3 nanoparticles. Mechanical swirling is employed to obtain homogenous particle dispersion in the internal resin. The mixed liquid is quickly placed in a mould and evaporated for 24 hours to prevent abrupt dryness. To reduce the internal moisture level, the produced specimen is separated from the moulds and put in a 100°C microwave.

Table 17.2 Process variables and their constraints.

Sl. No.	Symbols	Kenaf Fiber (wt.%)	Epoxy Matrix (wt.%)	Al_2O_3 Filler (wt.%)	Coal ash Filler (wt.%)
1.	A	10	80	–	10
2.	B	10	80	10	–
3.	C	10	80	5	5

17.2.3 Mechanical Characterisation

Tension testing was done to determine the stress distribution in the structure of a produced adhesive polymer hybrid mixture. According to the requirements of ASTM D 3039, the specimen was divided into a rectangular shape. Once the breaking or fracture was attained, the examination was conducted at an average speed of 1.2 mm per minute. In conformity with ASTM D2344, a three-point deformation study was done to confirm the flexural properties of the composites. The impact resistance of epoxy thermoplastic biocomposites has ability to absorb energy under impact pressure. The ASTM D 256 requirements were followed when conducting the Izod testing. Shoreline D hardness of epoxy nanocomposite polymeric layers was evaluated using technology that measures indentation toughness. When the bottom portion of the distortion fully interacted with the laminated material, the depression was forced towards the polymer laminate material specimen. The hardness indentation suggested that the material's durability may be influenced by its properties. Five distinct locations on the sample were tested and an average toughness was then calculated.

17.3 RESULTS AND DISCUSSION

17.3.1 Tensile Behaviour

As shown in Figure 17.1, nanofillers, coal ash, or both Al_2O_3 and coal ash significantly improved the tension properties of A, B and C type biocomposites. The insertion of an artificial aluminium oxide filler with a greater Young's modulus in a polymeric resin with a lower stiffness improves the tensile strength of the composite. Concrete-based polymers generally have higher mechanical characteristics and elasticity, followed by fly ash, which is consistent with the current research. Several researchers [13, 17, 18] have noticed that the elastic strength of a polymeric sandwich structure is not responsive to an interfacial contact between the polymers and the kenaf/nanofiller.

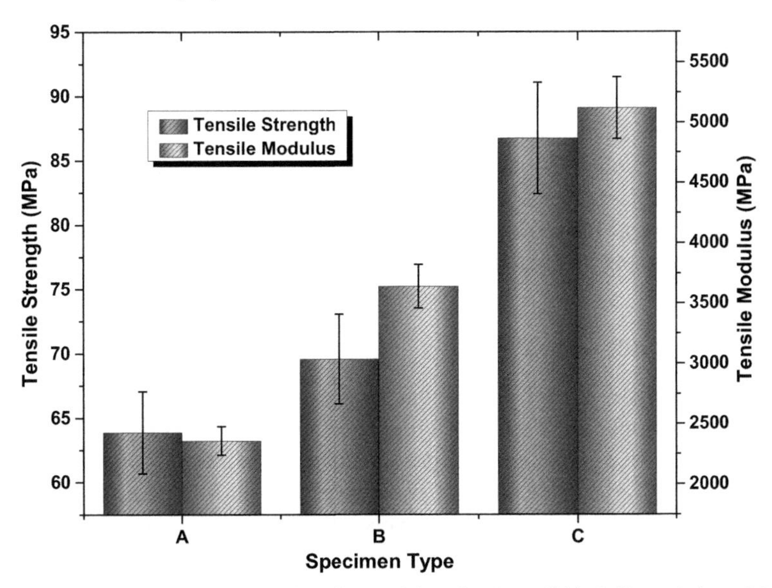

Figure 17.1 Tensile strength and their modulus for Kenaf/Al_2O_3/fly ash based hybrid composites.

Furthermore, because of the $CaCO_3$, there is an effective correlation between elastic strength and particle density. Because the configuration of concrete additives is complicated, it is anticipated that the tensile properties of A and C category fly ash ingredients would be better than Al_2O_3 nanomaterial incorporated adhesives [32–34]. Coal ash has been found to be strongly linked with polymers, as well as the breakage formed throughout the filler. The nanoparticle architecture is complicated enough to act as an impediment to fracture development and polymeric distortion. However, for some well-adhered Al_2O_3 fillers, the fracture may form through gaps in the Al_2O_3, displayed by a thin adhesive polymeric matrix adhering to a nanomaterial [21]. The failure mode is also uneven, with branches, bending cracking, or fissure fixing.

17.3.2 Bending Behaviour

Figure 17.2 displays an overview of the graph's findings. The elongation at break, whether with solitary reinforcements or both additives, appears to exceed five times the original magnitude. Adding ash, Al_2O_3 particles, or both to the epoxy matrix increased flexural strength up to 101.54 MPa. Flexural modulus was enhanced as well, although stress to breakage was reduced. The modulus of rupture and rigidity was increased in hybrid composites treated using ash and Al_2O_3 nanomaterials. The inclusion of both Al_2O_3 and coal ash reduced bending characteristics in comparison to the B-type composites. Bending characteristics could be depend on the efficient dispersion of fillers within an epoxy matrix, as well as the increased connectivity between polymer structures and the interface of filler particles [35]. The presence of a large interfacial area of micrometre filler that results in increased interfacial adherence between the polymeric matrix and the reinforcement is another feature that can be connected to a correlational study. Improving material properties is closely related to particulate factors like the interactions among polymeric matrixes and the granular contact, material behaviour, pore volume, concentration, and shape [17, 18]. The results are related to the specific structural, physical or molecular relationships of the polymer blend and the additives. The addition of coal ash and Al_2O_3 fillers to the kenaf fibre epoxy matrix leads to a deterioration in bending properties [28, 36].

17.3.3 Impact Behaviour

An electronic Izod impact assessment device measures the impact resistance of epoxy-blend composites. Figure 17.3 illustrates how adding fly ash, Al_2O_3 particles, or their mixture affects the strength of epoxy polymeric blended composites. Figure 17.3 shows that the impact resistance of B and C composites increased and reached the highest possible levels of 7.51 and 6.65 kJ/m^2 respectively. This is superior when compared to the hybrid in the A category. It has been determined that the stiffness of the polymer structure correlates with the polymer material's impact durability [37]. Fly ash and Al_2O_3 nanoparticles increase the stiffness of an epoxy hybridization laminate due to their ability to bind several separate resin polymers together via COOH and OH functional group interactions. Additionally, the impact toughness of the polymer hybrid laminate is decreased by the deposition of fly ash on additional kenaf microscopic particles [2, 4].

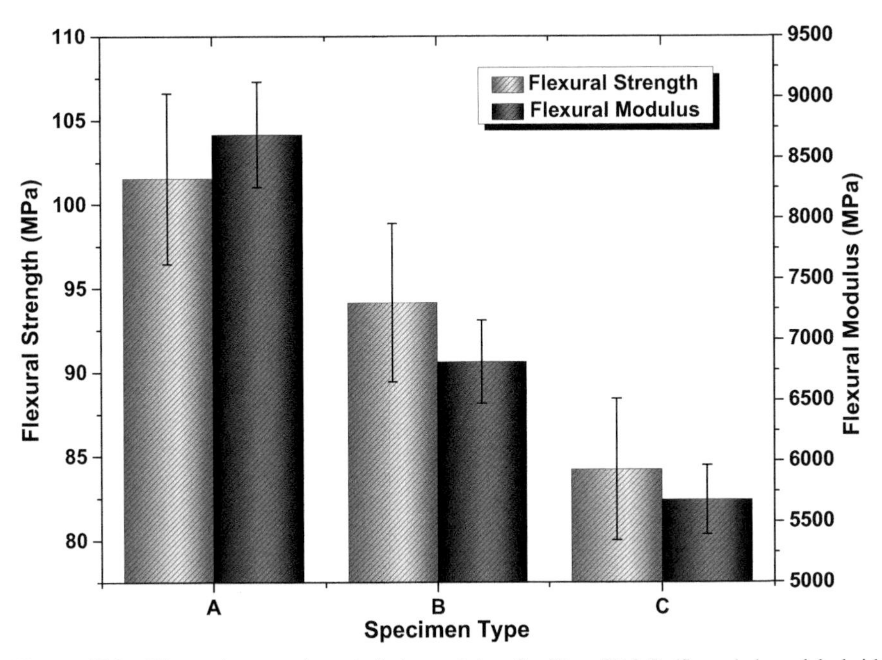

Figure 17.2 Flexural strength and their modulus for Kenaf/Al_2O_3/fly ash based hybrid composites.

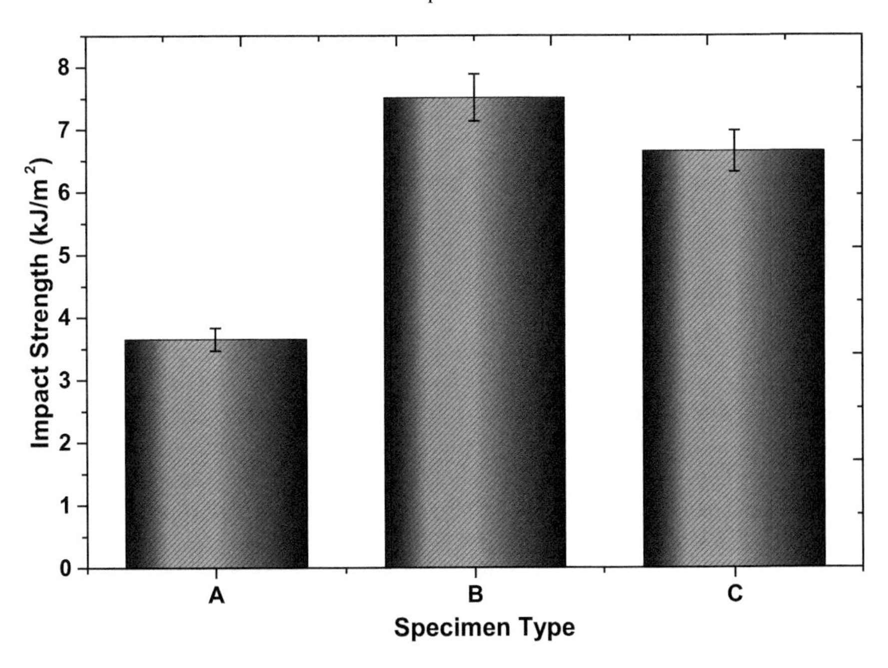

Figure 17.3 Impact strength of Kenaf/Al_2O_3/fly ash based hybrid composites.

17.3.4 Shore Hardness

The shoreline D hardness feature is the specimen's ability to resist crushing. Figure 17.4 displays the shoreline D hardness ratings of the created epoxy hybrid materials. Rigidity is a major factor in determining Shore D toughness. In the present study, the specimen B type's outside and inside became equally constructed of cellulose fibres, Al_2O_3 nanoparticles and the curing agent. Because of the less rigorous kenaf, coal ash, as well as Al_2O_3 fillers inside the external and internal stacks, sample B does have the greatest shore D toughness of 69.47, and sample A has the lowest shore D toughness at 55.28 [38, 39].

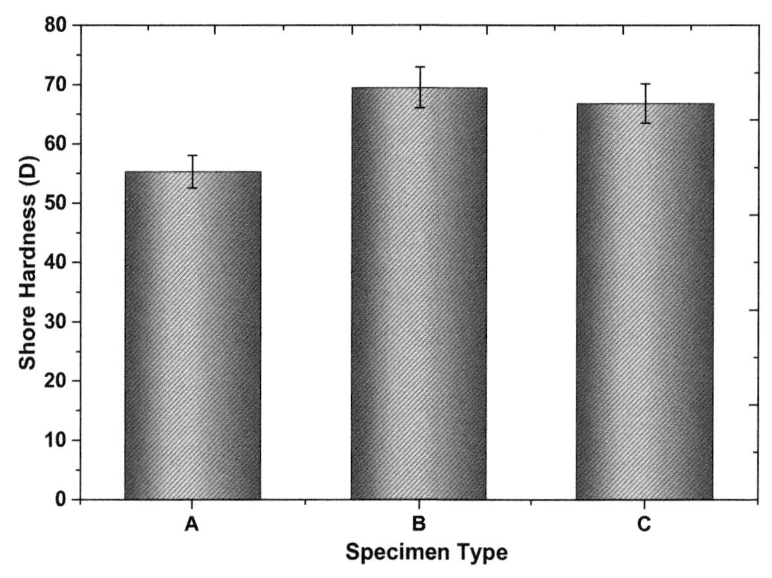

Figure 17.4 Shore hardness of Kenaf/Al_2O_3/fly ash based hybrid composites.

17.3.5 ILSS Behaviour

For example, creating gaps or interactions among additives and polymers might compromise the total ILSS of an adhesive-polymerized substrate. The manufactured materials have their ILSS resilience put to the test. All ILSS characteristics associated with the epoxy composites are shown in Figure 17.5. The sample type B among the manufactured polymer biocomposites seems to have better shear properties of 29.14 MPa, whereas the sample type A has inferior shear properties of 18.65 MPa. As the adhesion among compounds and polymers enhances, so does the ILSS. Mixtures of materials degrade more quickly when there are voids and flaws in the created substance, which lowers the overall durability of the materials [3, 4, 40]. Producing extra fractures and gaps in the C type composites reduced ILSS; other researchers have also observed similar reductions.

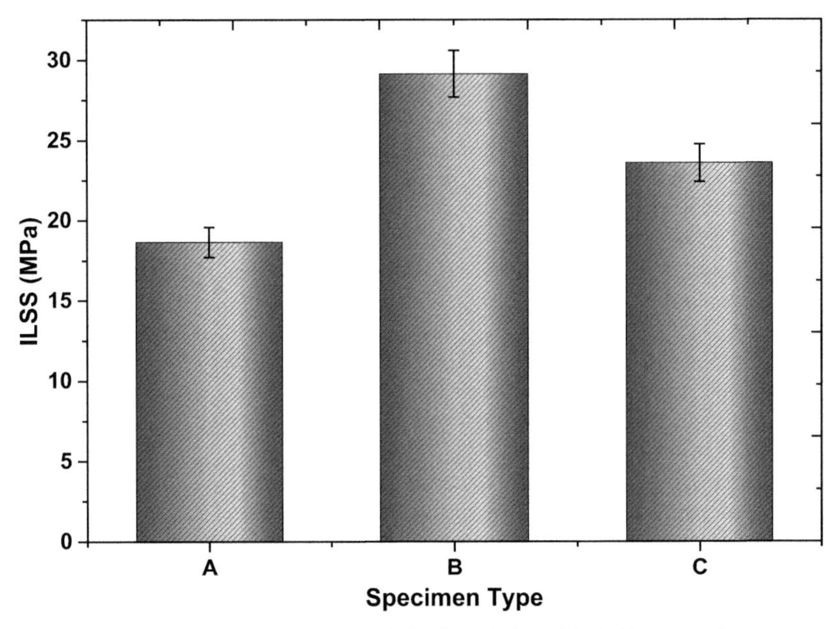

Figure 17.5 ILSS of Kenaf/Al_2O_3/fly ash based hybrid composites.

17.4 CONCLUSION

The results show that reinforcing with ash or Al_2O_3 nanomaterials improves the bending strength and the modulus of elasticity of the generated nanocomposites, especially when coal ash or Al_2O_3 nanomaterials are mixed with a polymer blend. The inclusion of ash and Al_2O_3 nanomaterials improves the tension strength and modulus of elasticity of all examined reinforcements, indicating that the elastic strength now indicates sensitivity to an interfacial filler-polymer interaction. In this regard, the gathered results may be attributed to the enhanced affinity of the augmented ash and Al_2O_3 nanomaterials for the polymer blend. Additionally, incorporating fly ash and Al_2O_3 nanoparticles improves an epoxy nanocomposite's impact strength and shoreline D toughness. The type C composites have an enhanced tensile strength (86.74 MPa) and modulus compared to other combinations (5120.78 MPa). When bent, the type A composite has the maximum bending strength (101.54 MPa) and modulus (8690.75 MPa). The type B composite has maximum impact, hardness, and ILSS values – 7.51 kJ/m^2, 69.47 D, and 29.14 MPa respectively.

REFERENCES

[1] Raju, J. S. N., Depoures, M. V., Shariff, J. and Chakravarthy, S. 2022. Characterization of natural cellulosic fibers from stem of *Symphirema involucratum* plant. J. Nat. Fibers. 19(13): 5355–5370.

[2] Lee, C.H., Khalina, A., Nurazzi, N.M., Norli, A., Harussani, M.M., Rafiqah, S.A., et al. 2021. The challenges and future perspective of woven kenaf reinforcement in thermoset polymer composites in malaysia: A review. Polymers. 13(9): 1390.

[3] Natarajan, E., Ramesh, S., Markandan, K., Saravanakumar, N., Dilip, A.A. and Batcha, A.R. 2023. Enhanced mechanical, tribological, and acoustic behavior of polyphenylene sulfide composites reinforced with zero-dimensional alumina. J. Appl. Polym. Sci. 140(16): e53748. https://doi.org/10.1002/app.53748.

[4] Giwa Ibrahim, S.A., Karim, R., Saari, N., Wan Abdullah, W.Z., Zawawi, N., Ab Razak, A.F., et al. 2019. Kenaf (*Hibiscus cannabinus* L.) seed and its potential food applications: A review. J. Food Sci. 84(8): 2015–2023.

[5] Ramesh, T., Sathiyagnanam, A.P., Poures, Melvin Victor De and Murugan, P., 2022. A comprehensive study on the effect of dimethyl carbonate oxygenate and EGR on emission reduction, combustion analysis, and performance enhancement of a CRDI diesel engine using a blend of diesel and Prosopis juliflora biodiesel. Int. J. Chem. Eng. 2022: 5717362. https://doi.org/10.1155/2022/5717362.

[6] Gurusamy, M., Vellaiyan, S., Kandasamy, M., Devarajan, Y. 2023. Optimization of process parameters to intensify the yield rate of biodiesel derived from waste and inedible *Carthamus lanatus* (L.) Boiss. seeds and examine the fuel properties with pre-heated water emulsion. Sustainable Chem. Pharm. 33: 101137. https://doi.org/10.1016/j.scp.2023.101137.

[7] Tharazi, I., Abu, B.S. and Farrahshaida, M.S. 2020. Application of response surface methodology for parameters optimization in hot pressing kenaf reinforced biocomposites. J. Mech. Eng. (JMechE). 17(3): 131–144.

[8] Natrayan L., Merneedi, A., Bharathiraja, G., Kaliappan, S., Veeman, D. and Murugan, P. 2021. Processing and characterization of carbon nanofibre composites for automotive applications. J. Nanomater. 1–7.

[9] Natarajan, E., Freitas, L.I., Santhosh, M.S., Markandan, K., Majeed Al-Talib, A.A. and Hassan, C.S. 2023. Experimental and numerical analysis on suitability of S-glass-carbon fiber reinforced polymer composites for submarine hull. Def. Technol. 19: 1–11. https://doi.org/10.1016/j.dt.2022.06.003.

[10] Prabhudass, J.M., Palanikumar, K., Natarajan, E. and Markandan, K. 2022. Enhanced thermal stability, mechanical properties and structural integrity of mwcnt filled bamboo/kenaf hybrid polymer nanocomposites. Materials. 15(2): 506.

[11] Ramesh, T., Sathiyagnanam, A.P., Melvin Victor De Poures and Murugan. P. 2022. Combined effect of compression ratio and fuel injection pressure on CI engine equipped with CRDi system using Prosopis juliflora methyl ester/diesel blends. Int. J. Chem. Eng. 1–12.

[12] Kumar, J.A., Amarnath, D.J., Jabasingh, S.A., Kumar, P.S., Anand, K.V., Narendrakumar, G., et al. 2019. One pot green synthesis of nano magnesium oxide-carbon composite: Preparation, characterization and application towards anthracene adsorption. J. Cleaner Prod. 237: 117691. https://doi.org/10.1016/j.jclepro.2019.117691.

[13] Ramaswamy, R., Gurupranes, S.V., Kaliappan, S., Natrayan, L. and Patil, P.P. 2022. Characterization of prickly pear short fiber and red onion peel biocarbon nanosheets toughened epoxy composites. Polym. Compos. 43(8): 4899–4908.

[14] Arjmandi, R., Yıldırım, I., Hatton, F., Hassan, A., Jefferies, C., Mohamad, Z., et al. 2021. Kenaf fibers reinforced unsaturated polyester composites: A review. J. Eng. Fibers Fabr. 16: 15589250211040184.

[15] Natarajan, E., Markandan, K., Sekar, S.M., Varadaraju, K., Nesappan, S., Albert Selvaraj, A.D., et al. 2022. Drilling-induced damages in hybrid carbon and glass fiber-reinforced composite laminate and optimized drilling parameters. J. Compos. Sci. 6(10): 310. https://doi.org/10.3390/jcs6100310.

[16] Nor, A.F.M., Hassan, M.Z., Rasid, Z.A., Aziz, S.A.A., Sarip, S. and Md. Daud, M.Y. 2021. Optimization on tensile properties of kenaf/multi-walled CNT hybrid composites with box-behnken design. Appl. Compos. Mater. 28: 607–632.

[17] Palanikumar, K., Natarajan, E., Markandan, K., Ang, C.K. and Franz, G. 2023. Targeted pre-treatment of hemp fibers and the effect on mechanical properties of polymer composites. Fibers. 11(5): 43. https://doi.org/10.3390/fib11050043.

[18] Choubey, G., Yadav, P.M., Devarajan, Y. and Huang, W. 2021. Numerical investigation on mixing improvement mechanism of transverse injection based scramjet combustor. Acta Astronautica. 188: 426–437. doi:10.1016/j.actaastro.2021.08.008.

[19] Chen, P., Chen, T., Li, Z., Jia, R., Luo, D., Tang, M., et al. 2020. Transcriptome analysis revealed key genes and pathways related to cadmium-stress tolerance in Kenaf (*Hibiscus cannabinus* L.). Ind. Crops Prod. 158: 112970.

[20] Vellaiyan S. 2023. Energy extraction from waste plastics and its optimization study for effective combustion and cleaner exhaust engaging with water and cetane improver: A response surface methodology approach. Environ. Res. 231: 116113. https://doi.org/10.1016/j.envres.2023.116113.

[21] Abbas, A.G.N., Aziz, F.N.A.A., Abdan, K., Nasir, N.A.M. and Huseien, G.F. 2023. Experimental evaluation and statistical modeling of kenaf fiber-reinforced geopolymer concrete. Constr. Build. Mater. 367: 130228.

[22] Nagappan, B., Devarajan, Y., Kariappan, E., Philip, S.B. and Gautam, S. 2020. Influence of antioxidant additives on performance and emission characteristics of beef tallow biodiesel-fuelled C.I engine. Environ. Sci. Pollut. Res. 28(10): 12041–12055. https://doi.org/10.1007/s11356-020-09065-9

[23] Chen, P., Li, Z., Luo, D., Jia, R., Lu, H., Tang, M., et al. 2021. Comparative transcriptomic analysis reveals key genes and pathways in two different cadmium tolerance kenaf (*Hibiscus cannabinus* L.) cultivars. Chemosphere. 263: 128211.

[24] Pragadish, N., Kaliappan, S., Subramanian, M., Natrayan, L., Satish Prakash, K., Subbiah, R., et al. 2022. Optimization of cardanol oil dielectric-activated EDM process parameters in machining of silicon steel. Biomass Conv. Bioref. 13: 14087–14096.

[25] Harussani, M.M. and Sapuan, S.M. 2022. Development of Kenaf biochar in engineering and agricultural applications. Chemistry Africa. 1–17.

[26] Elango, N. and Faudzi, A.A. 2015. A review article: Investigations on soft materials for soft robot manipulations. Int. J. Adv. Manuf. Technol. 80(5–8): 1027–1037. https://doi.org/10.1007/s00170-015-7085-3.

[27] Vellaiyan, S. 2023. *Bauhinia racemose* seeds as novel feedstock for biodiesel: Process parameters optimisation, characterisation, and its combustion performance and emission characteristics assessment with water emulsion. Process Saf. Environ. Prot. 176: 12–24. https://doi.org/10.1016/j.psep.2023.05.099.

[28] Natarajan, E., Santhosh, M.S., Markandan, K., Sasikumar, R., Saravanakumar, N. and Dilip, A.A. 2022. Mechanical and wear behaviour of Peek, PTFE and Pu: Review and experimental study. J. Polym. Eng. 42(5): 407–417. https://doi.org/10.1515/polyeng-2021-0325.

[29] Sathish, T., Gopal Kaliyaperumal, G. Velmurugan, S.J.A. and Nanthakumar, P. 2021. Investigation on augmentation of mechanical properties of AA6262 aluminium alloy

composite with magnesium oxide and silicon carbide. Mater. Today Proc. 46: 4322–4325.

[30] Santhosh, M.S., Sasikumar, R., Natrayan, L., Kumar, M.S., Elango, V. and Vanmathi, M., 2018. Investigation of mechanical and electrical properties of Kevlar/E-glass and Basalt/E-glass reinforced hybrid composites. Int. J. Mech.Prod. Eng. Res. Dev. 8(3): 591–598.

[31] Elango, N., Gupta, N.S., Lih Jiun, Y. and Golshahr, A. 2017. The effect of high loaded multiwall carbon nanotubes in natural rubber and their nonlinear material constants. J. Nanomater. 1–15. https://doi.org/10.1155/2017/6193961.

[32] Elango, N., Faudzi, A.A., Hassan, A. and Rusydi, M.R. 2014. Experimental investigations of skin-like material and computation of its material properties. Int. J. Precis. Eng. Manuf. 15(9): 1909–1914. https://doi.org/10.1007/s12541-014-0545-0.

[33] Veeman, D., Shree, M.V., Sureshkumar, P., Jagadeesha, T., Natrayan, L., Ravichandran, M., et al. 2021. Sustainable development of carbon nanocomposites: synthesis and classification for environmental remediation. J. Nanomater. 2021(1): 5840645.

[34] Vellaiyan S. 2023. Production and characterization of Cymbopogon citratus biofuel and its optimization study for efficient and cleaner production blended with water and cetane improver: A response surface methodology approach. Fuel. 351: 129000. https://doi.org/10.1016/j.fuel.2023.129000

[35] Balaji, N., Natrayan, L., Kaliappan, S., Patil, P.P. and Sivakumar, N.S. 2022. Annealed peanut shell biochar as potential reinforcement for aloe vera fiber-epoxy biocomposite: mechanical, thermal conductivity, and dielectric properties. Biomass Conv. Bioref. 14: 4155–4163.

[36] Prabhudass, J.M., Palanikumar, K., Natarajan, E. and Markandan, K. 2022. Enhanced thermal stability, mechanical properties and structural integrity of MWCNT filled bamboo/kenaf hybrid polymer nanocomposites. Materials. 15(2): 506. https://doi.org/10.3390/ma15020506.

[37] Asyraf, M.R.M., Syamsir, A., Bathich, H., Itam, Z., Supian, A.B.M., Norhisham, S., et al. 2022. Effect of fibre layering sequences on flexural creep properties of Kenaf fibre-reinforced unsaturated polyester composite for structural applications. Fibers Polym. 23(11): 3232–3240.

[38] Vellaiyan, S., Kuppusamy, S., Chandran, D., Raviadaran, R. and Devarajan, Y. 2023. Optimisation of fuel modification parameters for efficient and greener energy from diesel engine powered by water-emulsified biodiesel with cetane improver. Case Stud. Therm. Eng. 103129. https://doi.org/10.1016/j.csite.2023.103129

[39] Kaushik, D., Gairola, S., Varikkadinmel, B. and Singh, I. 2023. Static and dynamic mechanical behavior of intra-hybrid jute/sisal and flax/kenaf reinforced polypropylene composites. Polym. Compos. 44(1): 515–523.

[40] Elango, N., Mohd Faudzi, A.A., Muhammad Razif, M.R. and Mohd Nordin, I.N. 2013. Determination of non-linear material constants of RTV silicone applied to a soft actuator for robotic applications. Key Eng. Mater. 594–595: 1099–1104. https://doi.org/10.4028/www.scientific.net/kem.594-595.1099.

Chapter **18**

Impact of Fibre Hybridisation and Titanium Oxide Concentration on the Thermomechanical Properties of Sisal-reinforced Polymer Nanocomposites

Natrayan, L.

Department of Mechanical Engineering,
Saveetha School of Engineering,
SIMATS, Chennai, 602105, India
Email: natrayanphd@gmail.com

18.1 INTRODUCTION

The growing concern regarding biodiversity loss has sparked significant interest in leveraging abundant agricultural waste, which constitutes a substantial portion of the ecological crisis [1]. Agricultural leftovers are now readily generated in great quantities at a reasonable cost. A small portion of the vast quantities of garbage is used as a domestic fuel or a fertiliser, while the majority is burnt in the fields. As a result of the air pollution caused by them, these wastes hurt the ecosystem [2, 3]. Another option for addressing the aforementioned issue is to employ these leftovers as fluff in fibre composites. Certain construction company uses fibre-reinforced plastics rather than structural steel [4, 5]. Because of their inherent characteristics like toughness, low cost, and good biocompatibility, natural fabrics have lately found use in polymeric materials [6, 7]. Using renewable resources in polymeric composites reduces CO_2 emissions associated with plastic combustion. Despite the benefits, organic fibre materials have worse physicochemical and mechanical characteristics than artificial dispersion hybrids [8, 9]. To resolve these concerns, natural fabrics are mixed with a thicker inorganic or organic fibre to form a balanced composition in the polymer chain. This results in composite samples that fully use the greatest features of their component fibres, resulting in an optimised, better and cost-effective combination [10, 11]. Manufacturers may customise hybrid compositions and achieve attributes in a much more economically feasible manner with the hybridisation of fibre-reinforced polymeric materials, which is less feasible in binary mixtures including only one type of fibre distributed in the matrices [12]. The benefits of one fibre might supplement what is missing in another by using a polymer nanocomposite that combines multiple types of distinct fibres. As a result, enhanced material development might create a balance of quality and profitability [13].

Sisal, hardwood and coconut fibre have the greatest possibilities as polymer composite supplements of all plant fabrics. Sisal fibre is a natively plentiful substance that helps in generating ecologically beneficial supplies for various sectors, like vehicles, clothing and packaged food [14, 15]. Sisal is a rough fibre derived from the leaves of the agave tree and is among the most often used organic fibrous materials. Sisal is mostly grown in southern Mexico or other tropical regions [16]. Like some other natural fibres, Sisal fibres are primarily derived from renewable resources and are composed of lignocellulose, phenol, and trace quantities of paraffin and contaminants [17]. Sisal fibre has a microfibrillar inclination of 20, which is substantially greater than most other natural fibres. Sisal fibres are widely used to manufacture cables, matting, carpeting, cooking towels, and structural components with a high specific strength of up to 650 MPa [18].

A composite substance is made up of two separate components. A layered material indicates the integration of bonded layers of fabric and polymer. Incorporating materials, utilising active ingredients, or creating a protective coating can improve the structural qualities a composite [19, 20]. Applying resin to sisal gives it the strength to produce several basic products, including windows, sports goods, and water parks. Fibre content influences laminate qualities; adding fibre increases strength. The handling practices have an impact on a composite's qualities as well [21, 22].

Nanostructures are a viable method for creating improved substances for current as well as potential geotechnical engineering. Nanostructures comprise at least one nanometre-sized material that works as a filler in a matrix. Nano titanium oxide nanomaterials are stacked crystalline oxide particles [23, 24]. The usage of titanium oxide as a binding agent in polymers is significantly influenced by its form, particle density, size, surface morphology, and distribution level. Titanium oxide has high thermal conductivity and a porous structure, resulting in greater interface contact between the polymers and the nanofiller and a considerable improvement in polymeric characteristics [25]. Polymer nanocomposites outperform traditional micro-composites in terms of temperature, physical and biological composition [26].

A hybrid laminate is described as a matrix strengthened with more than two additives. The advantages of adopting hybrid composites emanate from the fact that the properties of a particular filler may be enhanced with another filler [16, 27, 28]. Several nano additives are employed today, including carbon nanotubes, micro mud, and graphene fibres. Several recent methods in batteries, detectors, ultracapacitors, etc. use nano titanium oxide as a nano filler. Because of its huge effective surface area, titanium oxide effectively transmits stress and enhances weight bearing capabilities [29].

Introducing titanium oxide nanoparticle fillers to composite materials improves their dynamic, thermodynamic, magnetic and biomechanical characteristics. Significant amount of research has shown that titanium oxide nanocomposites have stronger structural stability than porous carbon hybrids, while being less expensive [30, 31]. According to Experimental Modal Assessment, heat and dynamic performance are essential elements in mechanical systems. DMA is a popular technique for identifying the viscoelastic behaviour and morphologies of polymeric and fibre-reinforced composites. Bathroom walls made of epoxy coating interlaced with sisal are used in washing low-maintenance walls [32, 33]. Epoxy resin is a thermoplastic material hardened after being mixed with a curing agent. In the study discussed in this chapter, several combinations were investigated to determine the qualities of each. Tension experiments were also performed to investigate the laminated composite's tension performance [34, 35].

To analyse the dynamic characteristics of the composition, nonlinear dynamic experiments were performed on the DMA device. Numerous studies have discovered that adding nanofibres or granules to polymers can improve their temperature and dynamic qualities. Furthermore, the nanoscale filler plays a significant role in relaxing the complex molecular polymer molecules. The major aim of the study presented in this chapter was to investigate the influence of sisal fiber and the impact of adding a small quantity of nanoparticles to epoxy resin on the thermomechanical characteristics of multi-layered laminated composites.

18.2 Experimental Procedure

18.2.1 Materials

The materials used in the study include sisal fibre containing woven fibres and randomised fibres sourced from South Asian countries. The mean tenure of the randomised fibres employed in the study was 1.5 mm. Mowbray Chemical Industry in Bangalore, India supplied the epoxy resin and the hardener. The physical and the mechanical properties of the chosen fibre resins at room temperature are shown in Table 18.1. The resin-to-hardener proportion was approximately 10:1. Sisal was baked in a furnace for over 5 hours at 70°C to eliminate all residual water. The samples were manufactured manually by placing sheets within a steel mould in the desired combinations to make a rectangular panel, and then the board was dried at ambient temperature for 60 minutes at a hydrostatic fluid pressure of 40 kg/cm^2. In this research, a coating of fluid paraffin was placed on the mould to assist panel release. Every panel contained 10 stacks of fibres with a dimension of around 3 mm. Table 18.2 shows the specifics of the fibre mixture. A computerised scale was used to measure the adequate amount of randomised fibre to achieve the requisite sisal fibre content to create the random fibre layer. Laser cutting was used to slice the plates into specimens with desired shapes. The samples were dried at 80°C for 60 min to complete the bonding process.

Table 18.1 Physical and mechanical properties of sisal and epoxy matrix.

Sl. No.	Content	Sisal Fibre	Epoxy Matrix
1	Density (g/cm^3)	1.31–1.47	1.12–1.43
2	Tensile Strength (MPa)	550–720	67–79
3	Young's Modulus (GPa)	10–36	2.7–3.1
4	Moisture Absorption (%)	10.25	18.92
5	Elongation at Break (%)	2.1–2.76	1.2–6.3
6	Cellulose (%)	66–77	–
7	Hemicellulose (%)	11–13	–
8	Wax (%)	2.21	–

Titanium oxide (TiO_2) powders to be used as fillers were acquired from Arthy Chemical Industries in India. TiO_2 has a distinct shape and size that improves its thermomechanical characteristics. The TiO_2 in the matrices was dispersed in random directions. The TiO_2 percentages for every permutation were calculated (2, 4, 6 and 8 percent).

Table 18.2 Hybrid composites Parameters and their concentrations.

Trail No.	Sisal Fibre Type	TiO_2 (wt.%)
1	Woven	0
2	Chopped	0
3	Woven	2
4	Chopped	2
5	Woven	4
6	Chopped	4
7	Woven	6
8	Chopped	6
9	Woven	8
10	Chopped	8

18.2.2 Mechanical Characterisation

The influence of weight percent of TiO_2 and sisal fibre type on composite hardness was examined using Deeds hardness measurements. Hardness testing was carried out by D2583-87 ASTM. Tension tests were conducted on an Electronic Control Polymer UTM to investigate how the composite fibres break or deform as a function of the applied stress at ambient temperature. The Charpy test was used to investigate the effect of absorption qualities on a composite. The Charpy

Test Apparatus was used to conduct experiments on rectangle specimens of 14 × 30 × 3 mm dimensions and V-notch inclination of 60°. DSM was used to do heat transfer analysis on materials. The grafted polymer specimens were tested using a differential scanning calorimetric analyser. Temperatures ranged from 20 to 150°C and were measured with N_2 gas at a crosshead speed of 10°C/min.

18.3 RESULT AND DISCUSSION

18.3.1 Density Measurement

Figure 18.1 depicts the thickness of every specimen, and it is obvious from the picture that the kind of sisal and the nanoparticle weight fraction influence the composite thickness. The density of the laminated combination is primarily determined by the proportion of the filler material and the type of the fibre used. In comparison to random fibre specimens, woven fibre specimens have a greater density. The higher density housing of continuous fibres might be due to the voids between the fibre layers increasing the weight per unit surface area of the specimens. The figure clearly shows that the density of the composite sandwich reduces as the proportion of TiO_2 nanoplatelets increases. The density of epoxy composites using 2 wt% TiO_2 is considerably lower than that of fibres made using 4 wt% TiO_2. This is because nanoparticles having high densities have higher weights than nanoparticles with the same dimensions but lower densities [26, 36]. Since TiO_2 has a lower density compared to sisal and epoxy resin, the density decreases as the TiO_2 concentration increases.

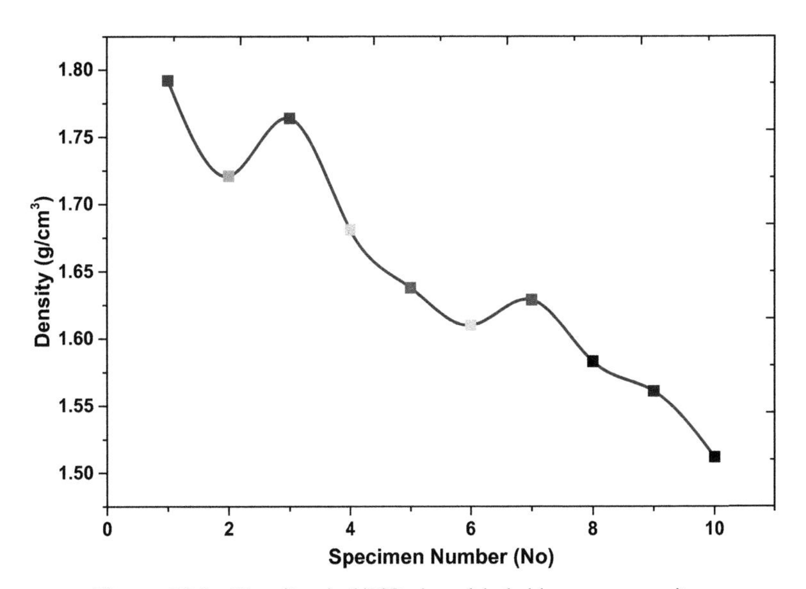

Figure 18.1 Density sisal/TiO_2-based hybrid nanocomposites.

18.3.2 Tensile Behaviour

The mechanical performance of the nanocomposite was evaluated to understand the behaviour of the combination with various types of sisal and varying weight percentages of TiO_2. The tensile strength readings of the various nanocomposites were determined and are depicted in Figure 18.2 to analyse how well the composite distorts or fractures as a function of the total load. Tensile strength improves by approximately 38 percent in continuous and randomised specimens when TiO_2 is used as a reinforcement, and tensile strength is generally higher in continuous fibre specimens [9, 12]. While the tensile properties of specimen 7 were examined compared to those of other specimens, the maximum strength was discovered in specimen 9 at 81.25 MPa.

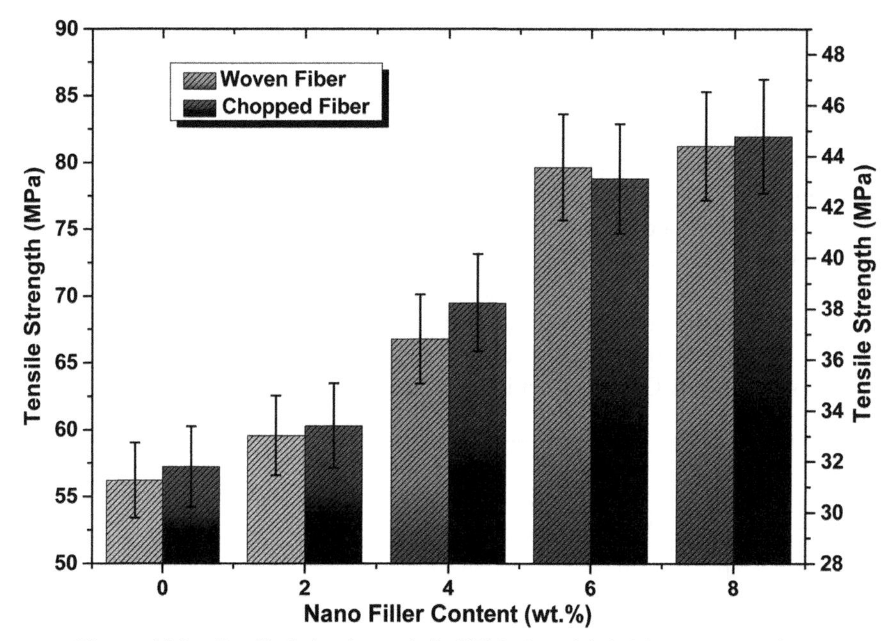

Figure 18.2 Tensile behaviour of sisal/TiO_2-based hybrid nanocomposites.

The interwoven section acts as a barrier for strain transmission from one site to the next, and the increased storing stiffness improves the structural properties of the substance due to the decrease in polymeric strategic sustainability under treatment. Poor binding among fibres and elastomeric resins in the randomised kind deteriorates the nanocomposite's mechanical properties. The fibre configuration in the grain direction is an important measure for boosting the nanocomposite's tensile strength [14, 25]. The rise in dry density and aggregation affects the tensile properties on a radial basis function configuration. The maximum tensile properties are shown in Figure 18.2 for 8 wt. percent TiO_2 continuous fibre. This might be due to the good TiO_2 dispersion.

18.3.3 Impact Strength

Figure 18.3 depicts the link between sampling and absorption energies. The results show that TiO_2 and sisal proportions substantially influence the absorption efficiency of the prepared specimens. The absorption energy in randomised sisal has a similar tendency; it is increased by 12.73, 18.24, 21.41 and 25.42 percent when 2, 4, 6 and 8 wt% TiO_2 are added, respectively. It has been determined that the impacting pressure increased for both types of sisal from 0 to 6 wt% and decreased by 8 wt% thereafter. Specimens with continuous fibre absorbed more power than random fibre specimens. This increase in impact resistance is because the capacity of a composite to absorb light is now determined by the characteristics of the ingredients as well as the strength of the connection between the fibre, resins, and TiO_2 fillers.

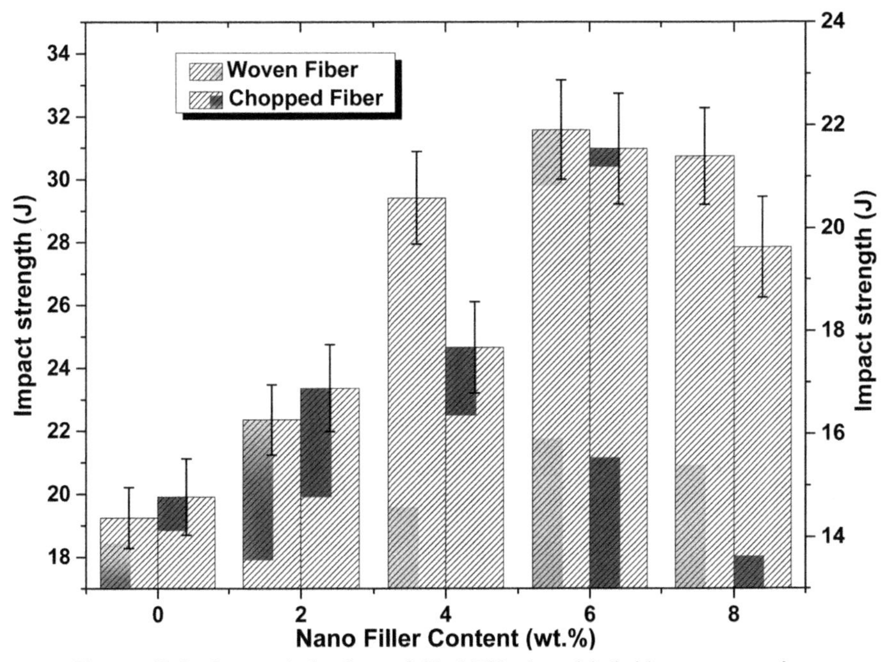

Figure 18.3 Impact behaviour of Sisal/TiO_2-based hybrid nanocomposites.

Due to the obvious poor link between the fibres and the resins in randomised sisal, it absorbed less power than continuous sisal. Furthermore, the excellent impact resistance can be attributed to the expectation of achieving a much more homogenous density of sisal in weaved construction vs. random structure [15, 24]. This lowers the influence of resinous enclaves, and the weaving structure's interlaced structure aids in distributing the high throughput, thus increasing the quantum of absorbed energy. The poor connections between the nanocomposites in random fibre specimens may promote fracture progression across the nanocomposite, absorbing considerably less power in random fibre specimens [23].

18.3.4 Hardness Measurement

Figure 18.4 depicts the toughness test results of a laminated combination supplemented by varying percentages of TiO_2. These findings are arithmetic mean measurements. As shown in the picture, the toughness of the laminated nanocomposites due to the incorporation of TiO_2 is significantly higher than that of the un-reinforced specimens. The resistance properties of the reinforced laminated hybrid reduce the tendency of deformation, in turn increasing the toughness of the material [24]. The figure illustrates that specimens containing continuous fibres have a higher significance than those with scattered fibres. The highest and the lowest fracture toughness of the laminate specimens were 36.52 and 19.65 respectively. This drop in hardness values may be attributed to holes or air pockets introduced in the random fibre specimens at the time of fabrication. The interwoven construction of continuous sisal fibre creates resin chambers at fusion sites that significantly increase the hardness values. Irregular fibres, on the other hand, have a poor distribution of fibres inside the resins. This poor distribution, caused by voids between among fibres and impurities formed throughout glue curing, reduces the hardenability [37, 38].

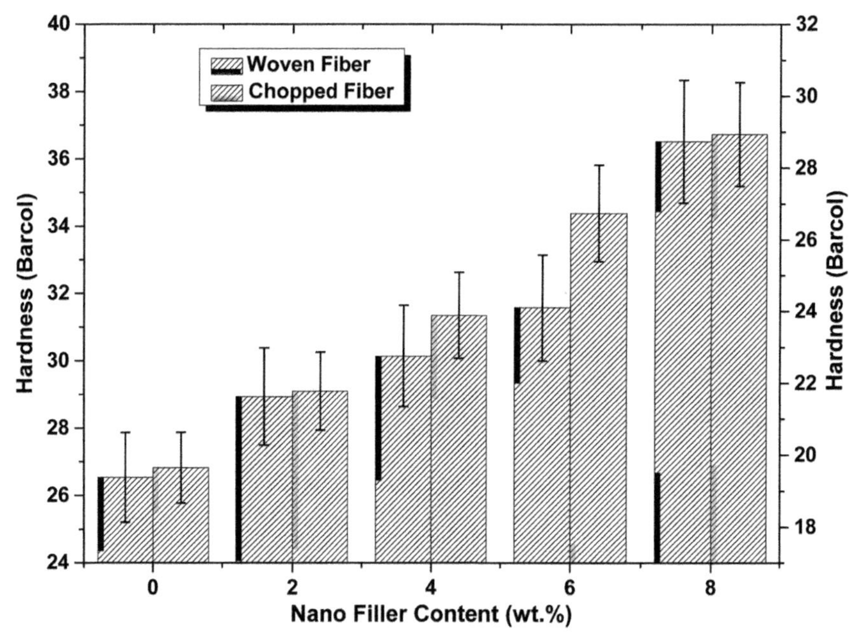

Figure 18.4 Hardness behaviour of Sisal/TiO_2-based hybrid nanocomposites.

18.3.5 Thermal Properties

According to the DSC data in Figure 18.5, the melting point for pure polymer resins is 90 degrees Celsius. Figure 18.5 depicts the effects of varying TiO_2 percentages and sisal type on TG values. It has been shown that the largest changes in TG delta occur whenever the composite transitions from a glass to an elastic range. Furthermore, the figure illustrates that when the wt percent of TiO_2 in the specimens increases, the TG divergence readings drop [39, 40].

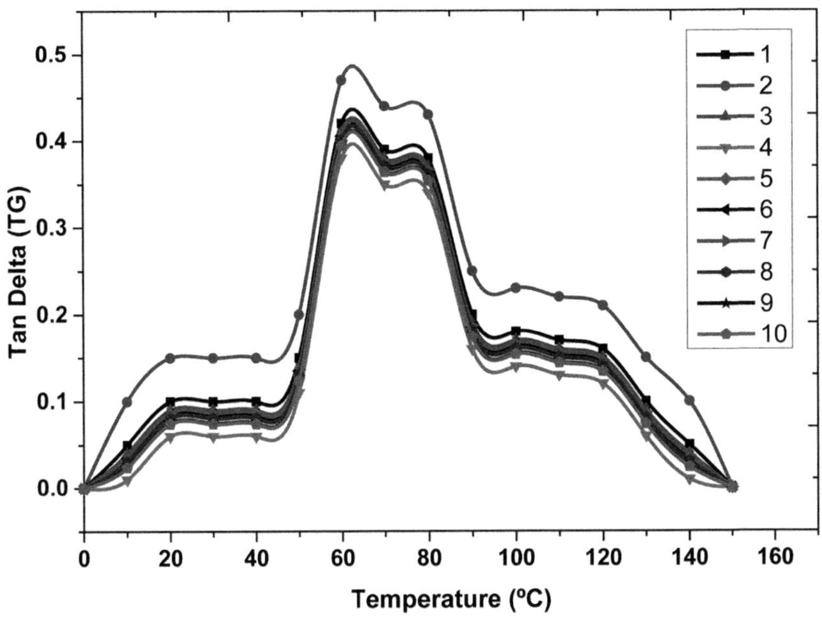

Figure 18.5 Tangent loss chart for all Sisal/TiO_2-based hybrid nanocomposite specimens.

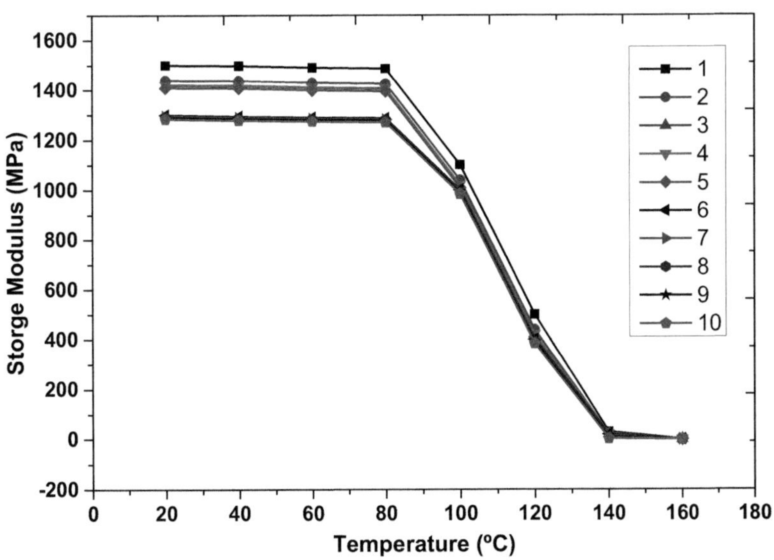

Figure 18.6 Storage Modulus chart for all Sisal/TiO_2-based hybrid nanocomposite specimens.

The thermoplastic filler interactions reduce the relaxing capability of polymeric chain fragments. Figure 18.6 depicts the storage stiffness charts for all specimens, indicating that weaved specimens had the highest compressibility, with specimen 1 having a compressibility of 1500.21 MPa at 20°C and specimen 2 having a compressibility of 1440.81 MPa at 20°C. This enhances adhesion with the epoxy matrix, effectively preventing vacuum formation at the interfaces between the fiber and the lattice [41].

18.4 CONCLUSION

In the study discussed in the chapter, hybrid composites were created using TiO_2 nanofillers and several kinds of sisal fibre. Post synthesis, the composites were mechanically and thermally tested. The following findings were drawn from the experimental assessments.

- Compared to other combinations, 8 wt.% of nanofiller with woven type fibre exhibits the highest tensile strength values, in the order of 81.25 MPa. Similarly, 6 wt.% of nanofiller with woven type sisal displays 31.58 J of impact and hardness properties.
- The density of epoxy composites using 2 wt% TiO_2 is dramatically lower than that of fibres made using 4 wt% TiO_2. This is because nanoparticles having high densities have higher weights than nanoparticles with the same dimensions but lower densities.
- From the thermal study, it can be observed that weaved specimens had the highest compressibility, with specimen 1 having a compressibility of 1500.21 MPa at 20°C and specimen 2 having a compressibility of 1440.81 MPa at 20°C.
- The highest and the lowest hardness of laminate specimens were 36.52 and 19.65 respectively. This drop in hardness values might be attributed to the holes or air pockets introduced into the random fibre specimens during fabrication. The interwoven construction of continuous sisal fibre creates resin chambers at fusion sites that significantly increase hardness values.
- Compared to chopped fibre, woven fibre shows higher strength. This is due to the obvious poor links between the fibres and the resins in randomised sisal; it absorbed less impact than continuous sisal. Furthermore, the excellent impact resistance can be owed to the expectation of achieving a much more homogenous density of sisal in weaved construction vs. random structure.

REFERENCES

[1] Idicula, M., Joseph, K. and Thomas, S. 2010. Mechanical performance of short banana/sisal hybrid fiber reinforced polyester composites. J. Reinf. Plast. Compos. 29: 12–29. https://doi.org/10.1177/0731684408095033.

[2] Venkateshwaran, N., Elayaperumal, A. and Jagatheeshwaran, M.S. 2011. Effect of fiber length and fiber content on mechanical properties of banana fiber/epoxy composite. J. Reinf. Plast. Compos. 30: 1621–1627. https://doi.org/10.1177/0731684411426810.

[3] Ary Subagia, I.D.G., Kim, Y., Tijing, L.D., Kim, C.S. and Shon, H.K. 2014. Effect of stacking sequence on the flexural properties of hybrid composites reinforced with carbon and basalt fibers. Compos. Part B Eng. 58: 251–258. https://doi.org/10.1016/j.compositesb.2013.10.027.

[4] Elango, N., Mohd Faudzi, A.A., Muhammad Razif, M.R. and Mohd Nordin, I.N. 2013. Determination of non-linear material constants of RTV silicone applied to a soft actuator for robotic applications. Key Eng. Mater. 594–595: 1099–1104. https://doi.org/10.4028/www.scientific.net/kem.594-595.1099.

[5] Sahari, J., Sapuan, S.M., Zainudin, E.S. and Maleque, M.A. 2013. Mechanical and thermal properties of environmentally friendly composites derived from sugar palm tree. Mater. Des. 49: 285–289. https://doi.org/10.1016/j.matdes.2013.01.048.

[6] Majeed, K., Jawaid, M., Hassan, A., Abu Bakar, A., Abdul Khalil, H.P.S., Salema, A.A., et al. 2013. Potential materials for food packaging from nanoclay/natural fibres filled hybrid composites. Mater. Des. 46: 391–410. https://doi.org/10.1016/j.matdes.2012.10.044.

[7] Vellaiyan, S., Kuppusamy, S., Chandran, D., Raviadaran, R. and Devarajan, Y. 2023. Optimisation of fuel modification parameters for efficient and greener energy from diesel engine powered by water-emulsified biodiesel with cetane improver. Case Stud. Therm. Eng. 103129. https://doi.org/10.1016/j.csite.2023.103129

[8] Bodros, E., Pillin, I., Montrelay, N. and Baley, C. 2007. Could biopolymers reinforced by randomly scattered flax fibre be used in structural applications? Compos. Sci. Technol. 67: 462–470. https://doi.org/10.1016/j.compscitech.2006.08.024.

[9] Morye, S.S. and Wool, R.P. 2005. Mechanical properties of glass/flax hybrid composites based on a novel modified soybean oil matrix material. Polym. Compos. 26: 407–416. https://doi.org/10.1002/pc.20099.

[10]. Veeman, D., Shree, M.V., Sureshkumar, P., Jagadeesha, T., Natrayan, L., Ravichandran, M., et al. 2021. Sustainable development of carbon nanocomposites: synthesis and classification for environmental remediation. J. Nanomater. 2021(1): 5840645.

[11] Mohammed, M., Betar, B.O., Rahman, R., Mohammed, A.M., Osman, A.F., Jaafar, M., et al. 2019. Zinc oxide nano particles integrated kenaf/unsaturated polyester biocomposite. J. Renewable Mater. 7(10): 967–982.

[12] Vellaiyan S. 2023. Production and characterization of cymbopogon citratus biofuel and its optimization study for efficient and cleaner production blended with water and cetane improver: A response surface methodology approach. Fuel. 351: 129000. https://doi.org/10.1016/j.fuel.2023.129000.

[13] Balaji, N., Natrayan, L., Kaliappan, S., Patil, P.P. and Sivakumar, N.S. 2022. Annealed peanut shell biochar as potential reinforcement for aloe vera fiber-epoxy biocomposite: mechanical, thermal conductivity, and dielectric properties. Biomass Convers. Biorefin. 14: 4155–4163.

[14] Kumar, M.S., Mangalaraja, R.V., Kumar, R.S. and Natrayan, L. 2019. Processing and characterization of AA2024/Al2O3/SiC reinforces hybrid composites using squeeze casting technique. Iran. J. Mater. Sci. Eng. 16(2): 55–67.

[15] Prabhudass, J.M., Palanikumar, K., Natarajan, E. and Markandan, K. 2022. Enhanced thermal stability, mechanical properties and structural integrity of MWCNT filled bamboo/kenaf hybrid polymer nanocomposites. Materials. 15(2): 506. https://doi.org/10.3390/ma15020506.

[16] Santhosh, M.S., Sasikumar, R., Natrayan, L., Kumar, M.S., Elango, V. and Vanmathi, M. 2018. Investigation of mechanical and electrical properties of Kevlar/E-glass and Basalt/E-glass reinforced hybrid composites. Int. J. Mech. Prod. Eng. Re. Dev. 8(3): 591–598.

[17] Elango, N., Gupta, N.S., Lih Jiun, Y. and Golshahr, A. 2017. The effect of high loaded multiwall carbon nanotubes in natural rubber and their nonlinear material constants. J. Nanomater. 1–15. https://doi.org/10.1155/2017/6193961.

[18] Nagappan, B., Devarajan, Y., Kariappan, E., Philip, S.B. and Gautam, S. 2020. Influence of antioxidant additives on performance and emission characteristics of beef tallow biodiesel-fuelled C.I engine. Environ. Sci. Pollut. Res. 28(10): 12041–12055. https://doi.org/10.1007/s11356-020-09065-9

[19] Elango, N, Faudzi, A.A., Hassan, A. and Rusydi, M.R. 2014. Experimental investigations of skin-like material and computation of its material properties. Int. J. Precis. Eng. Manuf. 15(9): 1909–1914. https://doi.org/10.1007/s12541-014-0545-0.

[20] EsmaeilpourShirvani, N., TaghaviGhalesari, A., Tabari, M.K. and Choobbasti, A.J. 2019. Improvement of the engineering behavior of sand-clay mixtures using kenaf fiber reinforcement. Transp. Geotech. 19: 1–8.

[21] Elango, N. and Faudzi, A.A. 2015. A review article: Investigations on soft materials for soft robot manipulations. Int. J. Adv. Manuf. Technol. 80(5–8): 1027–1037. https://doi.org/10.1007/s00170-015-7085-3.

[22] Vellaiyan, S. 2023. *Bauhinia racemose* seeds as novel feedstock for biodiesel: Process parameters optimisation, characterisation, and its combustion performance and emission characteristics assessment with water emulsion. Process Saf. Environ. Prot. 176: 12–24. https://doi.org/10.1016/j.psep.2023.05.099.

[23] Pragadish, N., Kaliappan, S., Subramanian, M., Natrayan, L., Satish Prakash, K., Subbiah, R., et al. 2022. Optimization of cardanol oil dielectric-activated EDM process parameters in machining of silicon steel. Biomass Convers. Biorefin. 13(15): 14087–14096.

[24] Natarajan, E., Santhosh, M.S., Markandan, K., Sasikumar, R., Saravanakumar, N. and Dilip, A.A. 2022. Mechanical and wear behaviour of Peek, PTFE and Pu: Review and experimental study. J. Polym. Eng. 42(5): 407–417. https://doi.org/10.1515/polyeng-2021-0325

[25] Sathish, T., Gopal Kaliyaperumal, G. Velmurugan, S.J.A. and Nanthakumar. P. 2021. Investigation on augmentation of mechanical properties of AA6262 aluminium alloy composite with magnesium oxide and silicon carbide. Mater. Today Proc. 46: 4322–4325.

[26] Paranthaman, V., Sundaram, K.S. and Natrayan, L. 2022. Influence of SiC particles on mechanical and microstructural properties of modified interlock friction stir weld lap joint for automotive grade aluminium alloy. Silicon. 14(4): 1617–1627.

[27] Vellaiyan, S. 2023. Energy extraction from waste plastics and its optimization study for effective combustion and cleaner exhaust engaging with water and cetane improver: A response surface methodology approach. Environ. Res. 231: 116113. https://doi.org/10.1016/j.envres.2023.116113

[28] Ramaswamy, R., Gurupranes, S.V., Kaliappan, S., Natrayan, L. and Patil, P.P. 2022. Characterization of prickly pear short fiber and red onion peel biocarbon nanosheets toughened epoxy composites. Polym. Compos. 43(8): 4899–4908.

[29] Natarajan, E., Markandan, K., Sekar, S.M., Varadaraju, K., Nesappan, S., Albert Selvaraj, A.D., et al. 2022. Drilling-induced damages in hybrid carbon and glass fiber-

reinforced composite laminate and optimized drilling parameters. J. Compos. Sci. 6(10): 310. https://doi.org/10.3390/jcs6100310.

[30] Palanikumar, K., Natarajan, E., Markandan, K., Ang, C.K. and Franz, G. 2023. Targeted pre-treatment of hemp fibers and the effect on mechanical properties of polymer composites. Fibers. 11(5): 43. https://doi.org/10.3390/fib11050043

[31] Raju, J.S.N., Depoures, M.V., Shariff, J. and Chakravarthy, S. 2022. Characterization of natural cellulosic fibers from stem of *Symphirema involucratum* plant. J. Nat. Fibers. 19(13): 5355–5370. https://doi.org/10.1080/15440478.2021.1875376.

[32] Natarajan, E., Ramesh, S., Markandan, K., Saravanakumar, N., Dilip, A.A. and Batcha, A.R. 2023. Enhanced mechanical, tribological, and acoustic behavior of polyphenylene sulfide composites reinforced with zero-dimensional alumina. J. Appl. Polym. Sci. 140(16): e53748. https://doi.org/10.1002/app.53748.

[33] Fajrin, J., Akmaluddin, A. and Femiana, G. 2022. Utilization of kenaf fiber waste as reinforced polymer composites. Results Eng. 13: 100380.

[34] Ramesh, T., Sathiyagnanam, A.P., Poures, Melvin Victor De and Murugan, P. 2022. A comprehensive study on the effect of dimethyl carbonate oxygenate and EGR on emission reduction, combustion analysis, and performance enhancement of a CRDI diesel engine using a blend of diesel and *Prosopis juliflora* biodiesel. Int. J. Chem. Eng. 2022: 5717362.

[35] Gurusamy, M., Vellaiyan, S., Kandasamy, M. and Devarajan, Y. 2023. Optimization of process parameters to intensify the yield rate of biodiesel derived from waste and inedible *Carthamus lanatus* (L.) Boiss. seeds and examine the fuel properties with pre-heated water emulsion. Sustainable Chem. Pharm. 33: 101137. https://doi.org/10.1016/j.scp.2023.101137

[36] Natrayan, L., Merneedi, A., Bharathiraja, G., Kaliappan, S., Veeman, D. and Murugan, P. 2021. Processing and characterization of carbon nanofibre composites for automotive applications. J. Nanomater. 1–7.

[37] Natarajan, E., Freitas, L.I., Santhosh, M.S., Markandan, K., Majeed Al-Talib, A.A., and Hassan, C.S. 2023. Experimental and numerical analysis on suitability of S-glass-carbon fiber reinforced polymer composites for submarine hull. Def. Technol. 19: 1–11. https://doi.org/10.1016/j.dt.2022.06.003.

[38] Ramesh, T., Sathiyagnanam, A.P., Melvin Victor De Poures and Murugan, P. 2022. Combined effect of compression ratio and fuel injection pressure on CI engine equipped with CRDi system using Prosopis juliflora methyl ester/diesel blends. Int. J. Chem. Eng. 1–12.

[39] Asumani, O. and Ratnam, P. 2021. Fatigue and impact strengths of kenaf fibre reinforced polypropylene composites: Effects of fibre treatments. Adv. Compos. Mater. 30(2): 103–115.

[40] Meikandan, M., Karthick, M., Natrayan, L., Patil, P.P., Sekar, S., Rao, Y.S., et al. 2022. Experimental investigation on tribological behaviour of various processes of anodized coated piston for engine application. J. Nanomater. 2022(1): 7983390.

[41] Guo, A., Zhihui, S. and Jagannadh, S. 2020. Impact of modified kenaf fibers on shrinkage and cracking of cement pastes. Constr. Build. Mater. 264: 120230.

Chapter 19

Perspectives on MXene Tribology

Kuhan Ganesan, Elango Natarajan,
Ang Chun Kit and Kalaimani Markandan*
Faculty of Engineering, Technology and Built Environment,
UCSI University, 56000, Malaysia

19.1 PROPERTIES OF MXENE

MXene is a family of two-dimensional (2D) materials consisting of transition metal carbides or nitrides with a general formula of $M_{n+1}X_nT_x$, where M represents a transition metal, X is carbon or nitrogen, T is a surface functional group, and n usually ranges from 1 to 3. Due to their distinctive characteristics such as high surface area, good mechanical properties, high electrical conductivity, and excellent thermal stability, MXene have become a subject of great interest and potential applications in various fields such as catalysis, energy storage, and electronics [1–4].

One of the notable properties of MXene is the high electrical conductivity. MXene exhibits high electrical conductivity, which makes the compounds ideal candidates for applications such as energy storage devices and electromagnetic interference shielding. For example, a study has shown that MXene films exhibit electrical conductivity up to 10,000 S/cm, which is higher than most metals [5]. Besides that, MXene has a high surface area, making it an excellent material for energy storage applications such as supercapacitors. The high surface area enables the electrode to store more charge, resulting in high energy density. A study has

*For Correspondence: kalaimani@ucsiuniversity.edu.my

shown that MXene electrodes can achieve energy densities of up to 38.6 Wh L^{-1}, which is higher than most other electrode materials [6]. In another study by Liu et al., it was reported that MXene exhibits good thermal stability, improving the reliability of the related devices effectively [7]. For instance, the thermal conductivity of $T_{i3}C_2T_x$ and $Ti_3C_2T_x$ – PVA composite reached as high as 55.8 and 47.6 W/m.K respectively, where it was also reported that PVA significantly improved the thermal stability of $Ti_3C_2T_x$ by reducing the thermal coefficient of Eg1 mode from –0.06271 to –0.03357 cm^{-1}/K due to the strong Ti–O bonds that were formed between MXene and the PVA polymer.

Figure 19.1 shows some scanning electron microscopy (SEM) images of as synthesized $Ti_3C_2T_x$ MXene as well as delaminated MXene as reported in literature. In terms of mechanical properties, it has been shown from first principle calculations that the M–C and/or M–N bond in the MXene structure is strong, which generates a high elastic constant exceeding 500 GPa [8]. Besides, the MXene structure can sustain up to 17% strain under uniaxial and biaxial tension; the value can be enhanced further, by 28%, after surface functionalization that prevents the surface atomic layers from collapsing. These properties can be highly appealing for 2D flexible devices. To summarize, the unique combination of properties exhibited by MXene makes them promising materials for a wide range of applications such as fuel cells, Li-on batteries and supercapacitors. Ongoing research is expected to uncover new and exciting properties of MXene and lead to the further development of their applications.

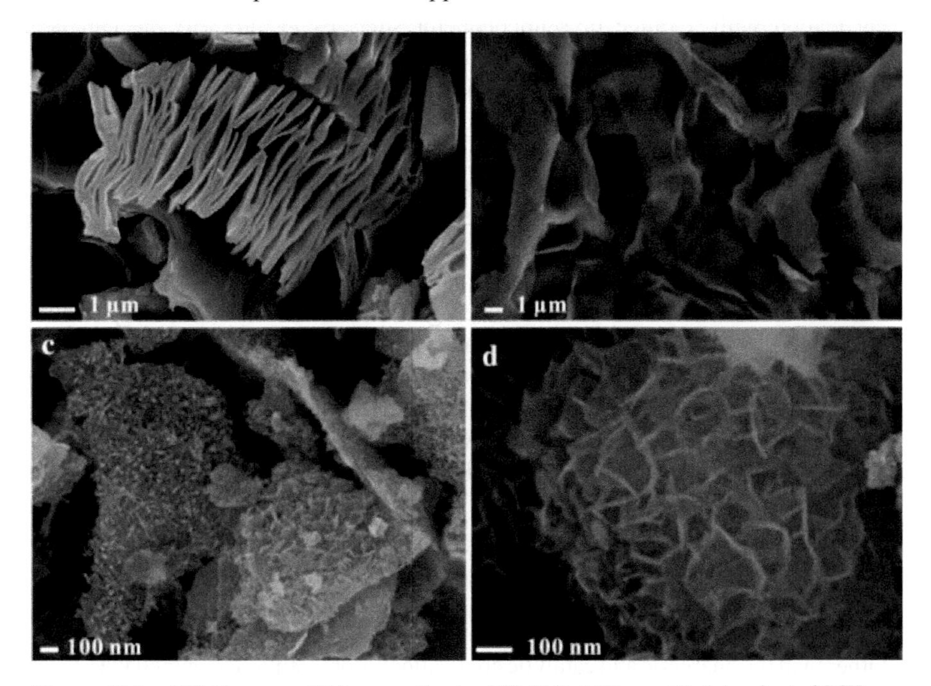

Figure 19.1 SEM images of (a) as-synthesized $Ti_3C_2T_x$ MXene, (b) delaminated MXene, and (c and d) MXene/HS composites [9].

19.2 RECENT ADVANCES IN APPLICATIONS OF MXENE

Today, MXene is being utilized in several fields such as energy storage, water purification, and biomedical applications. MXene is a 2D material that belongs to the family of transition metal carbides, nitrides and carbonitrides, with a typical formula of $M_{n+1}X_nT_x$. The elements M and X represent an early transition metal and carbon/nitrogen respectively, and T denotes surface functional groups. MXene has garnered considerable interest in recent times, owing to its exceptional characteristics, such as high surface area, good mechanical properties, high electrical conductivity, and excellent thermal stability [10, 11].

MXene has a wide range of potential applications, including energy storage devices. MXene has shown promising results as an electrode material for energy storage devices, including supercapacitors and batteries. A study showed that MXene electrodes can achieve high energy densities of up to 210 Wh/kg, which is higher than most other electrode materials [12]. Shielding from electromagnetic interference (EMI) has also been accomplished with MXene. MXene is a strong contender for EMI shielding applications due to its outstanding EMI shielding characteristics. According to a study, MXene films can have EMI shielding efficiencies of up to 60 dB – higher than most other EMI shielding materials [13].

MXene can be considered as the next generation high performance electrocatalyst for hydrogen generation due to the following reasons: (i) a large number of –OH and –O groups the on surface of MXene establish a strong correlation with a range of semiconductor surfaces [3, 14]; (ii) the enhanced electrical conductivity of MXene supports efficient charge-carrier transfer [15]; (iii) the metal sites exposed at the terminals of MXene allow strong redox activity compared to carbon materials [16]; and (iv) the hydrophilicity of MXene allows for sufficient contact with water molecules and stability in aqueous solution [17]. In a study by Zhang et al. platinum atoms were immobilized on MXene to serve as an efficient catalyst to enhance the latter's catalytic activity for HER [16]. The authors reported that the catalyst exhibited high catalytic ability with low overpotentials of 30 and 77 mV to achieve current densities of 10 and 100 mA/cm^2 with a mass activity that was 40 times greater than the commercial platinum-on-carbon catalyst. In a study by Yuan et al. highly active MXene nanofibers with high specific surface area was used as a catalyst for HER. It was reported that the MXene nanofibers enhanced HER activity, with a low overpotential of 169 mV at a current density of 10 mA/cm^2 [18]. The high specific surface area and exposed active sites were responsible for the enhanced hydrogen generation.

19.3 PROPERTIES OF MXENE FOR LUBRICATION APPLICATIONS

Other than the energetic and catalytic applications aforementioned, lamellar Ti_3C_2, as the most representative member of the MXene family, has gained significant interest for tribological purposes, given its graphite-like structure with self-lubricating properties and low shear strength [10, 11]. These findings were obtained from density functional calculations and molecular dynamic simulations [19]. Liu et al. reported that the addition of 0.8 wt.% Ti_3C_2 reduced the COF of PAO8 oil by 9.6% [20], while Gao et al. reported that $Ti_3C_2T_x$-based nano additives reduced the wear volume of base oil by 87%, with a high load carrying capacity (i.e. 500 N) [21]. In a more recent study, Markandan et al. reported that functionalization and hybrid MXene can improve the tribological behavior of engine oil [22]. For example, the coefficient of friction (COF) and wear scar diameter (WSD) reduced by 13.9% and 23.8% respectively with the addition of 0.05 wt.% MoS_2-Ti_3C_2, compared to base engine oil, due to the interlaminar shear susceptibility of MXene. To ascertain the findings, SEM images of the wear scar were obtained and it was found that ball bearings without MXene exhibited dark concentric groves, whereas the addition of unfunctionalized MoS_2-Ti_3C_2 hybrid nanoparticles did not show significant appearance of the dark concentric grooves, thus proving the lesser abrasive wear (Figure 19.2).

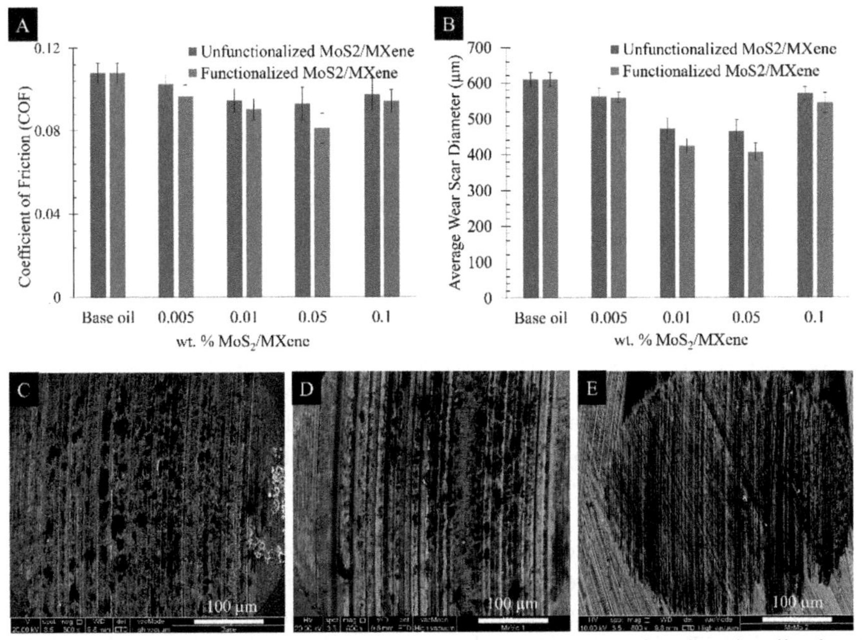

Figure 19.2 (A) COF and (B) average wear scar diameter of unfunctionalized and functionalized MoS_2-Ti_3C_2-based nanolubricant; (C–E) SEM images of wear surfaces of base oil without additive, base oil with unfunctionalized MoS_2-Ti_3C_2 and base oil with functionalized MoS_2-Ti_3C_2 respectively [22].

MXene materials have a high surface area due to their 2D layered structure, which provides more active sites for adsorption and reaction with lubricating molecules. The high surface area can also enhance the film-forming ability of lubricants and reduce the wear and friction of the surfaces. The protective layer formed by MXene on the surface of the moving parts reduces direct contact between the surfaces, thereby reducing wear and friction. For example, in a study by Gao et al. it was shown that commercial lubricating additive dialkyl dithiophosphate (DDP) functionalized MXene exhibited enhanced anti-wear and friction reduction abilities with a low coefficient of friction (not exceeding 0.11), reduction in the corresponding wear volume by 87%, as well as high load-carrying capacity up to 500 N [21]. MXene can form a protective tribofilm on the surface of the moving parts. This tribofilm acts as a physical barrier between the surfaces and reduces wear and friction. MXene has high thermal stability, which allows the tribofilm to withstand high temperatures without breaking down. This is because MXene has strong covalent bonds within the layers and weak van der Waals interactions between the layers, which allow it to maintain its structural integrity at high temperatures. As a result, MXene can maintain their lubrication properties even under high temperature and high load conditions.

19.3.1 Enhanced Oxidative Stability

Today, research on MXene has also proved to enhance the oxidative stability of biolubricants. These nanoparticles can act as antioxidants and prevent the formation of free radicals which are responsible for the degradation of lubricants. The addition of nanoparticles can also improve the thermal stability and wear resistance of biolubricants. For example, MXene are two-dimensional materials with a large surface area and high conductivity, which can act as electron reservoirs and facilitate charge transfer reactions. This property can effectively scavenge free radicals which are responsible for the oxidative degradation of lubricants.

Nagarajan et al. used advanced rapid microwave synthesis technique to synthesize nanoparticles of molybdenum disulfide (MoS_2)-MXene to be used as an additive in SAE 20W50 based diesel engine oil [23]. In comparison with base oil, the oxidation induction time (OIT) of the nanolubricant with 0.01 wt.% MoS_2 nanoparticles increased by 61.15%. Additionally, it was discovered that the anti-oxidant properties of the nanolubricant were enhanced as a result of the synergistic interaction between MoS_2 and zinc-dialkyldithiophosphate (ZDDP) [24, 25]. MoS_2 and ZDDP work together synergistically to encourage hydrogen donation which halts the spread of radicals. However, compared to 0.05 wt% and 0.01 wt%, the nanolubricants containing 0.1 wt%, 0.01 wt% and 0.005 wt% of MoS_2 nanoparticles showed lower OIT since the aforementioned concentrations were not optimal for generating a greater OIT in the nanolubricants. The significant increase in OIT of nanolubricants demonstrates the synergistic capacity of MoS_2 nanoparticles and ZDDP to exhibit superior oxidation stability, hence increasing the antioxidant characteristics of the nanolubricants.

19.3.2 Enhanced Thermal Conductivity

MXene is a type of two-dimensional material with high thermal conductivity by virtue of its layered structure and metallic nature. When added to any base oil, MXene can improve the lubrication performance by enhancing the heat dissipation and reducing the temperature rise in the lubricating system. In a recent study by Rubbi et al. a new class of nanofluids was formulated using soybean oil and MXene (Ti_3C_2) particles, to be utilized as a working fluid on a hybrid photovoltaic-thermal (PV/T) solar collector, for performance optimization [26]. At 55°C, the thermal conductivity of the SO-Ti_3C_2 nanofluid with 0.125 wt.% Ti_3C_2 was enhanced by 60.82%, compared to pure soybean oil. The authors reported 84.25% overall thermal efficiency at 0.07 kg/s discharge. Moreover, by substituting water-alumina with soybean oil-Ti_3C_2 nanofluids as the cooling fluid, the hybrid PV/T system showed optimized electrical output–increased by 15.44% at 0.07 kg/s of mass flow rate.

In a study by Rahmadiawan et al. 2021, MXene (Ti_3C_2) nanoflakes were combined with palm oil methyl ester (POME) to enhance the thermo-physical characteristics of nanofluids (POME/MXene) [27]. Five different concentrations of MXene (0.01, 0.03, 0.05, 0.08 and 0.1 wt%) were used to produce the POME/MXene nanofluid. The highest improvement in thermal conductivity was measured to be 176% compared to the base fluid at MXene concentration and temperature of 0.1 wt% and 65°C respectively. This discovery provides an exciting field of research to induce rapid cooling using MXene-based nanfluids which can be utilized as a promising heat transfer fluid. In conclusion, MXene has demonstrated superior thermal conductivity and oxidative stability compared to other nanoparticles such as graphene, maghemite or MoS_2. This makes MXene a promising candidate for improving the thermal and oxidative properties of various base oils, which can ultimately lead to improved performance and longer equipment lifespan in various industries.

19.3.3 Coefficient of Friction and Wear Scar Diameter

Nanoparticles can be added to biolubricants to improve the latter's performance in terms of reducing the coefficient of friction and wear scar diameter. The addition of nanoparticles to biolubricants can improve their tribological properties by providing a protective film that reduces the contact between the two surfaces in contact, reducing friction and wear (Table 19.1). A common additive that has been reported besides MXene is graphene. In a study by Alqahtani et al. the effect of adding graphene nanoparticles to Shell Helix 5W-30 diesel engine oil was investigated [28]. The authors reported that the addition of 0.12 wt.% graphene improved the lubrication performance, reducing engine wear and friction by 15% and 21% respectively.

Owing to the excellent tribological properties of MXene, recent studies have explored the potential of MXene as a additive in various base oils. For example, in a recent study reported in the year 2023, MXene and tetradecylphosphonic acid (TDPA) were used as additives in PA08 [29]. It was reported that the additive ensured friction reduction and wear resistance at contact pressures less than 1.4 GPa, owing to the iron-oxide and $FePO_4$ tribochemical reaction film as well as the lower interlaminar shear between MXene layers. In a study by Zhou et al. MXene-HS nanoparticles were added into PAO-10, where addition of 0.12 wt.% of the former resulted in 49.4% wear scar diameter reduction (Figure 19.3) [30]. Based on the literature findings above, the addition of MXene nanoparticles to various base oils has been shown to improve engine performance by a higher percentage compared to other nanoparticles, by reducing friction and wear and improving fuel efficiency.

Table 19.1 WSD and COF of various nanoparticles added to base oils.

Base Oil	Nanoparticle	Concentration of the Nanoparticle (wt %)	COF (% reduction)	WSD (% reduction)	References
Shell Helix 5W-30	Graphene	0.12	15	21	(19)
Vegetable Oil	Al_2O_3/TiO_2	0.1	20.51	44	(22)
Trimethylpropane ester (TMP)	TiO_2	–	15	11	(23)
Rice Bran Oil	SiC	0.03	15.6	5.7	(24)
SAE 20W50 Diesel Engine Oil	MoS_2	0.01	19.24	19.52	(20)
Sn 500 and Axle Oil	MoS_2	0.05	60	7	(25)
Sal Oil	CuO	O.5	8	4.2	(21)
Synthetic Lubricating Oil	ZrO_2/SiO_2	0.1	16.24	14.59	(26)
Rape seed Oil	CeO_2	0.1	54	–	(27)
Castor Oil	TDPA–MXene	0.1	27.9	–	(11)
PAO–10	MXene–HS	0.12	–	49.4	(10)

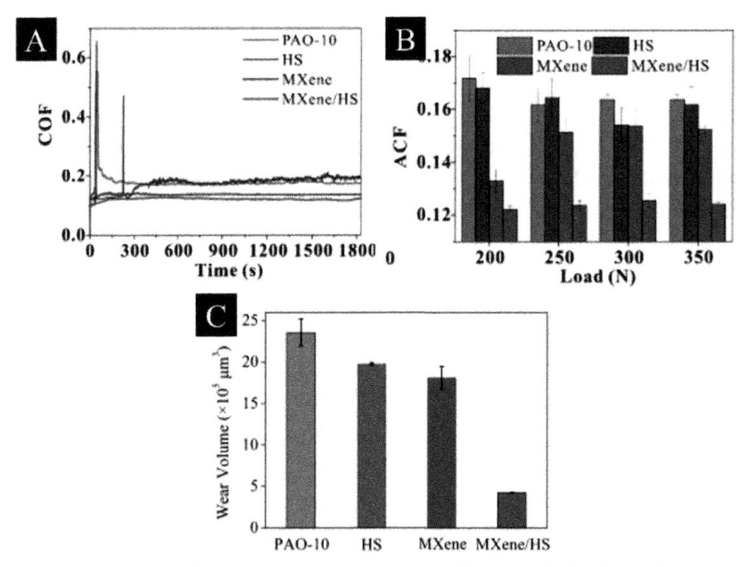

Figure 19.3 (A, B) Friction curves and average coefficient of friction using PAO-10 and PAO-10 containing 1.0 wt.% MXene, HS and MXene/HS at varying; (C) wear volumes of the different wear tracks [30].

19.4 LUBRICATION MECHANISM OF NANOPARTICLES

Lubrication by addition of nanoparticles can improve engine efficiency by reducing friction and wear between the moving parts. The nanoparticles in the engine oil can form a protective tribofilm, reducing direct metal-to-metal contact between the surfaces. Nanoparticles can also improve lubrication efficiency by enhancing the fluid film's thickness, reducing viscosity and improving surface quality. This can reduce frictional losses in the engine, thus improving fuel efficiency and reducing emissions. The selection and optimization of nanoparticles for engine oils is crucial for achieving the desired engine performance benefits. The mechanisms through which fillers in engine oils reduce COF and WSD are rolling, polishing and mending effects (Figure 19.4). The following sub-sections will discuss these mechanisms in detail.

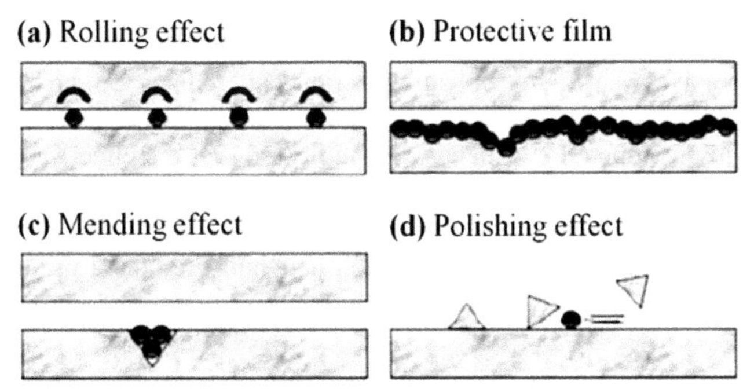

Figure 19.4 Lubrication mechanisms of nanoparticles: (a) rolling effect mechanism, (b) protective film, (c) mending effect mechanism, and (d) polishing effect mechanism [34].

19.4.1 Rolling Effect Mechanism

Nanoparticles added to an engine oil can improve its tribological performance through a rolling action mechanism. This mechanism involves the rolling of nanoparticles between the contacting surfaces, which can reduce friction and wear. When the nanoparticles are introduced into the engine oil, they can be dispersed uniformly throughout the lubricant. During operation, the nanoparticles are attracted to the surface of the contacting components due to their surface chemistry and van der Waals forces. As the components move relative to each other, the nanoparticles roll between the surfaces, creating a rolling action mechanism. The rolling action mechanism can reduce friction and wear through several mechanisms. First, the nanoparticles can fill in the surface defects and roughness, hence reducing the real contact area between the surfaces and decreasing friction by establishing a protective layer on the surfaces (the layer serves as a shield to minimize wear). Additionally, the movement of nanoparticles in a rolling manner can create pressure to produce a fluid film that improves hydrodynamic lubrication and consequently decreases friction. In the research conducted by Tao et al., it was discovered that when spherical diamond nanoparticles are utilized as an additive in paraffin oil, they can penetrate rubbing surfaces, support the load, and prevent direct contact between the two surfaces [31].

19.4.2 Polishing Effect Mechanism

The polishing action mechanism of nanoparticle-enhanced engine oil refers to the ability of nanoparticles to polish and smoothen the surface of the contacting components. The addition of nanoparticles to engine oil can enhance the lubricant's ability to fill in surface defects and irregularities, resulting in a smoother surface finish. When the nanoparticles in the lubricant come in contact with the surface of the component, they can interact with the surface material through tribochemical reactions. This can lead to the formation of a thin and uniform protective layer on the surface, which can further reduce the surface roughness and improve the tribological performance of the lubricant [32]. Moreover, the nanoparticles can also act as polishing agents by physically removing the surface asperities through abrasive wear. This polishing action can reduce the surface roughness and improve the contact between the components, leading to reduced friction and wear. A study conducted by Zhang et al. indicated that the effectiveness of copper nanoparticles functionalized with alkyl thiol in terms of anti-wear capability was dependent on their size [33]. The smaller particles exhibited a greater tendency to engage with the surfaces of the friction pair and produce a protective film on the surface, which was the primary reason for their better anti-wear performance.

19.4.3 Mending Effect Mechanism

The mending action mechanism in nanoparticle-enhanced engine oil refers to the ability of the nanoparticles to repair and regenerate the contacting surfaces during the lubrication process. The nanoparticles in the lubricant can fill in the surface defects and cracks on the contacting surfaces, thus reducing roughness and improving surface quality. This filling action can create a smoother surface, reducing the contact area and the friction between the surfaces. In addition, the nanoparticles can also react with the surface molecules to form a protective layer or coating. This coating can increase the hardness and reduce the surface roughness, leading to reduced wear and improved durability of the contacting surfaces. The mending action mechanism can also prevent further damage to the contacting surfaces by sealing off any loose or damaged material. The nanoparticles can bond to the surface and fill in any gaps or cracks, preventing further material loss and reducing the risk of catastrophic failure. Zhou et al. conducted a study that demonstrated the ability of nanoparticles to produce a physical tribofilm on a surface to compensate for the loss of mass due to friction [30].

In a study by Zhou et al. the lubrication mechanism on sliding contact surfaces lubricated by MXene/HS composite nanoparticles was studied [30]. By analyzing the surface wear track, it was possible to identify several factors contributing to the friction reduction and anti-wear effects of MXene/HS composites in PAO-10 base oil (Figure 19.5). First, ensuring the complete dispersion of MXene/HS composites in the oil allows for a continuous supply to the contact area, which is essential for their lubrication ability. Non-functionalized MXene have short-term stability, making them less effective in this regard. Second, HS nucleation and growth enlarges the interlayer space of MXene nanosheets, which weakens shear strength and counteracts some of the friction force during mechanical movements. Third, the composite additives can get spontaneously embedded in surface grooves, reducing roughness. During sliding, small HS nanosheets exfoliate from the composite and share the force while filling the grooves. Additionally, the high temperature generated during friction sliding promotes the adsorption and deposition of additives, creating a protective tribofilm on the worn surface. It is important to note that the formation of this tribofilm is not only due to physical adsorption and deposition, but also involves complex chemical reactions.

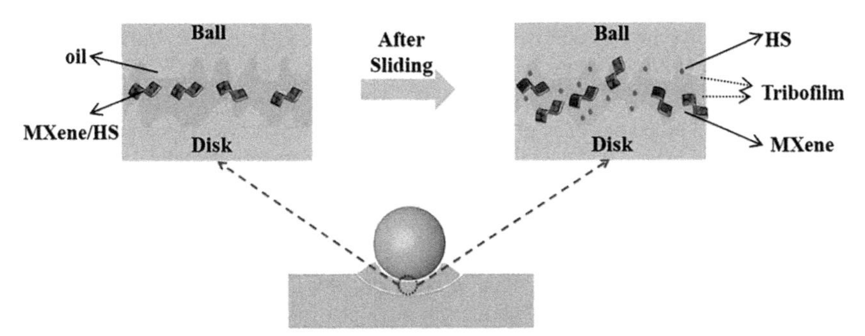

Figure 19.5 Lubrication mechanism of MXene/HS composites [30].

19.5 CONCLUSION

In conclusion, MXene has gained significant research interest in various fields such as catalysis, energy storage, and electronics, owing to its excellent mechanical, thermal, electrical and surface properties. When added to various base oils, MXene can improve their tribological behavior through rolling, polishing and mending effects, by acting as rolling elements, polishing agents and/or tribofilm precursors. These effects can reduce friction, improve wear resistance, repair worn surfaces and increase lubricant longevity. Optimizing nanoparticle size, shape, and surface chemistry can further enhance these effects, leading to improved lubrication performance.

REFERENCES

[1] Gogotsi, Y., Anasori, B. 2019. The rise of MXenes. ACS Nano. 13: 8491–8494. https://doi.org/10.1021/acsnano.9b06394.

[2] Gogotsi, Y. and Huang, Q. 2021. MXenes: Two-dimensional building blocks for future materials and devices. ACS Nano. 15: 5775–5780. https://doi.org/10.1021/acsnano.1c03161.

[3] Naguib, M., Mochalin, V.N., Barsoum, M.W. and Gogotsi, Y. 2014. 25th anniversary article: MXenes: A new family of two-dimensional materials. Adv. Mater. 26: 992–1005. https://doi.org/10.1002/adma.201304138.

[4] Rasheed, A.K., Khalid, M., Bin Mohd Nor, A.F., Wong, W.Y., Duolikun, T., Natu, V., et al. 2020. MXene-graphene hybrid nanoflakes as friction modifiers for outboard engine oil. IOP Conf. Ser. Mater. Sci. Eng. 834: 0–11. https://doi.org/10.1088/1757-899X/834/1/012039.

[5] Jia, Y., Pan, Y., Wang, C., Liu, C., Shen, C., Pan, C., et al. 2021. Flexible Ag Microparticle/MXene-based film for energy harvesting. Nano-Micro Lett. 13: 1–12. https://doi.org/10.1007/s40820-021-00729-w.

[6] Fan, Z., Wang, Y., Xie, Z., Wang, D., Yuan, Y., Kang, H., et al. 2018. Modified MXene/Holey graphene films for advanced supercapacitor electrodes with superior energy storage. Adv. Sci. 5. https://doi.org/10.1002/advs.201800750.

[7] Liu, R. and Li, W. 2018. High-thermal-stability and high-thermal-conductivity $Ti_3C_2T_x$ MXene/Poly(vinyl alcohol) (PVA) composites. ACS Omega. 3: 2609–2617. https://doi.org/10.1021/acsomega.7b02001.

[8] Yorulmaz, U., Özden, A., Perkgöz, N.K., Ay, F. and Sevik, C. 2016. Vibrational and mechanical properties of single layer MXene structures: A first-principles investigation. Nanotechnology. 27(33): 335702. https://doi.org/10.1088/0957-4484/27/33/335702.

[9] Zhou, C., Li, Z., Liu, S., Ma, L., Zhan, T. and Wang, J. 2022. Synthesis of MXene-based self-dispersing additives for enhanced tribological properties. Tribol. Lett. 70: 63. https://doi.org/10.1007/s11249-022-01605-3.

[10] Yan, H., Zhang, L., Li, H., Fan, X. and Zhu, M. 2020. Towards high-performance additive of Ti_3C_2/graphene hybrid with a novel wrapping structure in epoxy coating. Carbon. 157: 217–233. https://doi.org/10.1016/j.carbon.2019.10.034.

[11] Nguyen, H.T. and Chung, K.H. 2020. Assessment of tribological properties of Ti_3C_2 as a water-based lubricant additive. Materials (Basel). 13: 1–14. https://doi.org/10.3390/ma13235545.

[12] Yang, B., Liu, B., Chen, J., Ding, Y., Sun, Y., Tang, Y., et al. 2022. Realizing high-performance lithium ion hybrid capacitor with a 3D MXene-carbon nanotube composite anode. Chem. Eng. J. 429: https://doi.org/10.1016/j.cej.2021.132392. https://doi.org/10.1016/j.cej.2021.132392.

[13] Li, L., Cao, Y., Liu, X., Wang, J., Yang, Y. and Wang, W. 2020. Multifunctional MXene-based fireproof electromagnetic shielding films with exceptional anisotropic heat dissipation capability and joule heating performance. ACS Appl. Mater. Interfaces. 12: 27350–27360. https://doi.org/10.1021/acsami.0c05692.

[14] Wang, C., Wang, X., Zhang, T., Qian, P., Lookman, T. and Su, Y. 2022. A descriptor for the design of 2D MXene hydrogen evolution reaction electrocatalysts. J. Mater. Chem. A. 10: 18195–18205. https://doi.org/10.1039/d2ta02837a.

[15] Zong, H., Qi, R., Yu, K. and Zhu, Z. 2021. Ultrathin Ti2NTx MXene-wrapped MOF-derived CoP frameworks towards hydrogen evolution and water oxidation. Electrochim. Acta. 39. https://doi.org/10.1016/j.electacta.2021.139068.

[16] Zhang, J., Zhao, Y., Guo, X., Chen, C., Dong, C.L., Liu, R.S., et al. 2018. Single platinum atoms immobilized on an MXene as an efficient catalyst for the hydrogen evolution reaction. Nat. Catal. 1: 985–992. https://doi.org/10.1038/s41929-018-0195-1.

[17] Shi, L., Wu, C., Wang, Y., Dou, Y., Yuan, D., Li, H., et al. 2022. Rational design of coordination bond connected metal organic frameworks/MXene hybrids for efficient

solar water splitting. Adv. Funct. Mater. 32: 2202571. https://doi.org/10.1002/adfm.202202571.

[18] Yuan, W., Cheng, L., An, Y., Wu, H., Yao, N., Fan, X., et al. 2018. MXene nanofibers as highly active catalysts for hydrogen evolution reaction. ACS Sustain. Chem. Eng. 6: 8976–8982. https://doi.org/10.1021/acssuschemeng.8b01348.

[19] Zhang, D., Ashton, M., Ostadhossein, A., Van Duin, A.C.T., Hennig, R.G. and Sinnott, S.B. 2017. Computational study of low interlayer friction in $Ti_{n+1}C_n$ (n = 1, 2, and 3) MXene. ACS Appl. Mater. Interfaces. 9: 34467–34479. https://doi.org/10.1021/acsami.7b09895.

[20] Liu, Y., Zhang, X., Dong, S., Ye, Z. and Wei, Y. 2017. Synthesis and tribological property of $Ti_3C_2T_X$ nanosheets. J. Mater. Sci. 52: 2200–2209. https://doi.org/10.1007/s10853-016-0509-0.

[21] Gao, J., Du, C.-F., Zhang, T., Zhang, X., Ye, Q., Liu, S., et al. 2021. Dialkyl dithiophosphate-functionalized $Ti_3C_2T_x$ MXene nanosheets as effective lubricant additives for antiwear and friction reduction. ACS Appl. Nano Mater. 4: 11080–11087. https://doi.org/10.1021/acsanm.1c02556.

[22] Markandan, K., Nagarajan, T., Walvekar, R., Chaudhary, V. and Khalid, M. 2023. Enhanced tribological behaviour of hybrid MoS_2@Ti_3C_2 MXene as an effective anti-friction additive in gasoline engine oil. Lubricants. 11: 47. https://doi.org/10.3390/lubricants11020047.

[23] Nagarajan, T., Khalid, M., Sridewi, N., Jagadish, P., Shahabuddin, S., Muthoosamy, K., et al. 2022. Tribological, oxidation and thermal conductivity studies of microwave synthesised molybdenum disulfide (MoS_2) nanoparticles as nano-additives in diesel based engine oil. Sci. Rep. 12: 1–13. https://doi.org/10.1038/s41598-022-16026-4.

[24] Morina, A., Neville, A., Priest, M. and Green, J.H. 2006. ZDDP and MoDTC interactions and their effect on tribological performance—Tribofilm characteristics and its evolution. Tribol. Lett. 24: 243–256. https://doi.org/10.1007/s11249-006-9123-7.

[25] Thornley, A., Wang, Y., Wang, C., Chen, J., Huang, H., Liu, H., et al. 2022. Optimizing the Mo concentration in low viscosity fully formulated oils. Tribol. Int. 168: 107437. https://doi.org/10.1016/j.triboint.2022.107437.

[26] Rubbi, F., Habib, K., Saidur, R., Aslfattahi, N., Yahya, S.M. and Das, L. 2020. Performance optimization of a hybrid PV/T solar system using Soybean oil/MXene nanofluids as a new class of heat transfer fluids. Sol. Energy. 208: 124–138. https://doi.org/10.1016/j.solener.2020.07.060.

[27] Rahmadiawan, D., Aslfattahi, N., Nasruddin, N., Saidur, R., Arifutzzaman, A. and Mohammed, H.A. 2021. MXene based palm oil methyl ester as an effective heat transfer fluid. J. Nano Res. 68: 17–34. https://doi.org/10.4028/www.scientific.net/JNanoR.68.17.

[28] Alqahtani, B., Hoziefa, W., Moneam, H.M.A., Hamoud, M., Salunkhe, S., Elshalakany, A.B., et al. 2022. Tribological performance and rheological properties of engine oil with graphene nano-additives. Lubricants. 10: 1–11. https://doi.org/10.3390/lubricants10070137.

[29] Wen, G., Wen, X., Cao, H., Bai, P., Meng, Y., Ma, L., et al. 2023. Fabrication of Ti_3C_2 MXene and tetradecylphosphonic acid@MXene and their excellent friction-reduction and anti-wear performance as lubricant additives. Tribol. Int. 186: 108590. https://doi.org/10.1016/j.triboint.2023.108590.

[30] Zhou, C., Li, Z., Liu, S., Ma, L., Zhan, T. and Wang, J. 2022. Synthesis of MXene-based self-dispersing additives for enhanced tribological properties. Tribol. Lett. 70: 1–13. https://doi.org/10.1007/s11249-022-01605-3.

[31] Tao, X., Jiazheng, Z. and Kang, X. 1996. The ball-bearing effect of diamond nanoparticles as an oil additive. J. Phys. D. Appl. Phys. 29: 2932–2937. https://doi.org/10.1088/0022-3727/29/11/029.

[32] Talib, N. and Rahim, E.A. 2018. Performance of modified jatropha oil in combination with hexagonal boron nitride particles as a bio-based lubricant for green machining. Tribol. Int. 118: 89–104. https://doi.org/10.1016/j.triboint.2017.09.016.

[33] Zhang, Z.J., Simionesie, D. and Schaschke, C. 2014. Graphite and hybrid nanomaterials as lubricant additives. Lubricants. 2: 44–65. https://doi.org/10.3390/lubricants2020044.

[34] Yan, T., Ingrassia, L.P., Kumar, R., Turos, M., Canestrari, F., Lu, X., et al. 2020. Evaluation of graphite nanoplatelets influence on the lubrication properties of asphalt binders. Materials (Basel). 13. https://doi.org/10.3390/ma13030772.

Chapter **20**

Fly Ash—An Effective Lightweight Filler for Advanced Composites

Santhosh Mozhuguan Sekar*[1],
Elango Natarajan[1, 2] and Ang Chun Kit[1]

[1]Faculty of Engineering, Technology and Built Environment, UCSI University, Kuala Lumpur, Malaysia

[2]Department of Mechanical Engineering, PSG Institute of Technology and Applied Research, Coimbatore, Tamilnadu, India

20.1 INTRODUCTION

Composite materials offer enormous scope for fly ash (FA) to be utilized as fillers and reinforcements that can change the properties of the base matrix. FA is categorized as an organic material which is used for making lightweight insulating composites and hybrid composites. The hybrid composites can been made with carbides, oxides, borides, nitrides or ceramics to tailor the required characteristics of the material. The main ingredients of FA encompass SiO_2, Al_2O_3 and Fe_2O_3, while the supplementary elements consist of oxides of Mg, Ca, Na and K, among other similar grade reinforcements. The morphology of fly ash particulates is predominantly cylindrical, exhibiting a size distribution spanning from submicron dimensions to 100 micrometers as shown in Figure 20.1.

*For Correspondence: mozhuguan.santhosh@gmail.com

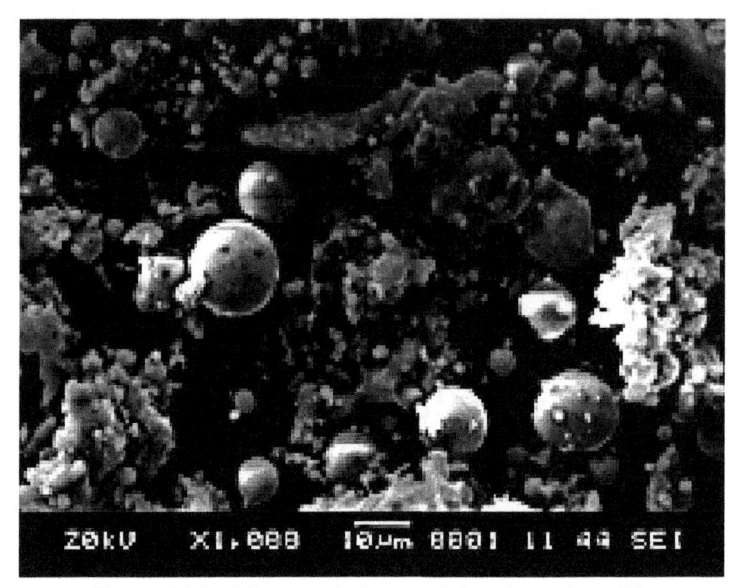

Figure 20.1 Microscopic image of fly ash particles [1].

These small particles possess a specific surface area that typically falls within the range of 250 to 600 square meters per kilogram. The specific weight of fly ash exhibits variations ranging from 0.6 g/cc to 2.8 g/cc, contingent upon its composition. Fly ash is an exhaust generated by thermal power stations where coal is burnt. The substance is gathered from the chimneys, in the form of a very lightweight granular material. The quantity of fly ash across the globe is approximately proportional to the amount of coal fired in power stations. Research done by Fortune Business Insights in 2022 revealed a 6.1% compound annual growth rate for the worldwide fly ash industry from 2022 to 2029, with a valuation of USD 12.70 billion in 2022 and USD 19.19 billion in 2029. In 2021, the industry worth was USD 12.25 billion [2].

Both the production and the use of fly ash are at their highest in the Asia-Pacific area. More than half of the fly ash produced across the globe is from China. India is the world's Third largest exporter of fly ash after the United States and Germany. As coal-fired power plants expand in emerging economies, fly ash will become more readily available in the years to come. However, reservations about the impact of coal-powered power plants on the planet are limiting the fly ash industries. Fly ash is in high demand because of its numerous practical uses such as making concrete, making bricks and blocks, building roads, stabilizing soil, farming, treating sewage and preventing erosion. It has been reported that the usage of or reinforcement with fly ash reduces the cost of commercial products by up to 40% [1, 4].

In order to conduct the study to be discussed in this chapter, Science Direct, Google Scholar, PubMed and Scopus databases were searched with keywords such as 'Fly ash', 'metal matrix composite' and 'polymer matrix composites', and articles published in the last five years were collected. The search resulted in a total of 15,030 articles: about 64 articles in PubMed, 3161 articles in Scopus, 1206 articles in Science Direct and 10600 articles in Google Scholar. An uptrend trend in the usage of fly ash fillers was noticed as shown in Figure 20.2. It was observed that there is a limited use of fly ash in medical applications as observed in the PubMed database; hence, future research may be focused on medical applications. Scientists and researchers worldwide have been keen working towards utilizing fly ash particles as reinforcements in metal and polymer matrices. The following section provides an overview of the available matrix materials for fly ash fortification and explores the suitability of a variety of metal matrices with fly ash.

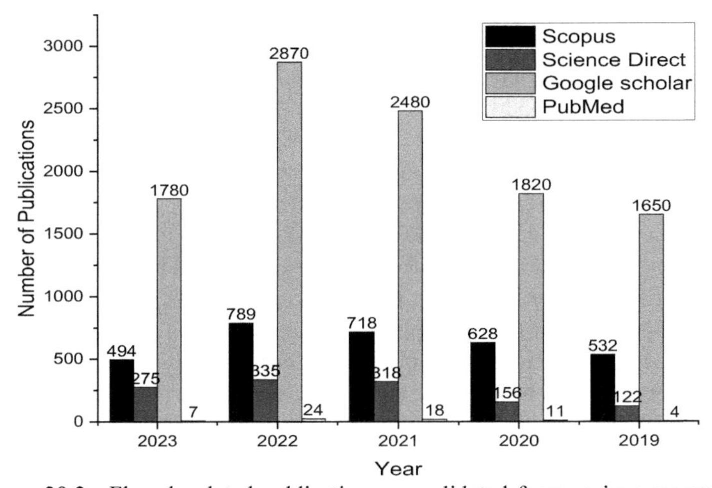

Figure 20.2 Fly ash related publications consolidated from various resources.

20.2 Prior Art Survey

Metal composite materials strengthened with fiber-reinforced polymers (FRPs) are produced using various methods, including stir casting, in situ deposition approach, press casting, and compo casting strategy. Many researchers have conducted investigations pertaining to the support of aluminum composites with various ceramic materials.

20.2.1 Improvement in Mechanical Properties

Vivpin et al. [3] dealt with the fabrication of aluminium fly ash composites (2 wt.%, 4 wt.% and 6 wt.%) through stir casting and studied their tribological behavior. AMMC with 6 wt.% of fly ash reinforcement showed a decreased wear of 0.32 g, which was 13.6% less compared to that with 2 wt.% reinforcement. The low coefficient of friction of 0.12 was noticed with 4 wt.% of reinforcement. Hence, it was concluded that the higher the weight percentage of FA, the higher the coefficient of friction. Prakash et al. [4] investigated stir cast hybrid composites of aluminium alloy A413, fly ash and Boron carbide (B_4C). They used 1.5 wt.%, 3 wt.% and 4.5 wt.% of both FA and B_4C, along with a small amount of magnesium for increasing the wettability of FA, and potassium hexa fluro titanate (K_2TiF_6) for B_4C. They performed machinability studies to investigate material removal rate and surface roughness and observed higher wettability and metal removal rate with fly ash reinforcement.

Rajendran and Suresh [5] developed SiC and fly ash reinforced LM30 Aluminium hybrid composites with different combinations: 10 wt.% SiC + 15 wt.% FA, 10 wt.% SiC + 5 wt.% FA, 5 wt.% SiC + 15 wt.% FA, 5 wt.% SiC + 10 wt.% FA, 15 wt.% SiC + 10 wt.% FA, and 10 wt.% SiC + 10 wt.% FA. They reported that LM30 + 5 wt.% SiC + 15 wt.% FA had the highest hardness value –87 HV. Furthermore, the authors mentioned that thermal conductivity can be improved with these fillers. The thermal conductivity of LM 30 + 10 wt.% SiC + 15 wt.% FA was 443.51 W/mK – the highest among all the tested samples. Kanth et al. [6] developed Al 7075 + 2.5 wt.% SiC + 2.5 wt.% FA and A17075 + 5 wt.% SiC + 5 wt.% FA composites through stir casting and reported that the densities of the composites decreased as fillers increased. A17075 + 5 wt.% SiC + 5 wt.% FA showed the lowest density (2.45 g/cm^3), compared to matrix material's density of 2.82 g/cm^3. The general reason for the low density could be voids and porosities. In spite of the decrease in density, they reported that other mechanical properties like hardness and tensile strength were enhanced as fly ash reinforcement was increased. This is an interesting topic for further studies – if we are able to obtain a high-strength low-density composite material, it can be deployed in several industrial applications.

Shekhawat et al. [7] produced composites reinforced with eggshell powder (ESP) and FA. Their investigations with three precursor ratios (30:70, 50:50 and 70:30 of ESP and FA respectively), and 0.5, 1 and 2% of Na_2SiO_3/NaOH activator revealed that 70 wt.% ESP + 30 wt.% FA possessed the highest maximum dry density (MDD). They also reported that the geopolymer with 50 wt.% ESP and 50 wt.% FA acquired a compressive strength of 2149 kPa, which they regarded as the optimum filler percentage for an environmentally friendly and cost-effective composite.

The research investigation conducted by Kumar et al. [8] reported the impact of acetone and ethyl alcohol solvents on epoxy laminated composites of teak wood dust and fly ash (FA). It was reported that the laminates that were prepared using ethanol as a solvent were subjected to heating without adding any hardener. Verma et al. [9] investigated the impact of incorporating fly ash (FA) and oxide particles on the mechanical and tribological properties of aluminum metal matrix composite (AMMC). From their studies on samples with 5 wt.% fly ash and varying amounts of aluminium oxide (6, 9 and 12 wt%), it was evident that the incorporation of fly ash and 12 wt.% of Al_2O_3 into the Al base metal matrix had the most superior mechanical characteristics. A very notable increase in hardness from 82 BHN to 162 BHN, tensile strength from 57 N/mm^2 to 156 N/mm^2, and flexural strength from 2.4 kN to 6 kN was also reported. In a study conducted by

Singh et al. [10], the mechanical characteristics of fly ash reinforced composite, including toughness, durability, elastic modulus, maximum tensile strength and wear resistance, were reported. In another study, Sudarshan et al. [11] investigated using fly ash of (53–106 μm), and (0.5–400 μm) and their research revealed that the reinforced material exhibited superior wear resistance compared to the unreinforced materials.

In a study conducted by Satapathy [12], the modulus of rupture (MOR), modulus of elasticity (MOE), hardness, and damage volume decrease from sand damage test was reported. Tiwari [13] carried out a study with aluminium (ADC 12) matrix reinforced with 6 wt% fly ash (60–100 μm in size), and magnesium (0.5 wt%). The reason for using Mg was to increase the wettability and reduce the surface tension of fly ash. The inclusion of fly ash particles into the base matrix may result in pitting rust in case of lower wettability. The said investigation on ageing temperatures of 130°C, 150°C, 175°C and 200°C showed an increase of approximately 45.25% in tensile strength and 42.99% in hardness at a temperature of 175°C. It was also observed that the highest impact strength was recorded at 150°C of aging temperature. Charles and Arunachalam [14] concluded that 10 wt% of SiC and a low concentration of fly ash are good for increasing the wear resistance and toughness of Al-alloy/Silicon carbide (SiC)/fly ash composite.

A patent by Rohatgi [15] reported the method of preparing fly ash reinforced monoaluminum phosphate solution (MAP solution), which could be further used to prepare aluminum-fly ash metal matrix composites from fly ash preforms. They demonstrated preform preparation from cenosphere fly ash and precipitator fly ash, along with the preparation of lead-fly ash composites.

Liu et al. [16] investigated the utilization of matrix as the primary phase change material (PCM) for energy storage purposes, with solid waste fly ash serving as the carrier material. The fabrication process involved the creation of FA combined phase change materials (PCMs) using an optimal mass ratio of 1.7:1. The chemical and physical features, anatomy, and thermal conductivity of the PCMs were analyzed and it was observed that fly ash inclusion significantly improved the morphology of the PCMs.

20.2.2 Effect of Processing Techniques

Zhiqiu Huang et al. [17] used the compo-casting technique to produce an AZ91D/fly ash composite with Mg_2Si and MgO reinforcements. They noted even dispersion of 100-micron fly ash materials in the AZ91D Mg metal matrix. Primarily in-situ Mg_2Si and MgO were observed on the surfaces of cenosphere particles, along with a small amount of Mg_2Si within the matrix. This observation indicated better wettability caused by Mg and was coherent with Tiwari [13]. Murthy et al. [18] attempted to strengthen AA7075 matrix using TiO_2 and fly ash, through stir casting followed by hot forging. They used 3 wt.% of fly ash, and varying wt.% of TiO_2 (2.5 to 10 wt.%), and reported that thermal conductivity is decreased by these fillers. Uju et al. [19] investigated the structural impact of fly ash fillers in Al–7Si–0.35Mg alloy. The authors reported that the integration of compo-casting and squeeze-casting techniques could produce a refined grain structure and lower porosity, resulting in improved compressive strength. Chand et al. [20] employed powder metallurgy to fabricate a metal matrix composite material consisting of fly ash and Mg.

Ravi Shankar [21] reported that 10 wt% of fly ash can be used along with SiC for the reinforcement and strengthening of Al alloys. Vinod et al. [22] used double stir casting process to fabricate A356 alloy/fly ash composite and reported that the incorporation of both inorganic and organic materials enhances the mechanical properties as they have low porosities. Kesavulu et al. [23] and Ramachandra et al. [24] used the stir-casting process—whereas the former used 5, 10, 15, and 20 wt.% of fly ash, the later restricted the filler percentage to 15 wt.%. Their research revealed that the incorporation of fly ash into aluminium alloys results in a notable enhancement of abrasive wear (approximately 20–30%).

Sobczake et al. [25] employed both squeeze casting technology and gravity casting methods, and evaluated these processing techniques subsequently. Also, they reported that squeeze casting demonstrated structural uniformity, reduced porosity and favorable interfacial bonding. Rohatgi et al. [26] examined the utilization of Al-fly ash composites in various industries, especially automobiles. After a comprehensive analysis, the authors reported that fly ash has ecological and power advantages and can be prospectively used with Al alloys to save cost, energy and pollution in terms of emission levels. In another study by the same authors [27], pressure penetration method was used for the fabrication of A356/fly ash chemosphere composites. It was found that this method resulted in a higher strength than other methods.

Mohankumar et al. [28] also employed stir casting to produce B4C and fly ash reinforced metal composites. They concluded that these fillers improved hardness and wear resistance. This improvement can be attributed to the obstructive effect of these particles on the movement of dislocations within the material. In another study, Dinaharan et al. [29] employed stir casting and friction stir processing (FSP) to fabricate fly ash reinforced magnesium alloy AZ31. They reported that microhardness and wear resistance of the stir cast composite was lower compared to the FSPed composite. Plastic deformation on the ridges and large craters on the worn surface were observed in stir cast composites, while these features were not found in FSPed composites at all.

Prasad et al. [30] and Kumaraswamy et al. [31] have also studied squeeze cast LM6 Al/fly ash composite and Al 7075/graphite composite respectively. The significant findings from various studies on fly ash reinforced composites are presented in Table 20.1.

The literature review reveals that fly ash would enhance the strength, abrasion resistance, and service life of the material, hence being a prospective filler for preparing composites in many distinct applications. Fly ash may also assist in reducing the density of the material, which is the most considerable feature and need of the modern automobile sector [32]. Because of its low cost and abundant availability, reinforcement of aluminium and other metals like Mg and Ti with fly ash may lead the next level research in waste management.

Table 20.1 Research on fly ash composites.

Authors	Matrix Material	Fly ash (wt.%)	Other Fillers	Fabrication Method	Parameters Investigated / Results	Outcome of the Research
Vivpin et al. [3]	Aluminium alloy	2, 4 and 6%	–	Stir casting	Wear behaviour	Low wear rate
Prakash et al. [4]	Aluminium alloy A413	1.5, 3, 4 and 5%	Boron Carbide (1.5, 3 and 4.5%)	Stir casting	Machinability study–material removal rate and surface roughness	Increased material removal rate and wettability
Rajendran et al. [5]	LM 30	5, 10, 15%	Silicon Carbide (5, 10 and 15%)	Stir casting	Heat transfer and thermal conductivity	Improved thermal conductivity
Kanth et al. [6]	Al 7075	2.5, 5%	Silicon Carbide (2.5 and 5%)	Stir casting	Hardness and tensile strength	Improved mechanical properties
Shekhawat et al. [7]	Polymer	3, 5, 7%	Egg shell powder (3–5 wt.%)	Geo polymerization	Compressive strength	Improved strength with 7 wt.% reinforcement
Kumar et al. [8]	Epoxy	2.5%	Teak Wood Powder	Compression method	Effect of Acetone and Ethyl Alcohol solvent coating	Improved flexibility
Verma et al. [9]	Aluminium	5%	Aluminium Oxide (6, 9 and 12%)	Stir casting	Tensile and compressive behaviours	Improved elongation and compression
Singh et al. [10]	Aluminium Alloy	5%	Al_2O_3, SiC	Stir casting	Tribological behaviour	Increased wear resistance
Sudarshan et al. [11]	Aluminium Alloy	6 and 12%	–	Stir casting	Wear rate	Increased wear resistance
Satapathy et al. [12]	Zirconia	5, 10, 15, 20 and 25%	–	Stir casting	Tensile and impact behaviours	Improved tensile and impact properties

(*Contd.*)

(*Contd.*)

Authors	Matrix Material	Fly ash (wt.%)	Other Fillers	Fabrication Method	Parameters Investigated / Results	Outcome of the Research
Tiwari et al. [13]	Al Alloy (ADC 12)	6%	–	Stir casting	Tensile, hardness and impact strength	Increased mechanical properties
Charles et al. [14]	Aluminium	2.5%	Silicon Carbide	Stir casting	Tensile, wear and hardness	Improved tensile, wear and hardness
Rohatgi et al. [15]	Aluminium	3–15%	–	Stir casting	Ductility of the composite	Reduced ductility at 15 wt.%
Liu et al. [16]	Sodium sulfate dehydrate	1%	–	Facile process	Chemical, thermal and morphology studies	Improved resistance towards chemicals
Huanga et al. [17]	AZ91D	5%	Mg_2Si and MgO	Compo casting	Mechanical properties	Increased mechanical properties
Murthy et al. [18]	AA7075	3%	Titanium oxide (2.5–10%	Stir casting	Thermal coefficient and thermal conductivity	Improved thermal conductivity
Uju et al. [19]	Aluminium	2.5%	SiC, MgO	Compo casting and squeeze casting	Mechanical behaviour study	Reduced porosity and enhanced compression strength
Chand et al. [20]	Magnesium	0.5 to 2 wt.%	–	Squeeze casting	Mechanical behaviour	Improved mechanical properties
Shankar et al. [21]	Magnesium	10%	SiC	Squeeze casting	Wear rate	Improved wear resistance
Vinod et al. [22]	A356 alloy	2.5%	–	Stir casting	Mechanical behaviour	Improved mechanical properties
Kesavulu et al. [23]	Aluminium	5, 10, 15 and 20%	–	Stir casting	Hardness	Improved hardness

Authors	Matrix Material	Fly ash (wt.%)	Other Fillers	Fabrication Method	Parameters Investigated / Results	Outcome of the Research
Radhakrishna et al. [24]	Aluminium	15%	12 wt.% Si	Stir casting	Sliding, erosive and corrosive behaviour	Improved corrosion resistance limitation towards erosion
Bienias et al. [25]	Aluminium	2.5%	–	Squeeze casting and gravity casting	Morphology study	Reduced porosity and good interracial bonding
Rohatgi et al. [26]	Aluminium	20%	–	Stir casting	Mechanical behaviour	Improved mechanical properties
Rohatgia et al. [27]	Aluminium A356	20 to 65%	–	Gas infiltration technique	Thermal studies	Composites' melting temperature, pressure, and particle size
Mohankumar et al. [28]	Aluminium A359	5, 10, 15, 20 and 25%	Boron carbide	Stir casting	Mechanical behaviour	Improved hardness, wear rate, coefficient of friction
Dinaharan et al. [29]	Magnesium alloy AZ31	10%		Stir casting and friction stir processing (FSP)	Sliding wear, worn surface and wear debris	Low wear rate with fly ash reinforcement
Prasada et al. [30]	LM 6	5–12.5%	–	Stir-squeeze casting	Comparison of hardness and wear rate	Low wear rate and improved hardness
Kumarasamy et al. [31]	Al 7075	10%	Graphite (2, 4 and 6%)	Stir casting	Mechanical behaviour studies	Increase in tensile and wear and decrease in density

20.3 CONCLUSION

Based on the short literature review on fly ash (FA) composites, some important conclusions can be drawn:

- Fly ash upto 25% by weight can be successfully added to commercially available aluminum and its alloys and also to metals/alloys such as magnesium, titanium and zirconia.
- For improving the wettability and retention of fly ash in the composite, magnesium can be added in the melt during production.
- Lower particle sizes up to 150 μm may result in the better dispersion of fly ash particles into the matrix.
- Finally, utilizing the most abundantly available industrial waste – fly ash – can effectively reduce environmental pollution, for the latter can be useful as a reinforcement/ingredient towards product quality improvement.

REFERENCES

[1] Norhaiza, G., Khairunisa, M. and Saffuan, W.A. 2019. Utilization of Fly Ash in Construction. IOP Conf. Series: Mater. Sci. Eng. 601: 012023.

[2] The global fly ash market is projected to grow from $12.70 billion in 2022 to $19.19 billion by 2029, at a CAGR of 6.1% in forecast period. 2022–2029. https://www.fortunebusinessinsights.com/industry-reports/fly-ash-market-101087.

[3] Vivpin K.S., Singh, R.S. and Rajiv, C. 2017. Effect of fly ash particles with aluminium melt on the wear of aluminium metal matrix composites. Eng. Sci. Technol. Int. J. 20: 1318–1323.

[4] Udaya Prakash, J., Moorthy, T.V. and Milton Peter, J. 2013. Experimental investigations on machinability of aluminium alloy A413 /Flyash/ B4C composites using wire EDM. Procedia Eng. 64: 1344 –1353.

[5] Rajendran, M. and Suresh, A.R. 2018. Characterization of aluminium, metal matrix composites and evaluation of thermal properties. Mater. Today Proc. 5: 8314–8320.

[6] Uppada, R.K., Putti, S.R. and Mallarapu, G.K. 2019. Mechanical behavior of fly ash/ SiC particles reinforced Al-Zn alloy –based metal matrix composites fabricated by stir casting method. J. Mater. Res. Tech. 8(1): 737–744.

[7] Poonam, S., Gunwant, S. and Rao, M.S. 2019. Strength behavior of alkaline activated egg shell powder and flyash geo ploymer cured at ambient temperature. Constr. Build. Mater. 223: 1112–1122.

[8] Nithin, K.N., Siddesh, C., Preran, R.H., Shivagiri, S.Y. and Revanasiddappa, M. 2018. Synthesis and characterization of fly ash/wooden fiber reinforced with epoxy resin polymer composites. Mater. Today Proc. 5: 501–507.

[9] Vikas, V, P.C. Tewari, Roshan, Z.A. and Syed, T.A. 2019. Effect of addition of fly ash and Al_2O_3 particles on mechanical and tribological behavior of Al MMC at varying load, time and speed. Procedia Struct. Integrity. 14: 68–77.

[10] Sugandha, S., Swati, G. and Sukriti, Y. 2017. A review on mechanical and tribological properties of micro/nano filled metal alloy composites. Mater. Today Proc. 4: 5583–5592.

[11] Sudarshan, M.K.S. 2008. Dry sliding wear of fly ash particle reinforced A356 Al composites. Wear. 265349–360.

[12] Satapathy, L.N. 2000. A study on the mechanical, abrasion and microstructural properties of zirconia–fly ash material. Ceram. Int. 26: 39–45.

[13] Sumit, K.T., Sanjay, S., Rana, R.S. and Alok, S. 2017. Effect of aging on mechanical behavior of ADC12–Fly Ash particulate composite. Mater. Today Proc. 4: 3513–3524.

[14] Charles, S. and Arunachalam, V.P. 2004. Property analysis and mathematical modelling of machining properties of aluminium alloy hybrid (Al-alloy/SiC/fly ash) composites Produced by liquid metallurgy and powder metallurgy techniques. Indian J. Engg. Mater. Sci. 11: 473–480.

[15] Rohatgi, P.K. 1998. Method of producing metal matrix composites containing fly ash. U.S. Patent. 5711362.

[16] Lei, L., Ben, P., Changsheng, Y., Min, G. and Mei, Z. 2018. Low-cost, shape-stabilized fly ash composite phase change material synthesized by using a facile process for building energy efficiency. Mate. Chem. and Phys. 222: 87–95.

[17] Zhiqiu, H. and Sirong, Y. 2011. Microstructure characterization on the formation of *in situ* Mg_2Si and MgO reinforcements in AZ91D/Fly ash composites. J. Alloys Compd. 509: 311–315.

[18] Shivananda Murthy, K.V., Girish, D.P.., Keshava Murthy, R., Temel Varol and Koppad, P.G. 2017. Mechanical and thermal properties of AA7075/TiO_2/Fly ash hybrid composites obtained by hot forging. Prog. Nat. Sci.: Mater. Int. 27: 474–481.

[19] Uju, W.A., Oguocha, I.N.A. 2012. A study of thermal expansion of Al-Mg alloy composites containing fly ash. Mater. Des. 33: 503–509.

[20] Raghu, C. and Swamy, R.P. 2016. Characterization of magnesium based fly ash reinforced composite using powder metallurgy. J. Eng. Res. Appli. 71–76.

[21] Ravi, S., Manivannan, A. and Vijayakumar, D. 2012. Effect of fly ash particles on the mechanical properties and microstructure on compacted magnesium reinforced with SiC particles. 2nd International Conference Manufacturing Engineering & Management. 155–162.

[22] Vinod, B., Ramanathan, S., Ananthi, V. and Selvakumar, N. 2019. Fabrication and characterization of organic and in-organic reinforced A356 Aluminium matrix hybrid composite by improved double-stir casting. Silicon. 11(2): 817–29.

[23] Kesavulu, A, Anand Raju, F., Dr. M.L.S. Deva Kumar. 2014. Properties of aluminium fly ash metal matrix composite. Int. J. Innovative Res. Sci. Eng. Technol. 3(11): 17160–17164.

[24] Radhakrishna, M. and Ramachandra, K. 2007. Effect of reinforcement of fly ash on sliding wear, slurry erosive wear and corrosive behavior of aluminium matrix composite. Wear. 262: 1450–1462.

[25] Bienias, J., Walczak, M., Surowska, B. and Sobczak, J. 2003. Microstructure and corrosion behavior of aluminum fly ash composites. J. Optoelectron. Adv. Mater. 5(2): 493–502.

[26] Rohatgi, P.K., Weiss, D. and Nikhil, G. 2006. Applications of fly ash in synthesizing low-cost MMCs for automotive and other applications. JOM. 71–76.

[27] Rohatgia, P.K., J. Kima K., Gupta N., Alaraj Simon and A. Daoud. 2006. Compressive characteristics of A356/fly ash cenosphere composites synthesized by pressure infiltration technique. Composites: Part A. 37: 430–437.

[28] Mohankumar, S., Aravind, Ra., SelvaKumar, G., Raja, P. and Selvam, V. 2019. Experimental investigation on the tribological-mechanical properties of B4C and fly ash reinforced Al 359 composites. Mater. Today Proc. 21: 748–754.

[29] Dinaharan, I., Vettivel, S.C., Balakrishna, M. and Akinlabi, E.T. 2019. Influence of processing route on microstructure and wear resistance of fly ash reinforced AZ31 magnesium matrix composites. J. Magnesium Alloys. 7155–7165.

[30] Prasada, K.N.P. and Ramachandra, M. 2018. Determination of abrasive wear behavior of Al–fly ash metal matrix composites produced by squeeze casting. Mater. Today Proc. 5: 2844–2853.

[31] Kumarasamy, S.P., Vijayananth, K., Thankachan, T. and Muthukutti, G.P. 2017. Investigation on mechanical and machinability behavior of aluminium/fly ash cenosphere/Gr hybrid composites processed through compo casting. J. Appl. Res. Technol. 15(5): 430–441.

[32] Natrayan, L., Kaliappan, S., Pravin, P., Patil, Y., Sesha, R., Suresh T.N.K., et al. 2022. Mechanical and wear behaviour of nano-fly ash particle-reinforced Mg metal matrix composites fabricated by stir casting technique. J. Nanomater. 2022: 5465771. https://doi.org/10.1155/2022/5465771.

Chapter 21

Green Nanocomposites: A Review on Extraction and Applications

S.P. Yeap, A.H. Abu Bakar,
C.S. Hassan* and F.A. Jamaludin

Faculty of Engineering, Technology and Built Environment,
UCSI University, 56000, Kuala Lumpur, Malaysia

21.1 INTRODUCTION TO GREEN NANOMATERIALS

Materials with at least one of the dimensions falling within the size range of 1 nm–100 nm are regarded as nanomaterials. Owing to their tiny size, nanomaterials exhibit a huge specific surface area and are known for their enhanced properties (such as mechanical strength, electrical, optical, magnetic, thermal and antibacterial properties, catalysis, etc.) which are superior to their bulk counterparts. More importantly, incorporating a small quantity of these nanomaterials into a bulk item produces a resultant composite with improved functionality [1, 2] or multi-functionalities [3]. For instance, nanomaterials as fillers have gained attention for their special properties and the ability to interact well with epoxy resin [4, 5]. The composite toughened by nanomaterials may exhibit improved elastic modulus, improved tensile strength, improved flexural strength, reduced impact duration, enhanced energy absorption, enhanced peak load, and a suppression in damage propagation [6, 7].

*For Correspondence: suhana@ucsiuniversity.edu.my

Enlightened by such promising application outcomes, researchers have been venturing into various strategies to synthesize large quantities of nanomaterials. Generally, nanomaterials can be synthesized through two approaches, namely the Top-Down and Bottom-Up approaches [8]. The Top-Down approach involves the processing of materials from bulk size to nano-size through mechanical processes [9]; on the other hand, the Bottom-Up approach involves nucleation and growth of nanoparticles from ions which undergo chemical reactions [10, 11]. The Bottom-Up approach is favorable for both lab-scale and industrial-scale production of nanomaterials as the process is less energy-intensive and time-consuming (does not require low hours of mechanical milling). Furthermore, the size, shape, and properties of a nanomaterial can be further tuned by changing the stoichiometric ratio of the chemical precursor, pH of the reaction medium, heating method, as well as the type of surfactant used during the Bottom-Up synthesis [12–15].

However, the Bottom-Up approach is not without limitation. In particular, some of the chemicals used for the synthesis process are hazardous. For instance, in the chemical reduction process to produce reduced graphene oxide (a 2-D nanomaterial known for superior electrical conductivity and mechanical strength), several reducing agents which include hydrazine [16], sodium borohydride ($NaBH_4$) [17, 18], and strong alkaline solutions are commonly used. Similarly, such reducing agents have been utilized for the chemical synthesis of other nanoparticles such as SiO_2 and AgO. Moreover, some nanomaterial chemical synthesis processes involve the usage of toxic solvents/precursors such as dimethyl formamide [19, 20], *N*-Methyl-2-pyrrolidone [20], trioctyl phosphine oxide [21], etc. These chemicals are hazardous, potentially produce toxic by-products, and may get retained on the surface of the nanomaterials [22]. Such events may cause adverse effects on human health and lead to environmental pollution, thus making its utilization for large-scale production of nanomaterials unsuitable [23, 24].

Realizing such limitations, scientists are exploring more green synthesis methods for the sustainable synthesis of nanomaterials [25]. Some of the key merits of green synthesis include: (1) avoiding the usage of toxic chemicals, (2) reducing material cost through the use of the same biological component as the reducing agent and the capping agent, (3) towards a more energy saving process by excluding the needs of high temperature/pressure conditions, and (4) suitable for large-scale nanomaterial production [11]. In this regard, replacing the aforementioned toxic reducing agents with natural compounds such as juice extracts [26], fruit peels extracts [27], plant extracts [28], and tea extracts [29] have been the center of research focus. This is attributable to the abundance of reducing species such as pectin, hemicellulose, cellulose [30] and antioxidant compounds [31, 32] in these natural products.

In addition to the replacement of toxic chemicals with green chemicals, extraction of nanomaterials from a more sustainable resource has gained much attention recently. For instance, SiO_2 nanoparticles have been successfully extracted from corn cobs husks [33], rice husk [34, 35], palm kernel shell [36], coir pith [37], bamboo leaves [38, 39], and other agricultural wastes. Utilization of agricultural wastes for nanomaterial synthesis is greener, sustainable, safer, and can reduce the cost of raw materials. A keyword search through

LENS.ORG implied that the publication of scholarly work on the topic related to green synthesis of nanomaterials has gained substantial attention over the past 30 years (see Figure 21.1). From the exponential increase in the publication trends, it can be concluded that this specific research area has yet to reach its saturation stage.

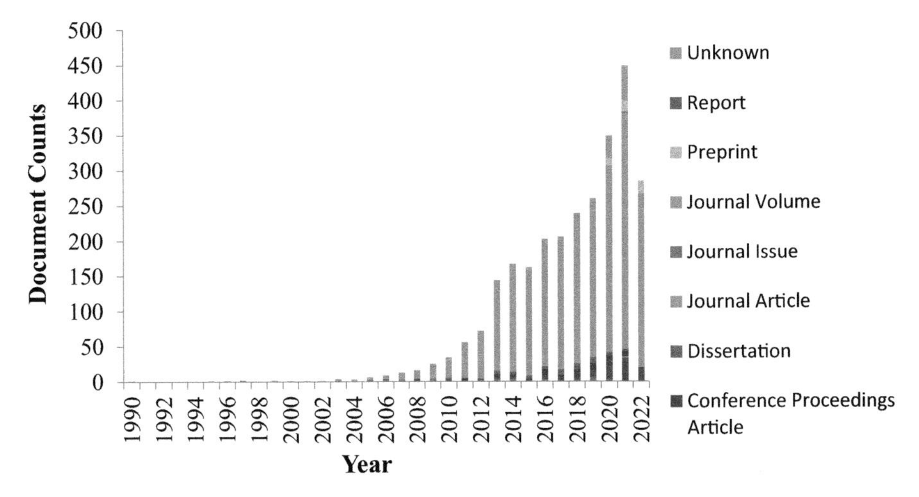

Figure 21.1 Scholarly work publication trend for "(Green Synthesis) AND (Nano)" recorded from year 1990 to year 2022 (as of 3rd Oct 2022). Data sources: https://www.lens.org/lens/. (*Note:* To increase the accuracy of the database search, the search was limited to TAC: Title, Abstract and Claim)

21.2 Extraction of Nanomaterials

Green nanomaterials/nanoparticles such as nanocellulose, nanosilica and nanolignin, are commonly extracted from biomass. Biomass is basically plant-based or lignocellulosic material which consists of three major components: (a) the fiber of the plant—cellulose and hemicellulose, (b) the wall of the plant—lignin, and (c) the inorganic composition such as silica and minerals. In order to extract green nanoparticles from the biomass, two main approaches are used: physical method (top-down) and chemical method (bottom-up) [50]. This chapter shall focus on nanosilica extraction methods and applications.

21.2.1 Nanosilica Extraction Methods

Most of the researchers have extracted nanosilica using the bottom-up approach. Adam et al. (2011) [51] produced nanosilica from rice husk by treating it with 1.0 M NaOH. They proceeded with titration using 3.0 M HNO_3 until pH 9.0 and the yellowish gel obtained was aged for 2 days, centrifuged, washed and dried. The final product was ground into fine powder. The nanosilica produced was in an amorphous state, with size between 15 to 91 nm.

A similar approach was used by Bhattacharya and Mandal (2018). They used alkaline treatment of rice straw with a liquid-to-solid ratio (LSR: ml/gm) of 28:1 for 155 minutes at 89°C. Sodium silicate present in the solution was neutralized with 2 v/v% sulfuric acid to pH 7. The precipitate was filtered and the filtered residue which consisted of $SiO_2.H_2O$ and Na_2SO_4 was washed with 5 w/v% $BaCl_2$ solution to remove the Na_2SO_4. The nanosilica thus produced had a particle size of 14.8 nm, with an amorphous structure. Equations (1) and (2) describe the reaction involved in the process [52]:

$$Na_2SiO_3 + H_2SO_4 \longrightarrow SiO_2.H_2O + Na_2SO_4 \quad (1)$$

$$BaCl_2 + SO_4^{-2} \longrightarrow BaSO_4 + 2Cl^- \quad (2)$$

Abd-Rabboh et al. (2022) used a similar method with a mixture of rice straw and husk, but concentrated HCl was applied during the precipitation process; then, the dried filter residue was burnt to remove all of the organic matter. In order to convert the sodium silicate to an amorphous structure, sodium silicate solution was mixed with a triblock copolymer, pluronic, which was dissolved in 80 g aqueous solution of HNO_3 at room temperature. After this process, the solid obtained was dried and calcined at 600°C for 4 h. From the SEM results, size of the silica particles obtained was between 0.35 and 45 μm; there was no detection of nano-size silica from Dynamic Light Scattering (DLS) analysis [53].

A different approach was applied on the alkaline hydrolysis part by Abu Bakar and Jia Ni Carey, (2020), wherein Anthraquinone (AQ) was incorporated as the catalyst to speed up the reaction at a lower temperature (75°C), with 3 hours of cooking time. The Soda-AQ solution worked well at a solution-to-solid ratio of 200:5 with 20 wt% NaOH and 0.1 wt% AQ. The silica extracted was confirmed with FTIR analysis and matched with the commercial silica gel, showing a strong IR spectrum of around 1054.8 cm^{-1} and 1040.6 cm^{-1}, indicating the presence of Si–O–Si asymmetric stretching group [54].

Another bottom-up approach was deployed by Singh et al. (2022), which involved thermal treatment in the first stage, followed by chemical treatment to extract the silica from rice straw ash. The combustion occurred at 600°C and 900°C to study the effect of different temperatures on the nanosilica structure. The ash was treated with 1 M NaOH at 100°C for 4 hours, followed by precipitation using 10% H_2SO_4 till pH 7. The solution was then centrifuged and dried. The different temperatures resulted in two different structures – amorphous at 600°C and crystalline at 900°C. TEM analysis showed that the size of the nanosilica varied between 27.47 nm (amorphous) and 52.79 nm (crystalline), with high agglomeration [55].

Thongma and Chiarakorn (2019) extracted silica from biomass ash produced from a biomass power plant. The rice husk biomass was burnt at 800–850°C [56]. The ash was subjected to alkaline treatment with 2 M NaOH, and precipitated with concentrated sulfuric acid into sodium silicate solution. Other researchers such as Worathanakul et al. (2009) and Khoshnood Motlagh et al. (2021) have applied similar methods to obtain nanosilica from biomass ash [57, 58]. However, Nagachandrudu et al. (2022) analyzed the effects of different types of acids such as HCl, H_2SO_4 and acetic acid on the silica extracted [59].

21.3 Application of Green Nanomaterials

21.3.1 Reinforcement Material

The ability of the structural features of nanomaterials to be tailored to achieve specific physical and chemical properties is one of the characteristics that has sparked interest in their use. Such enhanced properties promise a wide range of potential applications in science and engineering, such as serving as another imperative element used for reinforcements. Nanosilicon oxide, for example, has become an excellent filler in epoxy restorative materials due to its durability. The primary explanation is the presence of an oxide position in nanosilica material, which acts as a powerful adhesive throughout the matrices. Ponnusamy et al. (2022) [60] demonstrated that incorporating nanoparticles into composites improved the mechanical properties and microstructure of a Kenaf/Carbon fiber-reinforced epoxy composite (Figure 21.2a). The increased surface area and energy retention capacities, as well as the decreased void within the hybrid composite, were achieved as a result of the homogeneous scattering of nanosilica particles in the matrix, which ultimately resulted in improved fillers, matrix interactions, and fiber contact area; this caused efficient stress transfer within the composites when subjected to load. The authors reported that in unmodified samples with a nanosilica concentration of 1.5 wt%, tensile strength increased by 31%, flexural strength increased by 42.36%, and impact strength increased by 22.65% (Figure 21.2b). Subbiah et al. 2022 [61] investigated the role of nanosilica in improving load bearing, thermal stability, and antimicrobial properties of cardanol oil toughened epoxy composites reinforced with sisal fiber. According to the findings, nanosilica with sisal fiber in cardanol oil blended epoxy composites demonstrated improved mechanical properties, with tensile strength and modulus being 56% and 64% higher, respectively, than pure epoxy resin. The maximum impact resistance and hardness values of the composites were 6.48 J and 91 shore-D respectively. Furthermore, fatigue life counts improved for composites containing 1.0 vol% nanosilica, but decreased when the volume percentage was increased to 2.0. It was also reported that the antimicrobial properties were enhanced as the nanosilica and cardanol oil content increased.

The use of agricultural waste-derived nanosilica has piqued the interest of researchers due to its sustainable approach to addressing the problems of environmental degradation, and is being viewed as an economic substitute for industrial raw materials. Various agricultural sources of silica nanomaterials, such as rice husk, wheat husk, sugarcane bagasse, bamboo residues and maize stalk are available in abundant quantities, and can be used to synthesize silica nanomaterials [62–67]. Rahmawati et al. (2021) investigated the effects of nanosilica extracted from physically processed rice husk ash on the mechanical properties and fracture toughness of geopolymer cements. White rice husk ash was ground in a ball mill for 10 hours at 600 rpm to produce nanosilica with an average particle size of 339.09 nm [68]. It was reported that the nanosilica influenced mechanical strength, fracture toughness, and microstructure, as the addition of 2% nanosilica to geopolymer paste increased compressive strength by 22%, flexural strength by 82% and fracture toughness by 82%, while decreasing direct tensile strength

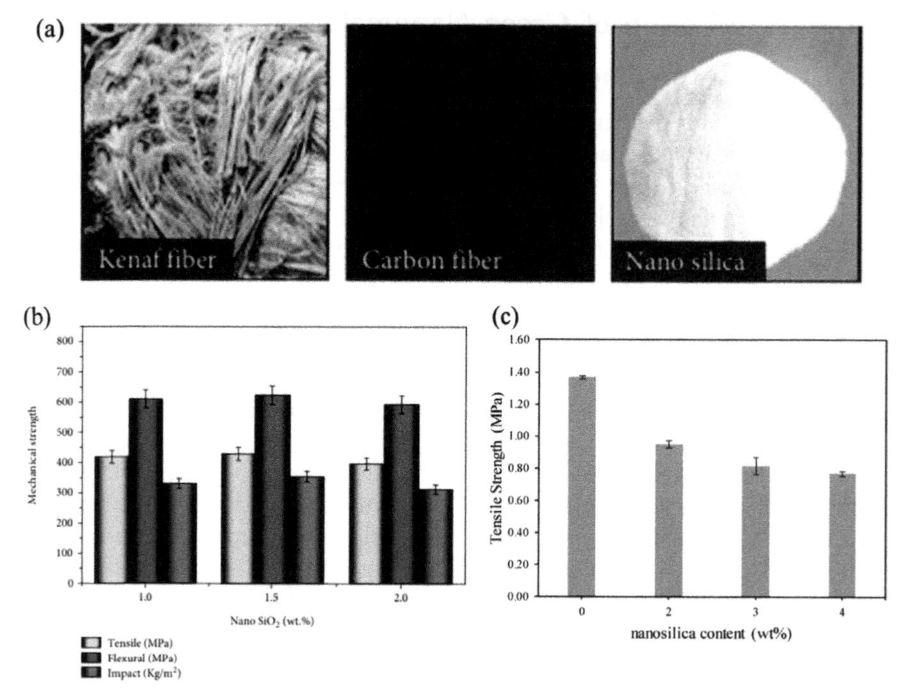

Figure 21.2 (a) Digital image of the Kenaf fibers, carbon fiber and nanosilica used to make the composite. (b) The tensile strength, flexural strength, and impact strength of the Kenaf/Carbon fiber-reinforced epoxy composite incorporated with different weight ratios of nanosilica [Reprinted with permission from [60], Copyright 2022, MDPI, Creative Commons Attribution License]. (c) Maximum tensile strength of geopolymer cement-based materials loaded with 0–4 wt% of nanoslica [Reprinted with permission from [68], Copyright 2021, MDPI, Creative Commons Attribution License].

by 31% (see Figure 21.2c). Vijayalakshmi et al. (2015) investigated the corrosion resistance behavior of rice husk ash nano-silica as a mild steel coating [69]. It was reported that the Nyquist plot obtained revealed that the corrosion inhibition was confirmed by the higher R_{ct} value of 5.882×10^5 Ohmfor the silica coating and 3.717×10^4 Ohm for the uncoated mild steel. In connection with this, the double layer capacitance value (C_{dl}) was found to be drastically reduced from 2.345×10^{-4} Farad for the uncoated to 7.145×10^{-6} Farad for the silica-coated substrate. The decrease in the C_{dl} value accompanied with the increase in the R_{ct} value indicates its corrosion-resistive behavior in an aggressive NaCl medium.

21.3.2 Agricultural Applications

Green-synthesized nanoparticles have also found numerous potential applications in agriculture due to their antimicrobial properties; they are typically used as pesticides [70], fertilizer agents [71, 72], assist in plant growth [73] and have many other applications. Nanoparticles have reportedly outperformed bulk silica in agricultural applications due to their microscopic size and high surface-to-volume ratio, which results in chemical and physical changes in the characteristics. Suriyaprabha et al. 2014 investigated the effects of green synthesized silica nanoparticles on conferring fungal resistance in maize, over bulk silica treatment. It has been reported that nanosilica-treated plants have lower levels of stress-responsive enzymes against fungi and higher levels of phenolic compounds, indicating that silica induces plant defense compounds [74]. The increased expression of stress-responsive compounds in maize after silica treatment promotes the bio-control activity of silica nanoparticles, indicating that nano silica improves fungal resistance by activating the damping-off mechanism more than its bulk counterpart. Furthermore, it has been reported that the hydrophobic potential and silica accumulation percentage that lead to leaf erectness in nano silica-treated maize are greater than those of bulk silica-treated maize, implying an increase in plant defense action against biotic and abiotic stresses, as well as fungal pathogens [75].

21.3.3 Electrical and Electronics Applications

Wide varieties of nanotechnologies have been implemented in various electrical and electronics applications such as energy conversion technologies, energy storage elements, insulation systems and semiconductor industries [40]. One of the key applications of green nanotechnology is in insulation systems since insulation systems are one of the most important elements or components in any electrical equipment/installations. Their role is to prevent undesired electric current flow. This is crucial for the safety of users and to prevent electric shocks. In high voltage systems, insulation coordination becomes one of the most important factors in designing the electrical systems, especially for the transmission and distribution of electricity, which involves varieties of insulation systems such as insulators which are used to separate the phase line conductors from each other and from the ground (tower), cables, electrical machines, and transformer insulations system [41]. On the other hand, the insulator application is also valid for semiconductor and electronic devices, in designing integrated circuit chip boards for instance [42]. Materials used for insulation do not allow the free flow of electric currents or charges. Typically, the characteristics of insulating materials include the ability to sustain high temperatures, non-conductivity, high dielectric strength depending on the type of application to prevent breakdown voltage, fire resistance, vapor permeability, and thermal expansion. Reviews of nanocomposite materials [76, 77] summarize the use of inorganic nanoparticles disseminated in an organic polymer matrix to form a class of hybrid organic-inorganic materials known as nanocomposites. The electrical, mechanical and thermal characteristics of polymers have improved in a promising way with nanoparticles used in place of micrometer-sized particles. Due to their enhanced properties, polymer nanocomposite systems can be used for a variety of purposes, including coatings, proton exchange membranes, catalysts, and packaging materials, as well as nanodielectrics in microelectronics. Epoxy is one of the most significant polymer foundation materials used in nanocomposites, which have recently attracted more attention as high voltage insulating materials.

Polymer blend is one of the best methods to make a new film with good properties which cannot be obtained by only one polymer. Metal nanoparticle polymer blend is receiving a lot of research attention these days. By incorporating gold nanoparticles (Au NPs) into the polymer mixture, samples are created with unexpected properties that significantly differ from those of the conventional material (See Figure 21.3) [43]. It has been reported that the finished product has excellent mechanical, electrical and optical qualities, making it suitable for various industries, including sensor design, photo imaging, and optics. Due to the quest for green technology, it has been observed that there is a focus on shifting to green materials such as jute [78]. Being 100% eco-friendly, accessible, and affordable, jute has been used as one of the bioresources in nanotechnology. In addition, jute is known to be the second-most popular natural cellulose fiber worldwide; this plant also generates a sizable amount of jute sticks as a byproduct. Jute fibers and sticks mostly consist of cellulose, hemicellulose and lignin, and contain a little amount of ash. Additionally, jute has served as a source for the development of nanomaterials that are used in a variety of applications. Hemicellulose and lignin, which can be extracted from jute fibers and sticks, can also be used as stabilizers or reductants when creating other nanomaterials. Past research focused on the position and future of jute in nanotechnology, with a detailed overview of the various research areas wherein jute can be used, including the preparation of nanocellulose, as scaffolds for other nanomaterials, in energy storage, sensors, coatings and electronics. In addition, green nanotechnology applications in insulating equipment can also be employed in the form of nano-fluid insulation such kernel oil [44], palm oil [45] or other possible vegetable oils as electrical insulation in transformers [46].

Apart from applications in insulating materials, the potential of green-nanotechnology has been explored in semiconductor industries in terms of applications in electronic/optoelectronic devices [47]. A lot of interest has been paid to the discovery of 2D organic semiconductors with atomically thin structures because of their developing optical, electrical, optoelectronic and mechatronic capabilities. An organic semiconductor, which comprises single molecules, short chains (oligomers), and polymers, is a substance that is predominantly composed of carbon and hydrogen and exhibits semiconducting properties. Recent advancements in these organic nanostructures have created new avenues for altering material properties through a variety of methods, including strain engineering, atomic doping, and 0D/1D/2D nanoparticle hybridization. Additionally, 2D organic nanostructures demonstrate a special trait of bio-functionality and are extraordinarily sensitive to bio-analytes. Highly effective bio-sensors can be created by utilizing the behavior of 2D organics. Moreover, in semiconductor—visible light communication applications, green semiconductor materials have been anticipated to be the best option due to the combination of novel semiconducting optical and electrical properties with simplified fabrication and tuning of properties associated with organic materials [48]. Contrary to inorganic semiconductors, organic semiconductors do not require highly organized crystals because their chemical structure primarily determines their electrical and optical properties. This simplifies the production of their devices—for instance,

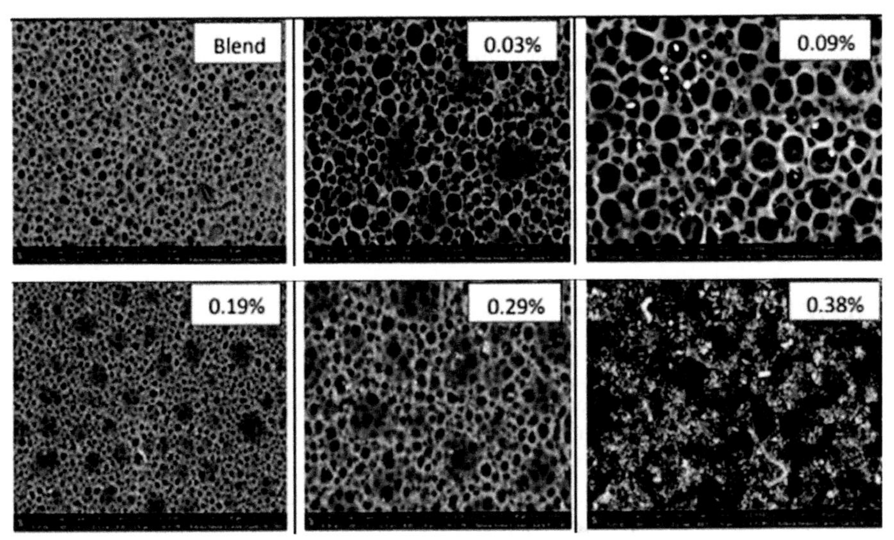

Figure 21.3 Polymer added with different amounts of Au nanoaprticles leading to changes in the surface structure of the polymer [Reprinted with permission from [43], Copyright 2019, Elsevier].

via deposition from solution—and is anticipated to result in lower device costs. Additionally, it makes it possible to fabricate many devices simultaneously, such as red, green and blue organic light-emitting diodes (OLEDs) used in televisions and mobile phone displays.

Rapid development of electrical and electronics-based technologies is aligned with IR4.0 needs where it features digitalization and promotes significant degrees of automation and autonomy in products or processes. Latest technological developments will benefit the mankind; however, impacts of technology on the current development of industries should not be neglected. The rapid technologies should consider their impact on climate change, environmental pollution, natural resources and sustainability. Therefore, the associated idea of 'green nanotechnology' seeks to take advantage of nano-innovations in materials science and engineering to produce products and processes that are energy efficient as well as economically and environmentally sustainable in order to enable society to create and sustain a green economy. It has been projected that these applications will have an impact on a number of economic areas, including clean technology, infrastructure development, and energy generation and storage [49].

21.4 CONCLUSION

In recent years, there has been a growing interest in the development of innovative materials for various applications. One notable advancement involves the creation of nanocomposites, where nano-sized fibers are used as reinforcement materials to produce lightweight structures while also offering a unique combination of strength. The use of green nanocomposites further signifies eco-friendliness, making them a promising alternative that highlights their potential across numerous applications while maintaining a reduced environmental impact. Further research is required to investigate nanoparticle concentration, nanoparticle type and nanoparticle orientation for modifying the mechanical properties and design of these materials. Additionally, factors such as cost and manufacturing reliability need to be considered to facilitate the implementation of green nanocomposites in a broader range of applications, including those with high-performance demands.

REFERENCES

[1] Ren, J., Li, Q., Yan, L., Jia, L., Huang, X., Zhao, L., et al. 2020. Enhanced thermal conductivity of epoxy composites by introducing graphene@boron nitride nanosheets hybrid nanoparticles. Mater. Des. 191: 108663.

[2] Long, J., Li, C. and Li, Y. 2022. Enhancement of mechanical and bond properties of epoxy adhesives modified by SiO_2 nanoparticles with active groups. Polymers (Basel). 14: 2052.

[3] Ali, A. and Andriyana, A. 2020. Properties of multifunctional composite materials based on nanomaterials: A review. RSC Adv. 10: 16390–16403.

[4] Tee, Z.Y., Yeap, S.P., Hassan, C.S. and Kiew, P.L. 2022. Nano and non-nano fillers in enhancing mechanical properties of epoxy resins: A brief review. Polym.-Plast. Technol. Mater. 61: 709–725.

[5] Natarajan, E., Freitas, L.I., Santhosh, M.S., Markandan, K., Majeed Al-Talib, A.A. and Hassan, C.S. 2023. Experimental and numerical analysis on suitability of S-Glass-Carbon fiber reinforced polymer composites for submarine hull. Def. Technol. 19: 1–11.

[6] Lal, L.P.J., Ramesh, S., Parasuraman, S., Natarajan, E. and Elamvazuthi, I., 2019. Compression after impact behaviour and failure analysis of nanosilica-toughened thin epoxy/GFRP composite laminates. Materials. 12: 3057.

[7] Natarajan, E., Ramesh, S., Markandan, K., Saravanakumar, N., Dilip, A.A. and Batcha, A.R.S. 2023. Enhanced mechanical, tribological, and acoustical behavior of polyphenylene sulfide composites reinforced with zero-dimensional alumina. J. Appl. Polym. Sci. 140: e53748.

[8] Abid, N., Khan, A.M., Shujait, S., Chaudhary, K., Ikram, M., Imran, M., et al. 2022. Synthesis of nanomaterials using various top-down and bottom-up approaches, influencing factors, advantages, and disadvantages: A review. Adv. Colloid Interface Sci. 300: 102597.

[9] Baig, N., Kammakakam, I. and Falath, W., Nanomaterials: a review of synthesis methods, properties, recent progress, and challenges, Mater. Adv. 2: 1821–1871.

[10] Thanh, N.T.K., Maclean, N. and Mahiddine, S. 2014. Mechanisms of nucleation and growth of nanoparticles in solution. Chem. Rev. 114: 7610–7630.

[11] Singh, J., Dutta, T., Kim, K.-H., Rawat, M., Samddar, P. and Kumar, P. 2018. 'Green' synthesis of metals and their oxide nanoparticles: applications for environmental remediation. J. Nanobiotechnol. 16: 84.

[12] Saragi, T., Permana, B., Saputri, M., Depi, B.L., Butarbutar, S.W., Safriani, L., et al. 2018. The effect of pH and sintering treatment on magnetic nanoparticles ferrite based synthesized by coprecipitation method. J. Phys. Conf. Ser. 1080: 012019.

[13] Jiang, W., Lai, K.-L., Hu, H., Zeng, X.-B., Lan, F., Liu, K.-X., et al. 2011. The effect of [Fe^{3+}]/[Fe^{2+}] molar ratio and iron salts concentration on the properties of superparamagnetic iron oxide nanoparticles in the water/ethanol/toluene system. J. Nanopart. Res. 13: 5135.

[14] Logaranjan, K., Raiza, A.J., Gopinath, S.C.B., Chen, Y. and Pandian, K. 2016. Shape- and size-controlled synthesis of silver nanoparticles using aloe vera plant extract and their antimicrobial activity. Nanoscale Res. Lett. 11: 520.

[15] Cobley, C.M., Skrabalak, S.E., Campbell, D.J. and Xia, Y. 2009. Shape-controlled synthesis of silver nanoparticles for plasmonic and sensing applications. Plasmonics. 4: 171–179.

[16] Chen, H., Ding, L., Zhang, K., Chen, Z., Lei, Y., Zhou, Z., et al. 2020. Preparation of chemically reduced graphene using hydrazine hydrate as the reduction agent and its NO_2 sensitivity at room temperature. Int. J. Electrochem. Sci. 15: 10231-10242.

[17] Yang, Z.-z., Zheng, Q.-b., Qiu, H.-x., Li, J. and Yang, J.-h. 2015. A simple method for the reduction of graphene oxide by sodium borohydride with $CaCl_2$ as a catalyst. New Carbon Mater. 30: 41–47.

[18] Rana, S., Sandhu, I.S., Chitkara, M. 2018 Exfoliation of graphene oxide via chemical reduction method. pp 54–57. *In*: 2018 6th Ed. Int. Conf. Wirel. Networks Embed. Syst. WECON 2018, IEEE.

[19] Zhang, Y., Shi, R. and Yang, P. 2014. Synthesis of Ag nanoparticles with tunable sizes using N, N-dimethyl formamide. J. Nanosci. Nanotechnol. 14: 3011–3016.

[20] Amgoth, C., Singh, A., Santhosh, R., Yumnam, S., Mangla, P., Karthik, R., et al. 2019. Solvent assisted size effect on AuNPs and significant inhibition on K562 cells. RSC Adv. 9: 33931–33940.

[21] Soosaimanickam, A., Jeyagopal, R., Hayakawa, Y. and Sridharan, M.B. 2015. Synthesis of oleylamine-capped $Cu_2ZnSn(S,Se)_4$ nanoparticles using 1-dodecanethiol as sulfur source. Jpn. J. Appl. Phys. 54: 08KA10.

[22] Malhotra, S.P.K. and Alghuthaymi, M.A. 2022. pp. 139–163. *In*: Agri-Waste and Microbes for Production of Sustainable Nanomaterials, (eds). K.A. Abd-Elsalam, R. Periakaruppan and S. Rajeshkumar. Elsevier. DOI: https://doi.org/10.1016/B978-0-12-823575-1.00011-1.

[23] Buasuwan, L., Niyomnaitham, V. and Tandaechanurat, A. 2019. Reduced graphene oxide using an environmentally friendly banana extracts. MRS Adv. 4: 2143–2151.

[24] Gour, A. and Jain, N.K. 2019. Advances in green synthesis of nanoparticles. Artif. Cells Nanomed. Biotechnol. 47: 844–851.

[25] Eissa, D., Hegab, R.H., Abou-Shady, A. and Kotp, Y.H. 2022. Green synthesis of ZnO, MgO and SiO_2 nanoparticles and its effect on irrigation water, soil properties, and Origanum majorana productivity, Sci. Rep. 12: 5780.

[26] Hou, D., Liu, Q., Cheng, H. and Li, K. 2017. Graphene synthesis via chemical reduction of graphene oxide using lemon extract. J. Nanosci. Nanotechnol. 17: 6518–6523.

[27] Wijaya, R., Andersan, G., Permatasari Santoso, S. and Irawaty, W. 2020. Green reduction of graphene oxide using kaffir lime peel extract (*Citrus hystrix*) and its application as adsorbent for methylene blue. Sci. Rep. 10: 667.

[28] Ismail, Z. 2019. Green reduction of graphene oxide by plant extracts: A short review. Ceram. Int. 45: 23857–23868.

[29] Vatandost, E., Ghorbani-HasanSaraei, A., Chekin, F., Naghizadeh Raeisi, S. and Shahidi, S.-A. 2020. Green tea extract assisted green synthesis of reduced graphene

oxide: Application for highly sensitive electrochemical detection of sunset yellow in food products. Food Chemistry: X. 6: 100085.

[30] Happi Emaga, T., Andrianaivo, R.H., Wathelet, B., Tchango, J.T. and Paquot, M. 2007. Effects of the stage of maturation and varieties on the chemical composition of banana and plantain peels. Food. Chem. 103: 590–600.

[31] Brewer, M.S. 2011. Natural antioxidants: Sources, compounds, mechanisms of action, and potential applications. Compr. Rev. Food Sci. Food Saf. 10: 221–247.

[32] Wolfe, K., Wu, X. and Liu, R.H. 2003. Antioxidant activity of apple peels. J. Agric. Food. Chem. 51: 609–614.

[33] Pieła, A., Żymańczyk-Duda, E., Brzezińska-Rodak, M., Duda, M., Grzesiak, J., Saeid, A., et al. 2020. Biogenic synthesis of silica nanoparticles from corn cobs husks. Dependence of the productivity on the method of raw material processing. Bioorg. Chem. 99: 103773.

[34] Phoohinkong, W. and Kitthawee, U. 2014. Low-cost and fast production of nano-silica from rice husk ash. Ad. Mater. Res. 979: 216–219.

[35] Nayak, P.P. and Datta, A.K. 2021. Synthesis of SiO_2-Nanoparticles from rice husk ash and its comparison with commercial amorphous silica through material characterization. Silicon. 13: 1209–1214.

[36] Imoisili, P.E., Ukoba, K.O. and Jen, T.-C. 2020. Green technology extraction and characterisation of silica nanoparticles from palm kernel shell ash via sol–gel, J. Mater. Res. Technol. 9: 307–313.

[37] Maroušek, J., Maroušková, A., Periakaruppan, R., Gokul, G.M., Anbukumaran, A., Bohatá, A., et al. 2022. Silica nanoparticles from coir pith synthesized by acidic sol-gel method improve germination economics. Polymers (Basel). 14: 266.

[38] Sarkar, J., Mridha, D., Sarkar, J., Orasugh, J.T., Gangopadhyay, B., Chattopadhyay, D., et al. 2021. Synthesis of nanosilica from agricultural wastes and its multifaceted applications: A review. Biocatal. Agric. Biotechnol. 37: 102175.

[39] Sethy, N.K., Arif, Z., Mishra, P.K. and Kumar, P. 2019. Synthesis of SiO_2 nanoparticle from bamboo leaf and its incorporation in PDMS membrane to enhance its separation properties. J. Polym. Eng. 39: 679–687.

[40] S Mallakpour, F.S. and C.M. Hussain. 2021. Green synthesis of nano-Al_2O_3, recent functionalization, and fabrication of synthetic or natural polymer nanocomposites: various technological applications. New J. Chem. 4885–4920.

[41] Mohammed Mostafa Adnan, E.G.T., Julia Glaum, Marit-Helen Glomm Ese, Sverre Hvidsten, Wilhelm Glomm, Mari-Ann Einarsrud. 2019. Epoxy-based nanocomposites for high-voltage insulation: A review. Adv. Electron. Mater. 1800505.

[42] Illarionov, Y.Y., Knobloch, T., Jech, M., Lanza, M., Akinwande, D., Vexler, M.I., et al. 2020. Insulators for 2D nanoelectronics: The gap to bridge. Nat. Commun. 3385.

[43] Abdelghany, A.O. 2019. Influence of green synthesized gold nanoparticles on the structural, optical, electrical and dielectric properties of (PVP/SA) blend. Physica B: Condensed Matter. 162–173.

[44] Oparanti, S.O., Khaleed, A.A. and Abdelmalik, A.A. 2021. Nanofluid from palm kernel oil for high voltage insulation. Mater. Chem. Phys. 259: 123961.

[45] N.S. Suhaimi, M.T.I., Rahman, A.R.A., Md. Din, M.F., Abidin M.Z.Z. and Khairi, A.K. 2022. A review on palm oil-based nanofluids as a future resource for green transformer insulation system. IEEE Access. 103563–103586.

[46] A. Heebah, A.N., M. Bara and O.G. Mrehel, Hammamet, Tunisia. 2021.

[47] Neupane, G.P., Ma, W., Yildirim, T., Tang, Y., Zhang, L. and Lu, Y. 2019. 2D organic semiconductors, the future of green nanotechnology. Nano Mater. Sci. 1(4): 246–259.

[48] Manousiadis, P.P., Yoshida, K., Turnbull, G.A. and Samuel, I.D. 2020. Organic semiconductors for visible light communications. Phil. Trans. R. Soc. A. 378(2169): 20190186.

[49] Ivo Iavicoli, V. L., Walter Ricciardi, Laura L Hodson and Mark D Hoover. 2014. Opportunities and challenges of nanotechnology. Environ. Health. 1–11.

[50] Akhayere, E., Kavaz, D. and Vaseashta, A. 2022. Efficacy studies of silica nanoparticles synthesized using agricultural waste for mitigating waterborne contaminants. Appl. Sci. (Switzerland). 12(18): 1–19.

[51] Adam, F., Chew, T.S., and Andas, J. 2011. A simple template-free sol-gel synthesis of spherical nanosilica from agricultural biomass. J. Sol-Gel Sci. Technol. 59(3): 580–583.

[52] Bhattacharya, M., and Mandal, M.K. 2018. Synthesis of rice straw extracted nano-silica-composite membrane for CO_2 separation. J. Cleaner Prod. 186: 241–252.

[53] Abd-Rabboh, H.S.M., Fawy, K.F., Hamdy, M.S., Elbehairi, S.I., Shati, A.A., Alfaifi, M.Y., et al. 2022. Valorization of rice husk and straw agriculture wastes of Eastern Saudi Arabia: Production of bio-based silica, lignocellulose, and activated carbon. Materials (Basel). 15(11): 3746.

[54] Abu Bakar, A.H. and Jia Ni Carey, C. 2020. Extraction of silica from rice straw using alkaline hydrolysis pretreatment. IOP Conf. Ser. Mater. Sci. Eng. 778(1): 012158.

[55] Singh, G., Dizaji, H.B., Puttuswamy, H. and Sharma, S. 2022. Biogenic nanosilica synthesis employing agro-waste rice straw and its application study in photocatalytic degradation of cationic dye. Sustainability (Switzerland). 14(1): 1–15.

[56] Thongma, B. and Chiarakorn, S. 2019. Recovery of silica and carbon black from rice husk ash disposed from a biomass power plant by precipitation method. IIOP Conf. Ser.: Earth Environ. Sci. 373(1): 012026.

[57] Worathanakul, P., Payubnop, W. and Muangpet, A. 2009. Characterization for post-treatment effect of bagasse ash for silica extraction. World Academy of Science, Engineering and Technology. 56: 398–400.

[58] Khoshnood Motlagh, E., Sharifian, S. and Asasian-Kolur, N. 2021. Alkaline activating agents for activation of rice husk biochar and simultaneous bio-silica extraction. Bioresour. Technol. Rep. 16: 100853.

[59] Nagachandrudu, S., Maheswari, S.T. and Jayaprakash, R. 2022. Effect of different acids on rice husk calcination and extraction of bio-silica. Asian J. Chem. 34(2): 371–375.

[60] Ponnusamy, M., Natrayan, L., Kaliappan, S., Velmurugan, G. and Thanappan, S. 2022. Effectiveness of nanosilica on enhancing the mechanical and microstructure properties of Kenaf/carbon fibre-reinforced epoxy-based nanocomposites. Adsorpt. Sci. Technol. 2022: 4268314.

[61] Subbiah, R., Arivumangai, A., Kaliappan, S., Balaji, V., Yuvaraj, G. and Patil Pravin, P. 2022. Effect of nanosilica on mechanical, thermal, fatigue, and antimicrobial properties of cardanol oil/sisal fibre reinforced epoxy composite. Polym. Compos. 43(11): 7940–7951.

[62] Sankar, S., Sharma, S.K., Kaur, N., Lee, B., Kim, D.Y., Lee, S., et al. 2016. Biogenerated silica nanoparticles synthesized from sticky, red, and brown rice husk ashes by a chemical method. Ceram. Int. 42(4): 4875–4885.

[63] Patel, K.G., Misra, N.M., Vekariya, R.H. and Shettigar, R.R. 2018. One-pot multicomponent synthesis in aqueous medium of 1, 4-dihydropyrano [2, 3-c] Pyrazole-5-carbonitrile and derivatives using a green and reusable nano-SiO_2 catalyst from agricultural waste. Res. Chem. Intermed. 44: 289–304.

[64] Rovani, S., Santos, J.J., Corio, P. and Fungaro, D.A. 2018 Highly pure silica nanoparticles with high adsorption capacity obtained from sugarcane waste ash. ACS Omega. 3(3): 2618–2627.

[65] Falk, G., Shinhe, G., Teixeira, L., Moraes, E. and Novaes de Oliveira, A.N. 2019. Synthesis of silica nanoparticles from sugarcane bagasse ash and nano-silicon via magnesiothermic reactions. Ceram. Int. 45(17): 21618–21624.

[66] Rangaraj, S. and Venkatachalam, R. 2017. A lucrative chemical processing of bamboo leaf biomass to synthesize biocompatible amorphous silica nanoparticles of biomedical importance. Appl. Nanosci. 7: 145–153.

[67] Adebisi, J.A., Agunsoye, J.O., Bello, S.A., Haris, M., Ramakokovhu, M.M., Daramola, M.O., et al. 2020. Green production of silica nanoparticles from maize stalk. Part. Sci. Technol. 38(6): 667–675.

[68] Rahmawati, C., Aprilia, S., Saidi, T., Aulia, T.B. and Hadi, A.F. 2021 The effect of nanosilica on mechanical properties and fracture toughness of geopolymer cement. Polymers (Basel). 13(13): 2178.

[69] Vijayalakshmi, U., Vaibhav, V., Chellappa, M. and Anjaneyulu, U. 2015. Green synthesis of silica nanoparticles and its corrosion resistance behavior on mild steel. J. Indian Chem. Soc. 92(5): 675–678.

[70] Ziaee, M. and Ganji, Z. 2016. Insecticidal efficacy of silica nanoparticles against rhyzopertha dominica F. and tribolium confusum jacquelin du Val. J. Plant Prot. Res. 56(3): 250–256.

[71] Janmohammadi, M., Amanzadeh, T., Sabaghnia, N. and Ion, V. 2016. Effect of nano-silicon foliar application on safflower growth under organic and inorganic fertilizer regimes. Botanica Lithuanica. 22: 53–64.

[72] Malik, M.A., Wani, A.H., Mir, S.H., Rehman, I.U., Tahir, I., Ahmad, P., et al. 2021. Elucidating the role of silicon in drought stress tolerance in plants. Plant Physiol. Biochem. 165: 187–195.

[73] Khan, Z.S., Rizwan, M., Hafeez, M., Ali, S., Adrees, M., Qayyum, M.F., et al. 2020. Effects of silicon nanoparticles on growth and physiology of wheat in cadmium contaminated soil under different soil moisture levels. Environ. Sci. Pollut. Res. 27: 4958–4968.

[74] Suriyaprabha, R., Karunakaran, G., Kavitha, K., Yuvakkumar, R., Rajendran, V. and Kannan, N. 2014. Application of silica nanoparticles in maize to enhance fungal resistance. IET Nanobiotechnol. 8(3): 133–137.

[75] Suriyaprabha, R., Karunakaran, G., Yuvakkumar, R., Prabu, P., Rajendran, V. and Kannan, N. 2012. Growth and physiological responses of maize (Zea Mays L) to porous silica nanoparticles in soil. J. Nanopart. Res. 14(12): 1–4.

[76] Mohammed Mostafa Adnan, E.G.-H.-A. 2019. Epoxy-based nanocomposites for high-voltage insulation: A review. Adv. Electron. Mater. 5: 1800505.

[77] Hesham Moustafa, A.M.-K. 2019. Eco-friendly polymer composites for green packaging: Future vision and challenges. Composites Part B: Engineering. 16–25.

[78] Shah, S.S. 2021. Present status and future prospects of jute in nanotechnology: A review. Chem. Rec. 1631–1665.

Index

D

E

F

G

H

I

L

M

R

S

T

U

V

W

Z

About the Editors

Dr. Elango Natarajan obtained doctoral degree in Mechanical Engineering from Anna University, Chennai, India in 2010. He worked as a post-doctoral research fellow at Center for Artificial Intelligence and Robotics in Universiti Teknologi Malaysia in 2013 and carried out research on soft actuators. He has served for engineering colleges/universities for about 25 years in various academic positions. He is currently attached to Faculty of Engineering, UCSI University, Malaysia as an Associate Professor in Mechanical and Mechatronic Engineering. Besides, he is the Deputy Director of Praxis, Industry and Community Engagement. He is a Chartered Engineer (CEng.) awarded by Engineering Council UK, Chartered member (CMEngNZ) awarded by Engineering Council NZ and Fellow awarded by Institution of Engineers India. He has published 130+ research articles with citations over 1500 and his h-index is 23 in Scopus. He has completed many research grant projects including projects supported by Ministry of Higher Education Malaysia worth of about RM 1 million. He has edited three Conference proceedings, and three books. He is an academic Editor of J. Advances in Materials Science and Engineering (Wiley) and Special issue editor at J. Welding International (T&F).

Ir. Dr. Kalaimani Markandan is an Assistant Professor and current Head of Programme at the Department of Chemical & Petroleum Engineering, UCSI University, Malaysia. She previously graduated with a first-class MEng (Hons) and PhD (Chemical Engineering) in the years 2013 and 2017 respectively from the University of Nottingham. Prior to joining UCSI University, she worked as a research scientist at Nanyang Technological University (NTU), Singapore. Her research area focusses on composite materials, functionally graded materials, additive manufacturing, mechanics of materials and MEMS. She is also a chartered member with the Institution of Chemical Engineers (IChemE), Chartered Engineer (CEng) with the Engineering Council UK (ECUK) and a Professional Engineer (PEng) with the Board of Engineers Malaysia (BEM).

Dr. Cik Suhana Binti Hassan is the Mechanical and Mechatronics Engineering Department Head at UCSI University, Malaysia. She earned her bachelor's and master's degrees in 2009 and 2011, respectively, from Universiti Teknologi PETRONAS, and her PhD in 2019 from Universiti Putra Malaysia. In 2011, she began her career as an Intellectual Property Executive at SIRIM Berhad, a corporate organisation wholly owned by the Malaysian government, and in May 2012, she began her career in academia as a lecturer in the Mechanical Engineering Department of UCSI University. Dr. Suhana is passionate about turning environmental waste into value-added products as part of her quest to live a more environmentally friendly life. Her research interests include bio-composites for automotive applications. She is also an active member of the materials community, having been named a Professional Member of the Institutes of Materials Malaysia and a Professional Technologist of the Malaysian Board of Technologists in the field of Material Science Technology.

Mr. Praveennath G. Koppad is currently a Ph.D. candidate at National Institute of Technology Karnataka, India in the Department of Mechanical Engineering and holds B.E and M.Tech degrees from Visvesvaraya Technological University, India. With 18 years of industrial and academic experience at various positions, his research is mainly focused on thermal spray coatings, non-ferrous metal casting, nanocomposites and 3D printing of polymeric materials. He has published more than 63 scientific contributions in the form of original articles, conference papers, editorials and book chapter with 1800+ citations and H-index of 25.

9781032423135